THE CODE OF CODES

THE CODE OF CODES

Scientific and Social Issues in the Human Genome Project

EDITED BY

DANIEL J. KEVLES AND LEROY HOOD

HARVARD UNIVERSITY PRESS
Cambridge, Massachusetts
London, England

First Harvard University Press paperback edition, 1993

Library of Congress Cataloging-in-Publication Data

The Code of codes : scientific and social issues
in the human genome project /
edited by Daniel J. Kevles and Leroy Hood.
p. cm.
Includes bibliographical references and index.
ISBN 0-674-13645-4 (cloth)
ISBN 0-674-13646-2 (paper)
1. Human Genome Project—Moral and ethical aspects.
2. Human Genome Project—Social aspects.
I. Kevles, Daniel J. II. Hood, Leroy E.
RB155.C575   1992
174'.25—dc20   91-38477
CIP

Designed by Gwen Frankfeldt

Contents

PART III   ETHICS, LAW, AND SOCIETY

Preface

The human genome comprises, in its totality, all the different genes found in the cells of human beings. The Nobel laureate Walter Gilbert has called it the "grail of human genetics," the key to what makes us human, what defines our possibilities and limits as members of the species *Homo sapiens*. What makes us human beings instead of chimpanzees, for example, is a mere 1 percent difference between the ape genome and our own. That distinction amounts to no more than a gross reckoning, however. The substance and versatility of the human genome lie in its details, in specific information about all the genes we possess—the number has been variously estimated at between 50,000 and 100,000—about how they contribute to the vast array of human characteristics, about the role they play (or do not play) in disease, development, and behavior.

The search for the biological grail has been going on since the turn of the century, but it has now entered its culminating phase with the recent creation of the human genome project, the ultimate goal of which is the acquisition of all the details of our genome. That knowledge will undoubtedly revolutionize understanding of human development, including the development of both normal characteristics, such as organ function, and abnormal ones, such as disease. It will transform our capacities to predict what we may become and, ultimately, it may enable us to enhance or prevent our genetic fates, medically or otherwise.

Unquestionably, the connotations of power and fear associated with the holy grail accompany the genome project, its biological counterpart. The project itself has raised professional apprehensions as well as high intellectual expectations. Undoubtedly, it will affect the way that much of biology is pursued in the twenty-first century. Whatever the shape of that effect, the quest for the biological grail will, sooner or later, achieve its end, and we believe that it is not too early to begin thinking about how to control the power so as to diminish—better yet, abolish—the legitimate social and scientific fears.

The project incorporates—indeed, is a product of—the development of genetics since the turn of the century, and perceptions of its social implications are strongly colored by the social uses of genetics in the past. In recognition of these facts, Part I of the book provides a historical introduction to acquaint the reader with the project's technical, social, and political background. Parts II and III explore the substance and implications of the project in relation both to genetics, technology, and medicine and to ethics, law, and society.

It is our conviction that the social and ethical issues of human genetics—which the project is not so much raising as intensifying—are analyzed most usefully when they are tied to the present and prospective realities of the science and its technological capacities. Science-fiction fantasies about the genetic future distract attention from the genuine problems posed by advances in the study of heredity. Many of the chapters examine or refer to a common set of technical ideas and methods that are fundamental to the mapping and sequencing of the human genome. To assist the reader, we have included a glossary of technical terms. We have also sought to minimize repetition of technical material from one chapter to the next, while permitting it to occur where it seems to facilitate comprehension.

Seven of the chapters in this book derive from lectures delivered at the California Institute of Technology during the 1989–90 academic year in a series on the human genome project that was jointly sponsored by the Program in Science, Ethics, and Public Policy, in the Humanities and Social Sciences Division, and the National Science Foundation Center for Molecular Biotechnology, in the Biology Division. We would like to express our thanks for the grants that made the series possible to President Thomas

Everhart, of Caltech, and to the National Science Foundation and the Program on Ethical, Legal, and Social Implications of the National Center for Human Genome Research. Also gratefully acknowledged is the support of the Andrew W. Mellon Foundation, which enabled one of us to contribute substantial time to the organization of the series, the editing of all the chapters, and the final preparation of the manuscript.

Our understanding of many of the issues covered in this book was enlarged by comments provided at the lectures at Caltech by Shirley M. Hufstedler, Leslie Steven Rothenberg, and Lucy Eisenberg; and by extensive post-lecture discussions that were made possible by Valerie Hood's opening her home and offering her dining table to the discussants. We would also like to thank the Audio Visual Department of Caltech for taping the lectures and discussions; Glenn Bugos for handling the equipment when necessary; Jane Dietrich for wrestling the raw lecture transcriptions into readable drafts; Rebecca Ullrich and Karen Thompson for assistance with editorial and administrative details; Bettyann Kevles for sharing her knowledge about special aspects of genetics and society; Gordon Lake for supplying documents on the European Community's genome project; Mark Cantley and his staff for facilitating use of the valuable BioDoc collection that he has created in the science section of DG-XII of the European Commission in Brussels; and Robert Cook-Deegan and Tracy Friedman for providing important information on the early development of the genome project in the United States.

Sheryl Cobb transcribed the original lecture tapes. We are deeply indebted to her and to Sue Lewis for dealing cheerfully and reliably with the endless administrative and secretarial details involved in mounting the lecture series and in preparing a book of this type. We are also grateful to Karen McCarthy for assistance with the final typing and preparation of the figures, and to Helga Galvan and Eloisa Imel for backup secretarial aid at a critical time. And we wish to thank Howard Boyer, our editor at Harvard University Press, who was quick to express interest in this book and has expedited its production, and Kate Schmit for her superb copyediting of the manuscript.

Neither of us necessarily agrees with everything in the chapters that follow. Our purpose in forging this book has not been to advance a uniform view of the human genome project and its implications but to stimulate thought about the diversity of issues that it provokes—and about the diversity of opinions and ideas that different people may hold about them.

Daniel J. Kevles
Leroy Hood

HISTORY, POLITICS, AND GENETICS

I

DANIEL J. KEVLES

Out of Eugenics: The Historical Politics of the Human Genome

1

The scientific search for the "Holy Grail" of biology dates back to the rediscovery, in 1900, of Gregor Mendel's laws of inheritance. Mendel had arrived at his law by studying the transmission of characters only in peas, but scientists quickly showed that his dominant and recessive factors of heredity—"genes," to use the term soon coined for them—governed inheritance in many other organisms. They also demonstrated that genes are located on chromosomes, the tiny, thread-like entities in the cell nucleus that color upon staining.* After 1910, they learned many of the details of Mendelian heredity from studies of fruit flies, which are advantageous subjects for genetic research because they breed rapidly and their breeding can be experimentally controlled. Human beings, who reproduce slowly, independently, and privately, are not good subjects for research. Nevertheless, since no creature fascinates us as much as ourselves, efforts began almost immediately after the rediscovery of Mendel's laws to test their applicability to human inheritance. By 1907 it had been shown convincingly that Mendelism could account for the transmission of eye color as well as of the inborn error of metabolism called alkaptonuria. (See Figure 1.)

*See Chapter 2 for a historical introduction to the key technical terms and concepts of genetics from Mendel to molecular biology.

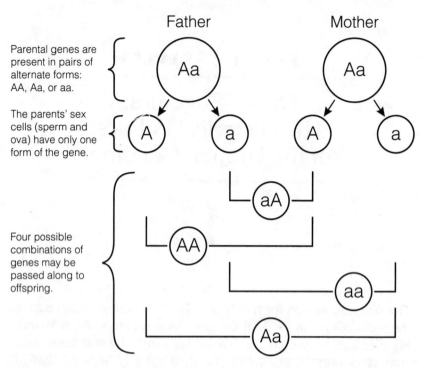

Figure 1 The simplest type of Mendelian inheritance. If *A* and *a* represent the domi-nant and recessive forms, respectively, of a gene encoding a particular trait, then the chances for each offspring are 3 out of 4 that the dominant trait will be expressed and 1 out of 4 that the recessive trait will be expressed. In human beings, blue eye color and alkaptonuria are examples of recessive traits. With a very large number of off-spring, the distribution of outcomes is *AA* + 2*Aa* + *aa*, but the ordinary human family is too small to exhibit this distribution.

In succeeding decades, a small number of scientists and physi-cians took it upon themselves to further the quest for the biological grail. Some were drawn to understand human heredity for its own sake, others were motivated by its relationship to medicine. However, perhaps most seekers were socially attracted and pro-fessionally nourished by its connection with eugenics—the cluster of ideas and activities that aimed at improving the quality of the human race through the manipulation of its biological heredity.

The goal of breeding better people goes back at least to Plato, but its modern version, eugenics, originated with Francis Galton, a younger first cousin of Charles Darwin's and a brilliant scientist in his own right. In the late nineteenth century Galton proposed that the human race might be improved in the manner of plant and animal breeding. It was Galton who named this program of human improvement "eugenics" (he took the word from a Greek root meaning "good in birth" or "noble in heredity"). Through eugenics Galton intended to improve human stock by getting rid of so-called undesirables and multiplying so-called desirables.[1]

Galton's eugenic ideas took popular hold after the turn of this century, developing a large following in the United States, Britain, Germany, and many other countries. The backbone of the movement was formed by people drawn from the white middle and upper-middle classes, especially professional groups. Its supporters included prominent laymen and scientists, particularly geneticists, for whom the science of human biological improvement offered an avenue to public standing and usefulness. Eugenicists declared that they were concerned with preventing social degeneration, which they found glaring signs of in the social and behavioral discordances of urban industrial society—for example, crime, slums, and rampant disease—and the causes of which they attributed primarily to biology—to "blood," to use the term of inheritable essence common at the turn of the century.[2]

To eugenically minded biologists, the biological roots of social degeneration had to be analyzed if they were to be extirpated—which made the study of human heredity essential to the eugenic program. Such biologists understood eugenics to be the application of human genetic knowledge to social problems and the development of that knowledge to be the basic branch of eugenic "science." As a result, the human genetics program included the study of medical disorders—for example, diabetes and epilepsy—not only for their intrinsic interest but because of their social costs. A still more substantial part of the program consisted of the analysis of traits alleged to make for social burdens—traits involving qualities of temperament and behavior that might lie at the bottom of, for example, alcoholism, prostitution, criminality, and poverty. A major object of scrutiny was mental deficiency—then commonly termed "feeblemindedness"—which was often identified by intelligence tests and was widely interpreted to be at the root of many varieties of socially deleterious behavior.

A large fraction—perhaps most—of research in human heredity was pursued in laboratories established to develop eugenically useful knowledge. In the English-speaking world, the most prominent of these institutions, both of which were created early in the century, were the Galton Laboratory for National Eugenics, at University College London, under the directorship of the statistician and population biologist Karl Pearson; and the Eugenics Record Office, which was affiliated with, and eventually became part of, the biological research facilities that the Carnegie Institution of Washington sponsored at Cold Spring Harbor, on Long Island, New York, under the directorship of the biologist Charles B. Davenport. Eugenic science was institutionalized in Germany beginning in 1918, with the establishment of what became the Kaiser Wilhelm Institute for Research in Psychiatry. The institutionalization continued with the creation, in 1923, of a chair for race hygiene at Munich, to which the biologist Fritz Lenz was appointed; and with the founding, in 1927, of the Kaiser Wilhelm Institute for Anthropology, Human Heredity, and Eugenics in Berlin, which was directed by the anthropologist Eugen Fischer, a conservative nationalist who then headed the Society for Racial Hygiene.[3]

Researchers at or affiliated with these laboratories gathered information bearing on human heredity by examining medical records or conducting extended family studies, often relying upon field-workers to construct trait pedigrees in selected populations—say, the residents of a rural community—on the basis of interviews and the examination of genealogical records. An important feature of German eugenic science was twin studies (the idea being that what is or is not genetic about human traits might be revealed by analysis of genetically similar or identical children raised in different family environments). By 1926, as a result of its surveys and studies, the Eugenics Record Office had accumulated about 65,000 sheets of manuscript field reports, 30,000 sheets of special traits records, 8,500 family trait schedules, and 1,900 printed genealogies, town histories, and biographies.

Karl Pearson, an adamant anti-Mendelian, sought to determine heritabilities by calculating correlations among relatives or between generations for the frequencies of occurrence of different diseases, disorders, and traits. Studies emanating from his laboratories typically explored the relationship of physique to intelligence; the resemblance of first cousins; the effect of parental occupation upon children's welfare or the birthrate; and the role of heredity in alcoholism, tuberculosis, and defective sight. How-

ever, the approach that dominated eugenic science in most laboratories was not correlational but Mendelian evaluation—the analysis of phenotypical and family data to account for the inheritance
of a variety of medical afflictions and social behaviors in genetic
terms.

Typical of Mendelian work in eugenic science were the studies
of Charles B. Davenport and his associates, which appeared in
his comprehensive 1911 book, *Heredity in Relation to Eugenics*, and
in later publications. Wherever family pedigrees seemed to show
a high incidence of a given character, Davenport concluded that
the trait must be biologically inheritable and he attempted to fit
the pattern of inheritance into a Mendelian frame. Although he
noted that single genes did not seem to determine important mental and behavioral characteristics, he did argue that patterns of
inheritance were evident in insanity, epilepsy, alcoholism, "pau-
perism," and criminality. The mental and behavioral characteristics of different races were a major concern for Davenport, who,
like eugenic scientists elsewhere, held different national groups
and "Hebrews" to represent biologically different races and to
express different racial traits. However, although he declared himself frequently on the subject, he actually did little research in it,
particularly of a Mendelian type, except for an inquiry into "race
crossing" between blacks and whites in Jamaica, the effects of
which, he concluded, were biologically and socially deleterious.[4]

Davenport helped introduce Mendelism into the studies of "feeblemindedness" conducted by Henry H. Goddard, the psychologist who introduced intelligence testing into the United States.
Goddard speculated that the feebleminded were a form of unde-
veloped humanity: "a vigorous animal organism of low intellect
but strong physique—the wild man of today." He argued that the
feebleminded lacked "one or the other of the factors essential to
a moral life—an understanding of right and wrong, and the power
of control," and that these weaknesses made them strongly susceptible to becoming criminals, paupers, and prostitutes. Goddard
was unsure whether mental deficiency resulted from the presence
in the brain of something that inhibited normal development or
from the absence of something that stimulated it. But whatever
the cause, of one thing he had become virtually certain: it behaved
like a Mendelian character. Feeblemindedness was "a condition
of mind or brain which is transmitted as regularly and surely as

color of hair or eyes."[5] According to later studies by Goddard and others, it also occurred with disproportionately high frequency among lower-income and minority groups—notably recent immigrants in the United States from eastern and southern Europe.

Eugenic research in Germany before the Nazi period was similar to that in the United States and Britain, and much of it remained similar after Hitler came to power. The Institute for Anthropology, Human Heredity, and Eugenics, for example, continued to press investigations into subjects such as the genetics of diabetes, tuberculosis, and brain disease; the heritability of criminality; the effects of race crossing (with no particular emphasis on Jews or Aryans). During the Hitler years, however, Nazi bureaucrats provided eugenic research institutions with handsome support and their research programs were expanded to complement the goals of the Third Reich. They exploited ongoing investigations into the inheritance of disease, intelligence, behavior, and race to advise the government on its biological policies.[6]

Davenport, Lenz, and eugenic scientists in other countries managed, in the end, to expose genuinely Mendelian patterns in the inheritance of traits that could be well specified—color blindness, for example—and were entirely physical. Their works showed that single genes might account for such abnormalities as brachydactyly, polydactyly, and albinism, and for such diseases as hemophilia, otosclerosis, and Huntington's chorea. Lenz, in particular, also developed important mathematical methods for overcoming ascertainment bias—the tendency in human genetic field surveys to find a higher frequency for a given trait among siblings than its true probability of occurrence. Some fraction of their work thus contributed usefully to the early study of human genetics.

But the fraction was rather small. Combining Mendelian theory with incautious speculation, eugenicists often neglected polygenic complexities—the dependence of a trait on many genes—in favor of single-gene explanations. They also paid far too little attention to cultural, economic, and other environmental influences in their accounts of mental abilities and social behaviors. Some of Davenport's trait categories were ludicrous, particularly in studies on the inheritance of what he called "nomadism," "shiftlessness," and "thalassophilia"—the love of the sea that he discerned in naval officers and concluded must be a sex-linked recessive trait because, like color blindness, it was almost always expressed in males.[7]

Class and race prejudice were pervasive in eugenic science. In northern Europe and the United States, eugenics expressed standards of fitness and social value that were predominantly white, middle-class, Protestant—and identified with "Aryans." In the reasoning of eugenicists, lower-income groups were not poor because they had inadequate educational and economic opportunities but because their moral and educational capacities, rooted in their biology, were inadequate. When eugenicists celebrated Aryans they demonstrated nothing more than their own racial biases. Davenport, indulging in unsupportable anthropology, found the Poles "independent and self-reliant though clannish"; the Italians tending to "crimes of personal violence"; and the Hebrews "intermediate between the slovenly Servians and the Greeks and the tidy Swedes, Germans, and Bohemians" and given to "thieving" though rarely to "personal violence." He expected that the "great influx of blood from Southeastern Europe" would rapidly make the American population "darker in pigmentation, smaller in stature, more mercurial . . . more given to crimes of larceny, kidnapping, assault, murder, rape, and sex-immorality."[8]

Eugenicists like Davenport knew little scientific self-doubt. Indeed, they displayed a high degree of scientific hubris that was compounded by a desire to be socially authoritative and useful. They urged the application of their putatively objective knowledge to the social problems of their day and offered their expertise to state and national governments for the formation of biologically sound public policy.

Overall, they advised interference in human propagation so as to increase the frequency of socially good genes in the population and decrease that of bad ones. The interference was to take two forms: one was "positive" eugenics, which meant manipulating human heredity and/or breeding to produce superior people; the other was "negative" eugenics, which meant improving the quality of the human race by eliminating biologically inferior people from the population. The elimination might be accomplished by discouraging biologically inferior human beings from reproducing or entering one's own population via immigration. In practice, little was done for positive eugenics, though eugenic claims did figure in the advent of family-allowance policies in Britain and Germany during the 1930s, and positive eugenic themes were certainly implied in the "Fitter Family" competitions that were sponsored at a number of state fairs during the 1920s in the United States. These competitions were held in the "human stock" sections of the fairs. At the 1924 Kansas Free Fair, winning families in the three categories—small, average, and large—were awarded a Governor's Fitter Family Trophy, which was presented by Governor Jonathan Davis, and "Grade A Individuals" received a medal that portrayed two diaphanously garbed parents, their arms outstretched toward their (presumably) eugenically meritorious infant. It is hard to know what made these families and individuals stand out as fit, but some evidence is supplied by the fact that all entrants had to take an IQ test—and the Wasserman test for syphilis.[9]

Much more was done for negative eugenics, notably the passage of eugenic sterilization laws. By the late 1920s, some two dozen American states had framed such laws, often with the help of the Eugenics Record Office, and enacted them. The laws were declared constitutional in the 1927 U.S. Supreme Court decision of *Buck v. Bell*, in which Justice Oliver Wendell Holmes delivered himself of the opinion that three generations of imbeciles were enough. The leading state in this endeavor was California, which as of 1933 had subjected more people to eugenic sterilization than had all other states of the union combined.[10]

The most powerful union of eugenic research and public policy occurred in Nazi Germany. Fischer's institute trained doctors for the SS in the intricacies of racial hygiene and analyzed data and specimens obtained in the concentration camps. Some of the material—for example, the internal organs of dead children and the skeletons of two murdered Jews—came from Josef Mengele, who had been a graduate student of Otmar von Verschuer's and was his assistant at the Institute for Anthropology, Human Heredity, and Eugenics. In 1942, Verschuer succeeded Fischer as head of the institute (and would serve postwar Germany as professor of human genetics at the University of Muenster).[11] In Germany, where sterilization measures were partly inspired by the California law, the eugenics movement prompted the sterilization of several hundred thousand people and helped lead, of course, to the death camps.

During the 1930s, in the United States and Britain at least, scientific opinion turned increasingly against eugenics, partly because of its association with the Nazis, partly because of the scientific shoddiness that colored its theories of human heredity. An assessment of the Eugenics Record Office pointed out, for example, that its vast records were worthless for the study of human genetics, not least because they fastened upon traits such as personality, character, sense of humor, self-respect, loyalty, holding a grudge, and the like, all of which could seldom be measured or honestly recorded if they were.[12] Eugenic science was also indicted for its distortions of race and class bias and for its neglect of how social and cultural environment might shape social behavior, not to mention performance on intelligence tests.

In the United States, where eugenic doctrine had been virulently deployed against minority groups, plant and animal geneticists were discouraged from having anything to do with human genetics because of its associations with racism, sterilizations, and scientific poppycock. The field was also unappealing because the techniques and skills of plant and animal genetics, in which most geneticists were trained, did not readily transfer to the study of human heredity, which depended upon at least some medical knowledge as well as mathematical methods like those necessary to overcome difficulties such as ascertainment bias. One American geneticist recalled (in an interview with the author in 1982) having been warned that it was just too difficult to get the necessary reliable information on human heredity: "The records are poor; classification is poor . . . Let's work with experimental organisms. The only thing you can do with human genetics is develop prejudices. And anyone who went into human genetics was immediately classified as a person of prejudice."

However, the eugenic idea remained tantalizing to some scientists and drew an exceptionally talented cadre into human genetics, including the British scientists Ronald A. Fisher, J. B. S. Haldane, Lancelot Hogben, and Julian Huxley and the American Hermann J. Muller. One might call them "reform eugenicists" because, unlike their predecessors, they held that any eugenics must be free of racial and class bias and must also be made consistent with what was known about the laws of heredity. In this latter regard, they had important allies among physicians, such as the British authority on mental deficiency Lionel Penrose, an anti-eugenicist, who thought that genetics might be advantageously deployed in preventive or therapeutic medicine. What bound Penrose together with reform eugenicists like Haldane was a deep-seated belief in the need to develop a sound science of human genetics.

Partly to emancipate the field from a prejudicial eugenics, the new students of human heredity preferred to search for well-defined, sharply segregating traits as immune as possible both to uncertainty in identification and to environmental influence. They thus welcomed with particular enthusiasm the rapidly increasing knowledge of the human blood groups, seven of which were known by the early 1930s. The blood groups displayed patterns of inheritance that seemed to conform to Mendel's laws. Being readily identifiable, they also might provide precise and universal genetic markers, presumably located at the same chromosomal place in most individuals and relative to which, it was hoped, the genes for other traits might be located. Lancelot Hogben, in his influential *Genetical Principles in Medicine and Social Science*, published in 1931, noted that if such unambiguous markers could be found for every chromosome, then there would be a set of socially unbiased benchmarks in connection with which the human genome could be catalogued—or, as a contemporary human geneticist might say, genetically mapped.[13]

Gene mapping depends on linkage analysis, a technique that had been pioneered by fruit-fly geneticists shortly before World War I for traits that occur in discernibly alternate forms—for example, eye color or wing type. If one or the other form of each of two such traits tend to be inherited together, their respective genes likely lie on the same chromosome and are said to be linked. Genes that lie close to each other will be jointly inherited with high frequency. Genes that are far apart will be jointly inherited with low frequency. The reason is that chromosome pairs often randomly cross over—that is, they reciprocally exchange segments—a process that can separate two linked genes, leaving one on the original chromosome and placing the other on the partner in the recombination. Thus, determining the frequency with which traits are jointly inherited provides a measure of the linear separations on a chromosome between the genes for the traits. (Two genes will undergo recombination 1 percent of the time for every one million bases of separation—a genetic distance of 1 centimorgan.) In principle, a type of genetic map may be drawn from the frequency determination: a delineation of relative distances between genes on the chromosome and a specification of the order in which they occur.

Although linkage maps had been drawn for fruit flies, which have only four pairs of chromosomes, they were decidedly difficult to accomplish with human beings. The normal human cell was known to contain two sex chromosomes and was thought to include twenty-three pairs of the chromosomes, called *autosomes*, that function independently of sex. The two sex chromosomes could be identified; indeed, they were termed X and Y after their shapes. However, laboratory techniques of the day were inadequate to separate the autosomes sharply; even counting them was a challenge (there are actually twenty-two pairs of human autosomes), and distinguishing one from the other was highly problematic. It was clear that the genes for sex-linked characters in males, such as hemophilia, resided on the X chromosome; it was unclear how to identify any one autosome as the locus of a particular marker gene, including the gene for a blood group. In the 1930s, the idea of a human genetic map was a vision ahead of its time.

Seemingly more realizable was the expectation of Hogben and others that linkage studies might hold promise for eugenic prognosis. Eugenicists had long been stymied by the problem of identifying the carriers of single genes for recessive traits, which were not expressed until—too late from a eugenic point of view—they joined homozygously in offspring. Linkage studies might reveal that a deleterious recessive gene occurred on the same chromosome as did one of the blood groups; it would not be necessary to know which chromosome was involved to spotlight someone found to have that blood group as a probable carrier of the recessive. Similarly, if the gene was a dominant, the identification of an infant's blood group would enable one to predict the probability—it would depend upon the degree of linkage—that the disease resulting from the dominant would be expressed in the child. Appropriate steps might then be taken to prevent the expression—or at least to mitigate the effects—of the disease itself. If the disease came on late in the child-bearing years, people fated to contract it could be advised before they had children of the chance of transmitting it to their offspring and they might then refrain from reproduction.

A good deal of effort was expended, especially in England, on the search for linkages. None was found between, on the one side, the blood groups or any universal, non-sex-linked character and, on the other, any type of genetic disease or disorder. But J. B. S. Haldane and Julia Bell, a collaborator at University College London, had better luck. Focusing on male sex-linked characters because the genes for them obviously resided on the X chromosome, they found in 1936 what, in Bell's report, Haldane called "a 'sensational' pedigree showing linkage of Haemophilia and color-blindness"—the first certain demonstration of linkage in human beings.[14]

In 1945, with Haldane's influential support, Lionel Penrose became Galton Professor and head of the Galton Laboratory of National Eugenics at University College London. Still a ferocious anti-eugenicist, he had the title of the laboratory's journal changed, in 1954, from *Annals of Eugenics* to *Annals of Human Genetics,* and in 1961 he managed to have his chair renamed the Galton Professorship of Human Genetics. Penrose focused the Galton research program away from eugenically oriented subjects and toward human and medical genetics as such, particularly on the study of hereditary phenomena that could be objectified quantitatively or otherwise. During the quarter-century after World War II, the Galton was one of the major centers in the English-speaking world for human genetics, a mecca for the growing number of scientists and physicians who were eager to master the methods of the field as well as to rid it of eugenic prejudice. In the United States in 1950, a corps of such enthusiasts formed the American Society of Human Genetics, and in 1954 they established the *American Journal of Human Genetics.*

In the United States during the postwar years, a key figure in liberating human genetics from its eugenic attachments was James V. Neel, who had started out in fruit-fly genetics, taken an M.D., and in 1948 assumed a joint faculty appointment in the Medical School and the Laboratory of Vertebrate Biology of the University of Michigan.[15] Neel recalled, "When I came into human genetics, I had one, I guess absolute, guiding principle: Try to be as rigorous as I would have been had I remained with *Drosophila*. That meant picking problems carefully, problems where we could get solid scientific evidence about inheritance in man." Neel's search for solid scientific evidence—and for indicators of deleterious genetic carriers—focused his attention, like that of others before him, on human blood. "You can spread it out, you can look at it, you can treat it objectively," he remarked.

Turning an objective light on blood disorders, Neel showed, in 1948, that sickle cell anemia, which at the time was thought to be the result of a dominant gene but whose symptoms ranged from the harsh to the benign, was the product of a single recessive gene. That year, independently, Linus Pauling and several collaborators at the California Institute of Technology discovered that the hemoglobin molecule in sickling cells differed physically from that in the normal type. The Pauling group, reinforcing Neel's conclusion, interpreted their results to mean that the trait and the disease derived from a particular recessive gene involved in the synthesis of the hemoglobin molecule.[16]

During the 1950s, human geneticists in both the United States and Britain drew upon the results of work then under way in the molecular biological and biochemical branch of plant, animal, and, increasingly, bacterial genetics and benefited from the rapid growth of knowledge concerning the biochemistry of the human body. The chief molecular biological breakthrough was, of course, the determination by James Watson and Francis Crick, in 1953, that genes are double-helical strands of deoxyribonucleic acid (DNA), the two strands running anti-parallel to each other and joined at periodic intervals by rungs formed of one of two pairs of bases—adenine and thymine or cytosine and guanine. Within a decade, scientists had recognized that the four bases form the alphabet of the genetic code. Variations in the linear ordering of the letters spell out units of genetic information—the sequences of code that are called genes.

By the mid-1960s, a large number of clearcut biochemical variations were known, including more than a dozen inborn errors of metabolism arising from probable enzyme deficiencies, and so were numerous hemoglobin and blood-serum protein variants. Also ascendant were methods for studying human chromosomes.

It was in 1956, in Lund, Sweden, that Joe-Hin Tjio and Albert Levan deployed several techniques in combination to demonstrate that the human genome contained twenty-two rather than twenty-three pairs of autosomes, which added to the two sex chromosomes put the total number of chromosomes in the normal human genome at forty-six. (Figure 2 shows the relative sizes of the human chromosomes and their banding patterns.) And in early 1959, almost simultaneously in France and England, it was shown that Down's syndrome resulted from a chromosomal anomaly, trisomy 21—that is, the possession of three copies of chromosome 21 instead of the normal two.[17]

The biochemical and cytogenetic advances boosted the new field of genetic counseling, which provided prospective parents with advice about what their risk might be for bearing a child with a genetic or chromosomal disorder. In the early years of counseling, some geneticists had sought to turn the practice to eugenic advantage—to reduce the incidence of genetic disease in the population, and by extension to reduce the frequency of deleterious genes in what population geneticists were coming to call the human gene pool. To that end, some claimed that it was the counselor's duty not simply to inform couples about the possible genetic outcome of their union but also to instruct them whether or not to bear children at all. Through the 1950s, however, the standards of genetic counseling had turned strongly against eugenically oriented advice—that is, advice aimed at the welfare of the gene pool rather than that of the family. It became standard that no counselor had the right to tell a couple not to have a child, even for the sake of the couple's own welfare.[18]

The revelations of the Holocaust had turned "eugenics" into virtually a dirty word. Besides, the more that was revealed about the complexity of heredity in human beings, the less did eugenics appear defensible in principle, or even scientifically within reach. Most human geneticists could agree with what Lionel Penrose declared in 1966: "our knowledge of human genes and their action is still so slight that it is presumptuous and foolish to lay down positive principles for human breeding."[19] The search for the biological grail, by then emancipated from its eugenic antecedents, had become an independent and respected scientific pursuit. The exploration of human heredity was valued for its own sake and

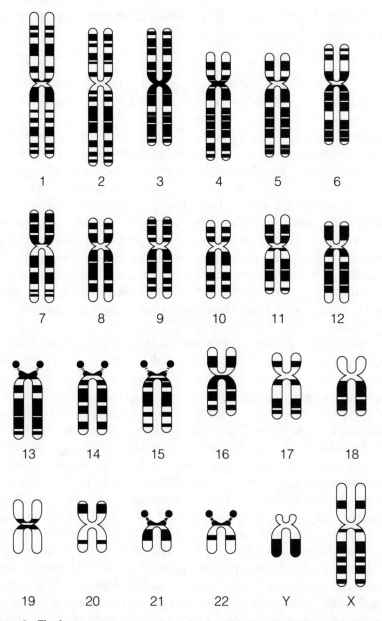

Figure 2 The human genome comprises 22 pairs of autosomes plus one pair of sex chromosomes (two X's for females, one X and one Y for males). After dye staining, each human chromosome displays a unique pattern of bands (areas darkened by inter-action with the dye) and can be identified by its individual pattern. Human chromo-somes range in size from 50 million to 250 million base pairs of DNA.

as a means to improve understanding, diagnosis, and treatment of disease.

———————

The human genome project originated largely from initiatives taken in the mid-1980s by Robert Sinsheimer and Charles DeLisi. Sinsheimer, a distinguished molecular biologist, had declared in 1969 that molecular biology, by enabling scientists to create new genes and new qualities, was opening a new and boundless human prospect: "For the first time in all time, a living creature understands its origin and can undertake to design its future." In 1977, Sinsheimer became chancellor of the comparatively young Santa Cruz campus of the University of California. The new human prospect remained on his mind, and to it was added an eagerness to put the institution on the world scientific map. Undaunted by a hairline-thin failure to obtain a large telescope for the campus, he had the idea sometime in 1984 to establish a major Santa Cruz project to determine the details of the human genome.[20]

Charles DeLisi, a physicist by training and formerly chief of mathematical biology at the National Institutes of Health, was director of the Office of Health and Environment at the Department of Energy (DOE), in Washington, D.C. The department, whose roots went back to the wartime Manhattan Project and the invention of the atomic bomb, had long sponsored research in the biological effects of radiation, especially genetic mutations. It maintained a Life Sciences Division at the Los Alamos National Laboratory, in Los Alamos, New Mexico, and in 1983 it had established there a major data base—"Genbank"—for DNA sequence information. DeLisi had thought hard about how such data might be analyzed to reveal the genetic bases of human disease. In October 1985, he found himself thinking hard about that problem again while reading a draft report on the detection of heritable mutations in human beings. He later recalled that he suddenly looked up from the report with the thought of a marvelous way to expose mutations: compare the genome of a child with that of its parents, DNA base pair by base pair. The thought led DeLisi to consider whether it might be feasible to obtain the base-pair sequence in an entire human genome.[21]

In May 1985 Sinsheimer had brought a dozen leading American and European molecular biologists to Santa Cruz for a workshop on the technical prospects of a human genome project, and in March 1986 DeLisi convened a similar workshop on the same subject at Los Alamos. Several of the participants—Walter Gilbert, for example—attended both gatherings (it was at the Los Alamos conclave that Gilbert declared the complete human genome to be the grail, adding that it was also the ultimate response to the commandment, "Know thyself").[22] Like Gilbert, most of the participants were leading practitioners of the methods and technologies necessary to seek the biological grail. In their collective view, since the late 1960s a variety of innovations had made the technical prospects for achieving it excellent.

Perhaps the most dramatic advance was the invention, in 1973, of recombinant DNA, the technique whereby a fragment of DNA could be snipped out of one genome and spliced into— recombined with—another. The snipping scissors are proteins called restriction enzymes, which bind to and cut DNA at specific sites dictated by the sequence of base pairs there.[23] Recombinant DNA opened an immense range of scientific possibilities, including the isolation of single human genes and determinations of their function. During the 1970s, Gilbert and Allan M. Maxam, at Harvard University, and Fred Sanger, at Cambridge University, in England, devised technologies to determine the sequence of base pairs in a stretch of DNA. In the early 1980s, scientists at the California Institute of Technology led by Leroy E. Hood invented a highly promising new technology that would automate and speed up the sequencing process.[24]

Identifying the location of a given sequence would depend in part upon having a physical map of the genome—that is, a linear array of DNA fragments spanning the length of each chromosome. With recombinant techniques, each human chromosome could be cleaved into fragments. The fragments could then be segregated by pulsed-field gel electrophoresis—a technology invented in the early 1980s, at Columbia University, by Charles Cantor and collaborators that could segregate relatively large fragments of DNA.[25] Once separated, the fragments could be inserted into genetic elements, such as plasmids, that are capable of replicating as recombinant molecules in suitable host cells—for example, bacteria. The process would achieve in the host cells a library of all the different human fragments, any "volume" of which could be taken out and sequenced.

The new techniques also promised to transform into reality Hogben's vision of obtaining a genetic map of the human genome. A chemical staining technique had been developed that made rigorously clear which autosome is which: a given chromosome displays a pattern of fluorescent bands unique to it. Particular genes could now be traced to a particular chromosome by special cell-culture methods. Most important, restriction enzymes could be employed to establish genetic markers on the human chromosomes that were rather more common and useful than, say, the blood-group genes. Some of the positions of a particular DNA restriction cutting site vary between any two individuals. In consequence, the DNA fragments created in two individuals by one restriction enzyme applied to the same chromosomal region tend to differ in length. These DNA markers are thus many-formed, or polymorphic in the terminology of genetics. Biologists call them restriction fragment length polymorphisms—or RFLPs (pronounced "riflips") for short. (See Figure 3.)

At the end of the 1970s, the biologist David Botstein and several colleagues had recognized that, since RFLPs are scattered across all chromosomes, they constitute a framework of genetic markers with reference to which any gene might be genetically mapped.[26] As of 1980, when the fundamental papers on RFLP gene mapping were published, 450 human genes had been mapped, mainly by cytological methods; by the mid-1980s, following the employment of RFLP methods, the total had tripled to almost 1,500 genes.[27]

RFLP mapping could also do what Hogben had futilely hoped the blood groups might—detect genes for disease. This power of RFLPs arises from their polymorphic feature. The RFLP marker may occur in one form on a normal chromosome but in another form on a chromosome containing the disease gene. If the latter form is also tightly linked to the gene, then finding that RFLP signals the gene's presence. The next step is to calculate its approximate distance from the RFLP marker, which opens the door to tracking down and analyzing the gene. In 1983 James Gusella, at the Harvard Medical School, Nancy Wexler, of Columbia University, and several collaborators announced that they had successfully employed RFLP mapping to detect the presence of the

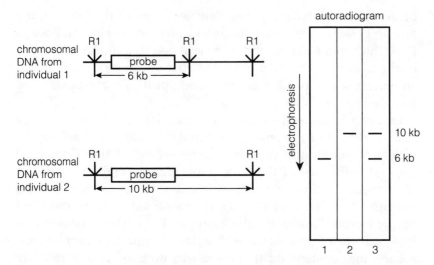

Figure 3 Restriction fragment length polymorphisms (RFLPs) are genetic markers. A restriction enzyme will cut chromosomal DNA only at a particular base-pair sequence; wherever that sequence occurs (here marked *R1*), the DNA chain will be broken. In individual 1, the DNA fragment containing the probe sequence is 6 kilobases (kb) in length; in individual 2, it is longer, 10 kb, because one of the restriction sites has been lost owing to a DNA base mutation. This difference in length is visible in the autoradiogram at right, which shows a sample from a third individual as well. The DNA fragments, labeled by radioactive probes, are separated by size by electrophoresis, and the probes are recorded on X-ray film. Individual 1 is homozygous for the 6 kb fragment, individual 2 is homozygous for the 10 kb fragment, and individual 3 is heterozygous.

gene for Huntington's disease. They located it on chromosome 4—an electrifying report that boosted RFLP mapping efforts already under way and stimulated new ones for the study of putatively genetic diseases and disorders.[28]

By the mid-1980s, discoveries about the role of genes in disease were coming along at a dizzying pace. Potential killers such as familial hypercholesterolemia, a cause of heart disease, had been traced to a recessive gene, and cancer had been identified as arising in part from the play of what were called oncogenes. Oncogenes are normal cellular genes that go haywire as a result of deregulation or mutation—a fact that stimulated Renato Dulbecco, a Nobel laureate in physiology or medicine, to declare in an editorial in *Science* magazine for March 7, 1986, three days after the end of the Santa Fe conference, that science had reached a turning point in cancer research and that further progress would be expedited by having a complete sequence of the DNA in the human genome. Dulbecco, though uninvolved in either the Santa Fe or the Santa Cruz meetings, proclaimed that the United States ought to commit itself to obtaining that sequence, by mounting an endeavor comparable in effort and spirit to the program that had "led to the conquest of space."[29]

Dulbecco's editorial joined with informal reports of the Santa Cruz and Santa Fe meetings to stimulate widespread talk in the biological community about a genome project. Walter Gilbert, an avowed enthusiast of the idea, persuaded several key scientists of its merits, including James D. Watson, the codiscoverer of the structure of DNA and the highly influential head of the Cold Spring Harbor Biological Laboratory, on Long Island. Gilbert tirelessly advocated proceeding with a crash program for human gene sequencing, contending that there was no good reason to delay because the technology was already available for the task. In June 1986, at a Cold Spring Harbor meeting, Gilbert declared that the project would be vastly accelerated by putting several thousand people to work on it, estimating that, at a sequencing cost of one dollar per base pair, the complete human sequence could be obtained for three billion dollars.[30] It would be a big project for biology. Still, Walter Bodmer, who had come to the meeting from London, where he was Director of Research at the Imperial Cancer Research Fund, observed that while it might resemble the gargantuan efforts of particle physics or space flights, its payoffs would be more certain. It did no good to get a man "a third or a quarter of the way to Mars . . . However, a quarter or a third . . . of the total human genome sequence . . . could already provide a most valuable yield of applications."[31]

Charles DeLisi, who had no qualms about big scientific tasks, allocated $4.5 million for the genome project from the Department of Energy's fiscal 1987 appropriation and moved to transform the project into a major department program. The department was not only interested in the health effects of radiation; as the principal supporter of particle accelerators in the United States, it was accustomed to Big Science projects, especially those involving sophisticated technologies. The volatility of national defense and energy policy since the early 1970s had also given it a sense of budgetary insecurity; it was always eager to embrace worthwhile new research projects that would help maintain the vitality of its national weapons laboratories, like those at Los Alamos and in Livermore, California.[32]

The department backed DeLisi when, in 1986, he advanced a plan for an ambitious five-year DOE human genome program that would comprise physical mapping, the development of automated high-speed sequencing technologies, and research into the computer analysis of sequence data. In September 1987, the secretary of energy ordered the establishment of human genome research centers at three of the department's national laboratories: Los Alamos, Livermore, and the Lawrence Berkeley Laboratory.[33] The department's move into genome work found enthusiastic support from Senator Pete Domenici, of New Mexico, a staunch supporter of the national weapons laboratories in his state who was himself worried about the fate of these institutions should peace break out. Domenici put the human genome project on the congressional agenda by holding hearings on the issue in the same month as the DOE's order and introducing a bill designed to advance it in connection with a general revitalization of the national laboratories.[34]

DOE's push toward a major program in the biological and medical sciences decidedly bothered a number of practitioners in those fields. The principal federal agency in the life sciences, including genetics, was the National Institutes of Health (NIH). It was dominated by biomedical scientists and, while it maintained a number of large laboratories for research into particular diseases, it spent roughly half its budget on support of external grants. NIH generally did not tell its grantees what kind of research to do; by tradition it fostered small-scale investigations and local initiative. In contrast, DOE was run by physical scientists and its research programs, symbolized by the national laboratories, tended to be big, bureaucratic, and goal-oriented. To many biomedical scientists, its entry into the genome sweepstakes threatened to divert funds from the NIH and subject mapping and sequencing research to centralized, czar-like control. At Cold Spring Harbor in June 1986, David Botstein, authoritative as one of the key inventors of RFLP mapping, earned applause when he cautioned biomedical scientists against letting themselves become "indentured" to mindless, Big Science sequencing.[35]

James Wyngaarden, the distinguished physician who headed the National Institutes of Health, had been reluctant to enter the sweepstakes, fearing that it might not be a good scientific bet and that the cost would jeopardize his agency's other programs. Key biomedical scientists insisted that NIH should get into the game, however, not least to take the principal control of it away from the Department of Energy. They brought Wyngaarden around. They also mobilized several reliable congressional friends of NIH against Domenici's bill, notably Senator Edward M. Kennedy of Massachusetts, who chaired the Senate Committee on Labor and Human Services, where NIH got its authorizations, and Senator Lawton Chiles of Florida, a member of the Senate Appropriations Committee who chaired the subcommittee responsible for the agency.[36] The outcome was that early in 1987 Wyngaarden endorsed the genome project in congressional testimony; in the fall, Domenici's measure was absorbed into an omnibus biotechnology bill that died in committee; and in December, Congress appropriated funds to both NIH and DOE—the health agency receiving $17.2 million, almost 50 percent more than the amount given the energy department—for human genome research in fiscal 1988.[37]

However, the increasing NIH commitment to the genome project by no means killed off the opposition to it; if anything, it intensified the dissent, which in 1987 spread more widely in the biomedical scientific community and grew increasingly heated. The project might now be largely in the friendly hands of NIH, but it suffered from the image that Walter Gilbert had given it— a three-billion-dollar Big Science crash program, built around a few large, bureaucratized centers that would be given over to DNA sequencing and accomplish their task within several years. The work would be tedious, routinized, and intellectually unrewarding, the critics contended. In their view, sequencing the entire human genome would amount to bad and wasteful science. Only 5 percent of the base pairs in human DNA are estimated to code for genes. These coding regions, called "exons," are interspersed among extensive noncoding regions, long stretches of DNA formally termed "introns" and informally referred to as "junk DNA." From the perspective of the MIT biologist Robert Weinberg, an authority in oncogenetics, "a gene appears as a small archipelago of information islands scattered amid a vast sea of drivel." It made no sense to Weinberg to spend time and resources to obtain data that would, for the most part, reveal little or nothing about human disease or development.[38]

Nor did it make sense to many others, especially since NIH would have to bear a large fraction of the enormous cost. The molecular biologist David Baltimore, a Nobel laureate and head of the Whitehead Institute for Biomedical Research at MIT, expressed a common apprehension: "The belief that new money is going to appear for a sequencing effort and that this project will not compete with other priorities is naive . . . A huge, low-priority project in biology would undercut the efforts of those who argue that very high priority science is not being funded today."[39]

Nevertheless, in February 1988 a committee of the National Research Council (NRC), which is an arm of the National Academy of Sciences, issued a surprisingly favorable report on the human genome project. It was surprising because the committee included David Botstein and several other scientists opposed to "big biology" and initially skeptical of the genome project. The report found considerable merit in genome research so long as it served a broad biological interest and did not operate as a crash program. The report urged proceeding in a phased, long-term fashion and suggested funding the project for fifteen years with an annual $200 million of new money—that is, money not to be taken from existing biomedical research. The money would at first be devoted primarily to physical and genetic mapping of the DNA of human and other organisms, which would speed the search for genes related to disease (and was a type of research that many biologists wanted to pursue anyway). Part of the money would also be invested in the development of technologies that would make sequencing rapid and cheap, cheap enough to be accomplished in many ordinary-sized laboratories rather than in just a few large facilities. In the view of the committee, the technological development could be done at up to ten large, multidisciplinary centers around the country. The biological research could be pursued in the usual way, with money granted on a competitive, peer-review basis to able investigators wherever they might be.[40]

Wyngaarden's thinking matched that in the report. Early in 1988, he had explained his preferences to David Baltimore, who agreed to chair a high-level scientific advisory meeting on the genome project that Wyngaarden arranged to convene at the beginning of March 1988, in Reston, Virginia. There, Baltimore, Botstein, Watson, and the rest of the advisory group closed ranks behind a project that would proceed along the lines recommended by the NRC committee's report. Wyngaarden, on his part, announced that he would create an NIH Office for Human Genome Research. In October 1988, Watson agreed to head the office. (Wyngaarden later said, "I had an A list and a B list, and Watson was the only name on the A list.") Watson's appointment was a de facto decision in favor of NIH over the nagging issue of which federal agency would lead the biological side of the project. About the same time, under pressure from Congress, the two agencies drafted a concordat that formed the basis of their working relationship for the next five years; it allocated mapping mainly to NIH and sequencing, particularly the development of technologies and informatics, to DOE, but it allowed collaboration in overlapping areas.[41]

Watson, Mr. DNA if anyone was, had already proved himself an influential spokesman for the project on Capitol Hill, and he was joined in his advocacy by forceful allies, including biomedical scientists as well as representatives of the pharmaceutical and biotechnology industries. The biomedical scientists tended to stress that the project promised high medical payoffs. The industrial spokesmen contended that it would be essential to national prowess in world biotechnology, especially if the United States expected to remain competitive with the Japanese.[42]

To be sure, by several measures, the United States led Europe and was far ahead of Japan in molecular biology and biotechnology in general and human genome research in particular. Of 1,000 firms found by a United Nations survey to be mainly involved in biotechnology, almost half were based in the United States, almost a third in Britain. The United States and Europe together accounted for 80 percent of funds spent on human genome research, Japan for only 5 percent. Between 1977 and 1986, the United States had originated more than 42 percent of 10,000 articles published on human genome research—about twice the fraction that had come from Britain, France, and Germany combined and ten times the fraction published by Japanese scientists.[43] However, the Japanese appeared to be mobilizing for a major push into molecular biology. In 1987, at the Venice Economic Summit of the leading industrial nations known as the Group of Seven (G7), the Japanese government announced that it was establishing a Human Fron-

tiers Scientific Programme, an international enterprise of cooperative basic research into neurobiology and molecular biology in
which the other G7 countries were invited to participate. The Japanese declared that they would provide the bulk of funding for a
three-year trial period, a welcome source of new research money
but one that meant giving the Japanese further access to the molecular biological know-how of Europe and the United States.[44]

The Japanese were also moving toward a major genome project
of their own and, since the early 1980s, had been pressing the
development of automated sequencing technologies. The effort
was spearheaded by a biophysicist at the University of Tokyo
named Akiyoshi Wada and it included Fuji Film, Hitachi Limited,
Mitsui Knowledge Company, and Seiko, corporations of manifest
ability to forge technologies low in unit cost and high in quality.
In 1986, Wada had declared that the automation of complex laboratory procedures "could well turn out to be the equivalent of the
Industrial Revolution in biological and biochemical laboratories."
Wada's group expected its automated sequencers to spit out the
identification of one million base pairs a day by the early 1990s—
more sequencing than was then being annually accomplished in
the entire world—and was reported to have already reduced the
sequencing cost to seventeen cents per base pair.[45]

In Europe during 1987–88, genome research and sequencing
efforts were gathering momentum in several countries, including
Britain, France, Italy, West Germany, the Netherlands, Denmark,
and even the Soviet Union. Some of the European enthusiasts of
the project had been stimulated by the American initiative. Some
had helped launch it—for example, Renato Dulbecco, who was
coordinating a project in his native Italy to sequence a fragment
of the X chromosome known to be a cause of mental retardation.
The British venture was led by Walter Bodmer, as co-director of
a special group in the Medical Research Council (the British equivalent of NIH), which considered it a high-priority program.[46] In
France, the prime genome promoter was the biologist Jean Dausset, who had won a Nobel prize in 1980 for his identification and
analysis of the human leukocyte antigens, proteins on the surface
of white blood cells that play a key role in immune response. HLA
proteins are polymorphic: widely diverse in composition—much
more so than the blood groups—and as specific to every individual as the complement of HLA genes that code for them. Untan-

gling the complexities of the HLA system had required scrutiny of thousands of different blood sera, systematic tests of populations, and computerized correlation studies, not to mention institutional backing and, eventually, international cooperation—the kind of organized attack that human genome research demanded.[47]

In 1984, Dausset had established a key facility for human genome mapping: the Centre d'Etude du Polymorphisme Humain, known as CEPH, at the Collège de France in Paris. CEPH had accumulated the DNA of a fixed sample of forty human families, on the assumption, as Dausset later explained, that the genetic linkage map would be efficiently achieved by "collaborative research on DNA from the *same* sample of human families." In 1987, largely supported by the Howard Hughes Medical Institute, CEPH was making clones of its DNA available to some three dozen investigators in Europe, North America, and Africa. They tested the clones for the presence of RFLPs of interest to them and donated the results of these tests to the CEPH data base, thus filling in points on a standard human genetic map. Prime Minister Jacques Chirac identified human genome research as a new priority for the nation, and by May 1988 the French government had agreed to devote eight million francs ($1.4 million) to the work, the money to be distributed by a committee chaired by Dausset.[48]

Enough was happening in human genome research around the world to prompt Sydney Brenner, a prominent molecular biologist who was involved in gene mapping at Cambridge University, to think that it might be advantageous to have an international organization in the field. In April 1988, Brenner advanced the idea to a responsive genome gathering in Cold Spring Harbor and in September 1988, at a meeting in Montreux, Switzerland, a founding council formally established the Human Genome Organization. HUGO, as it promptly became known, was privately funded, mainly by the Howard Hughes Medical Institute and the Imperial Cancer Research Fund. It elected its own members and was intended to help coordinate human genome research internationally, to foster exchange of data, materials, and technologies, and to encourage genomic studies in organisms other than human beings, such as mice. The American biologist Norton Zinder called HUGO a "U.N. for the human genome."[49]

In Europe as in the United States, the accelerating momentum of human genome research disquieted a number of biologists.

The French newspaper *Le Figaro* reported, in mid-1988, that many scientists considered human gene sequencing premature—somewhat as if one were "to list the millions of letters in an encyclopedia without having the power to interpret them, ignoring practically all vocabulary and syntax." When several months earlier eleven prominent French biologists met in Paris for a roundtable debate on the merits of the project, they expressed clear interest in it but also an apprehension that its novelty and size would take biology in the direction of the kind of Big Science characteristic of particle physics and space programs.[50]

In France, molecular biologists were particularly resistant to bigness and centralization, a powerful tradition in science as in all other areas of French public life. Most preferred the model of the Institut Pasteur, a private institution that had long insulated itself from state control, pursued a small-scale, artisanal mode of research, and achieved remarkable scientific distinction, including several Nobel prizes, in the process. To many Pasteurians, human genome research threatened to put a premium on managerial and technological skills, smother Little Science, and take away its resources.[51] In Italy, in contrast, thirty different laboratories appeared to be content to work on the part of the X chromosome that had been parceled out to them, but elsewhere even many enthusiasts of genome research were disinclined to subordinate themselves to any centralized division of intellectual labor—for example, an allocation of a chromosome to a country. Sydney Brenner himself would declare in 1989, "We [in Cambridge] do not intend to be assigned a part of a chromosome by some Politburo somewhere. That's no way to do genetics."[52]

Still, the misgivings were offset in Europe by recognitions of reality. Jean-Michel Claverie, one of the few genome advocates at the Institut Pasteur, where he was the head of the group on scientific computing, noted at the 1988 Paris debate that exploring the genome of the mouse or rabbit would doubtless be scientifically more useful and interesting in the short term, but he had to add that "man is the sole species that will pay to have its genome sequenced."[53] The most compelling reality was the consequences of remaining out of the human genome sweepstakes—the disadvantage at which Europe would likely find itself in medical and biological science and technology, including diagnostic methods and therapeutic materials. John Tooze, the executive secretary of the European Molecular Biology Organization, and Lennart Philipson, director general of the European Molecular Biology Laboratory, in Heidelberg, had admonished in 1987, "We in Europe cannot afford simply to stand aside and first watch, helping to harvest some of the benefits without contributing to their production . . . The 'Book of Man', some 3,500 million base pairs long, may well be available on compact discs by the year 2000; it should have some European authors."[54]

An increasing number of scientific statesmen were beginning to agree—and to go even further. The participants in the Paris debate acknowledged that the discipline and size required for a human genome project placed it beyond the capacities of ordinary academic science; they saw in it a European dimension. Ernst Winnacker, the vice-president of the Deutsche Forschungsgemeinshaft, one of West Germany's leading research agencies, and the director of the Gene Center in Munich, contended that if Europe were both to "collaborate scientifically and compete technologically with the United States and Japan" in the genome area, its genome effort had to be coordinated at a European level.[55]

Precisely such reasoning led the European Commission, the Brussels-based executive arm of the European Community— the term was coming to replace the phrase European Communities, meaning the European Economic Community, the European Coal and Steel Community, and the European Atomic Energy Community—to propose in July 1988 the creation of its own human genome project.[56] Called a health proposal, it was entitled "Predictive Medicine: Human Genome Analysis." Its rationale rested on a simple syllogism—that many diseases result from interactions of genes and environment; that it would be impossible to remove all the environmental culprits from society; and that, hence, individuals could be better defended against disease by identifying their genetic predispositions to fall ill. According to the summary of the proposal: "Predictive Medicine seeks to protect individuals from the kinds of illnesses to which they are genetically most vulnerable and, where appropriate, to prevent the transmission of the genetic susceptibilities to the next generation."[57]

In the view of the Commission, the genome proposal, which it found consistent with the Community's main objectives for research and development, would enhance the quality of life by decreasing the prevalence of many diseases distressful to families and expensive to European society. Over the long term, it would make Europe more competitive—indirectly, by helping to slow the rate of increase in health expenditures; and directly, by strengthening its scientific and technological base. (The Commission noted that informed estimates of the potential annual market for DNA diagnostic kits alone amounted to 1,000 to 2,000 million European Currency Units (ECUs), each equal in 1989 to about $1.11. To the end of fostering European prosperity by creating a "Europe of health," the Commission proposed to establish a modest human genome project, providing it with 15 million ECU (about $17 million) for the three years beginning January 1, 1989.[58]

The Community's rules of governance had been revised in the Single Act, of February 1986, to compel the Council of Ministers, the policymaking authority to which the European Commission is responsible, to share a small degree of power with the European Parliament. On August 16, 1988, in accord with a "cooperation procedure" called for by the revision, the Council submitted its genome proposal to the Parliament. The Parliament would assess it in a "first reading" and could, if it chose, suggest amendments to the Council. The Council would then seek to form a "common position" on the text, one mutually agreeable to itself and the Parliament.[59] In the Parliament, primary responsibility for evaluating the genome proposal was given, on September 12, 1988, to the Committee on Energy, Research and Technology, which considered it in several meetings and, by late January 1989, was ready to vote on a report concerning the matter.[60]

The drafting of committee reports in the European Parliament is guided by a member—a rapporteur—who is designated for the purpose and who can exercise enormous influence over the position that the committee eventually adopts. The rapporteur appointed for the genome proposal was Benedikt Härlin, a Green Party member from West Germany. Opposition to genetic engineering was widespread there, and it was especially sharp among the Greens, a disparate coalition united mainly by a common interest in environmental protection. The Greens' desire to preserve nature was suffused with distrust of technology and suspicions of human genetic manipulations. The Greens had helped impose severe restrictions on biotechnology in West Germany and raised objections to human genome research on grounds that it might lead to a recrudescence of Nazi biological policies. As James Burn, a Scottish expert on biotechnology and a longtime resident of West Germany, once told a reporter, "Germans have an abiding and understandable fear of anything to do with genetic research. It is the one science that reminds them all of everything they want to forget."[61]

The Härlin report, insisting that the European Community remember the past, raised a red flag against the genome project as an enterprise in preventive medicine. It reminded the Community that in the past eugenic ideas had led to "horrific consequences" and declared that "clear pointers to eugenic tendencies and goals" inhered in the intention of protecting people from contracting and transmitting genetic diseases. The application of human genetic information for such purposes would almost always involve decisions—fundamentally eugenic ones—about what are "normal and abnormal, acceptable and unacceptable, viable and non-viable forms of the genetic make-up of individual human beings before and after birth." The Härlin report also warned that the new biological and reproductive technologies could make for a "modern test tube eugenics," a eugenics all the more insidious because it could disguise more easily than its cruder ancestors "an even more radical and totalitarian form of 'biopolitics.'" Holding that the primary function of a European health and research policy must be "to block any eugenic trends in relation to human genome research," the report judged the proposed program in predictive medicine "unacceptable" as it stood.[62]

Härlin actually wished to make it acceptable, not to reject it. ("You can't keep Germany out of the future," he later said about his own country's involvement in genome research.)[63] On January 25, 1989, the energy committee voted twenty to one to adopt the Härlin report. It thus urged Parliament's endorsement of the European Commission's proposal as it was modified by thirty-eight amendments contained in the report, including the complete excision of the phrase "predictive medicine" from the text. Collectively, the modifications were mainly designed to exclude a eugenically oriented health policy; to prohibit research seeking to modify the human germ line; to protect the privacy and anonymity of an individual's genetic data; and to ensure ongoing debate into the social, ethical, and legal dimensions of human genetic research.[64]

In mid-February 1989, the Härlin report whisked through a first reading in the European Parliament, drawing support not only from the Greens but also from conservatives on both sides of the English Channel, including German Catholics.[65] The Parliament's action prompted Filip Maria Pandolfi, the new European Commissioner for Research and Development, in early April 1989 to freeze indefinitely Community human genome monies. The move was believed to be the first by a commissioner to block one of Brussels' own technological initiatives. Pandolfi explained that time for reflection was needed, since "when you have British conservatives agreeing with German Greens, you know it's a matter of concern."[66]

The reflection produced, in mid-November, a modified proposal from the European Commission that accepted the thrust of the amendments and even the language of a number of them. The new proposal called for a three-year program of human genome analysis as such, without regard to predictive medicine, and committed the Community in a variety of ways—most notably, by prohibiting human germ-line research and genetic intervention with human embryos—to avoid eugenic practices, prevent ethical missteps, and protect individual rights and privacy. It also promised to keep the Parliament and the public fully informed via annual reports on the moral and legal basis of human genome research.[67] On December 15, 1989, the modified proposal was adopted by the European Community Council of Ministers as its common position on the genome project. On June 29, 1990—the Parliament having raised no objection—the common position was promulgated by the Council as the human genome program of the European Community, authorized for three years at a total cost of 15 million ECU, 7 percent of which was designated for ethical studies.[68]

All the while, human genome programs at the national level in Europe had been prospering. Already in 1989, European gene analysis was being carried out in eighteen countries with the support of fifty funding agencies. That year, the British government committed itself to a formal human genome program funded at a level of 11 million pounds for three years and 4.6 million pounds a year subsequently. In France during 1990, genome funding received 100 million francs and a promise from Hubert Curien, the minister of research and technology, of substantial increases by

1992. In the Soviet Union, the Politburo approved a 1989 human genome budget of 25 million rubles plus $5 million in hard currency, a sizable amount by the standards of Soviet civilian research. In 1990, the European Community announced that it would join the Human Frontiers Scientific Programme that the Japanese had initiated and recently agreed to fund as a research-granting agency headquartered in Strasbourg. In fiscal 1991 the Community spent $34 million on genome research: roughly half, taken from its various general scientific research programs, supported projects in member states—for example, CEPH—and half was slated for the specific genome effort of the Community itself, which had been melded into its Program in Biomedicine and Health.[69]

The specter of eugenics loomed over consideration of the human genome project in the United States, too. In mid-1990, the journalist Robert Wright noted in *The New Republic* that "biologists and ethicists have by now expended thousands of words warning about slippery eugenic slopes, reflecting on Nazi Germany, and warning that a government quest for a super race could begin anew if we're not vigilant." Jeremy Rifkin, the sleepless critic of genetic engineering, was quick to point to the eugenic possibilities raised by the project, and apprehensions of ethical dangers infiltrated the Congress, expressing themselves across the right-to-left spectrum—from the liberal Democrat Albert Gore of Tennessee, who had long been concerned about governmental intrusion in private genetic matters, to the conservative Republican Orrin Hatch, who worried that the human genome project might foster increased practice of prenatal diagnosis and abortion.[70]

Yet among the Americans most sensitive to the eugenic hazards and the ethical challenges inherent in the project were a number of its leading scientific enthusiasts, particularly its chief official advocate, James Watson. No Johnny-come-lately to such questions, Watson had published an article in 1971, in *The Atlantic*, entitled "Moving Toward Clonal Man," warning that society ought not to leave to scientists alone decisions about new reproductive technologies such as test-tube conceptions, that it had better foster broad debate about the social implications of science or face the possibility that "our having a free choice will one day suddenly be gone." Although Watson did not fancy himself an authority on ethical issues, he considered it not only appropriate but imperative that the NIH genome program stimulate study and debate about its social, ethical, and legal implications. To that end, at the 1988 press conference announcing his appointment to head the new Office of Genome Research, he declared that NIH should spend some of its genome funds to address the ethical implications of the work; subsequently, he announced that such activities would be eligible for roughly 3 percent of the NIH genome budget.[71]

The commitment of NIH resources to the provocation of ethical debate was unprecedented, as was making bioethics an integral part of an NIH biological research program. By no means all biologists—some estimated less than a majority—concerned with the genome project supported Watson's policy, but Watson, undaunted, defended it at a 1989 scientific conference on the genome: "We have to be aware of the really terrible past of eugenics, where incomplete knowledge was used in a very cavalier and rather awful way, both here in the United States and in Germany. We have to reassure people that their own DNA is private and that no one else can get at it."[72]

Watson was not only undaunted in his commitment to ethics but also, it would appear, shrewd. His policy undoubtedly helped defuse anxieties about the prospect of a genome project indifferent to or unrestrained by ethical considerations. Whatever the attention Gore and Hatch gave to such matters, it was discussion of the project's medical and economic ramifications that dominated congressional panels and hearings. In 1988, Congress awarded NIH and DOE together some $39 million for the genome project for the following year.[73] In October 1989, Secretary of Health and Human Services Louis Sullivan elevated Watson's office in NIH to a National Center for Human Genome Research. In 1990, federally sponsored research in the human genome operated at an appropriation of about $88 million, with the National Center receiving about two-thirds of the total, the Department of Energy the rest. Watson announced that the Center would use up to half its budget to establish and operate several genome centers around the country, each to work on special aspects of the project and each to be funded for five years at $2 to $3 million a year. He also allocated several percent of the Center's share to conferences and research on ethical issues and he established a working advisory group on ethics that included five scientists, Nancy Wexler among them, as well as a lawyer and an ethicist.[74]

In 1991, the year it was inaugurated as a formal federal program, the human genome project received roughly $135 million and moved into high gear, its infrastructure solidly in place.[75] The NIH centers, seven at first, were in operation—five focused on human gene mapping, one on mouse gene mapping, and one on yeast chromosome sequencing.[76] The genome installations at the Lawrence Livermore, Lawrence Berkeley, and Los Alamos national laboratories were occupied with forging gene mapping and sequencing technologies and informatics. Four additional projects, jointly funded by DOE and NIH, were engaged in large-scale sequencing efforts and innovations. Gene mapping and sequencing activities were also to be found in dozens of other laboratories, each of them an investigator-initiated activity supported by NIH.

Genomic data were pouring out of laboratories on both sides of the Atlantic (though not on the western side of the Pacific: the Japanese, having grossly overestimated their capacity to develop supersequencing machines, had scaled down their goals to one hundred thousand base pairs a day and their genome budget to an annual $8 million).[77] Human gene mapping information was being entered into a centralized data base at Johns Hopkins University. (By October 1990, almost 2,000 human genes had been mapped.)[78] Gene sequencing data was going into a data base at the European Molecular Biology Laboratory and at GenBank, at Los Alamos, which now contained entries on 60 million DNA base pairs from a number of species, including about 5 million from human beings. The human genome project was steadily gathering the technology, techniques, and experience to obtain the biological grail. The first complete human sequence was expected to be that of a composite person: it would have both an X and a Y sex chromosome, which would formally make it a male, but this "he" would comprise autosomes taken from men and women of several nations—United States, the European countries, and Japan. He would be a multinational and multiracial melange, a kind of Adam II, his encoded essence revealed for the twenty-first century and beyond.

HORACE FREELAND JUDSON

A History of the Science and Technology Behind Gene Mapping and Sequencing

2

Genetics—the explanation of how organisms pass on traits of anatomy, physiology, and behavior to their descendants and of how each individual expresses those traits in its formation and throughout its life—is the central problem of biology. Which is to say, as we gain fuller understanding of the transmission and expression of hereditary characters we progressively unlock understanding of whole classes of other problems in biology, as well. Thus, genetics underlies all of the biology of cells, including developmental biology, or embryology, but also enzymology and the study of cancers and many other diseases; all of immunology and of endocrinology; neurobiology generally as well as many disorders of the nervous system and of the mind; and, ultimately, the processes of evolution. Though no one supposes that genetic analysis will be sufficient to explain everything we want to know in these vast scientific domains, every biologist recognizes that genetics is necessary to all such explanation. Necessary in two ways: genetics supplies essential components of the accounts biologists seek and provides crucial methods of access to the rest.

Central to genetics, of course, is the concept of the gene: but genes act only in concert, which is to say that almost from the beginning the concept of the gene included the map of genes, their locations and relations, and later the sequences of their

chemical subunits. Modern genetics began in 1900, with the rediscovery of Mendel's rules and Mendel's paper of thirty-five years
earlier. The word *gene* first appeared about 1909. The first demonstration that a particular gene had a locus—that it could be assigned to a particular chromosome—was published in 1910; the
first genetic map, which showed the relative locations of six genes
on one chromosome, appeared in 1913. In the three-quarters of
a century since, as the experimental systems of organisms and
techniques that geneticists employ have grown more sophisticated, the concept of the gene has changed and deepened—and
with it, integrally, the idea of gene maps and sequences.

Now, biologists propose to map and sequence the entire gene
set of a creature as complex as any—man—whose genes number
at least fifty thousand and perhaps several times that. This is, of
course, the human genome project. What this proposal signifies,
how it can be carried out, what gains it could bring—all these
emerge from the historical development of genetics from 1900 to
the present day. Despite the sophistication and the sometimes
stupefying intricacy of genetics today, the main line of its history
is straightforward and casts a brilliant light upon the recent proliferation of methods and ambitious aims.

Two preliminary observations will clarify that main line. First,
the fundamental discoveries of genetics emerged as a series of
approximations. One can think of these as successive models of
the processes of heredity; they were also successive redefinitions
of the gene and of the nature of gene maps and gene sequences.
Each model, each redefinition, in order to incorporate more of the
evident diversity of inheritance, refined and made more complex
the version preceding it. Technically, too, genetic analysis has
proved conservative: even today we see the earliest methods of
mapping adapted and readapted.

Second, heredity necessarily has that dual aspect, as the transmission of characters from one generation on to the following and
as the expression of characters in the developmental process by
which an individual organism builds itself. The transmission of
characters and their expression were triumphantly united in the
elucidation, in 1953, by James Watson and Francis Crick, of the
three-dimensional molecular structure of the genetic substance—
deoxyribonucleic acid, DNA, the celebrated double helix.

The double helix of DNA is a conception of astonishing parsi-
mony in the relation of structural detail to functional necessity.

DNA takes the form of two strands, twining coaxially, clockwise, the one running up, the other down. Each strand is a string of chemical subunits called nucleotides. The nucleotides are of four kinds, differing only in the shape of a flat bit, called a base, that sticks out from the side. The bases are adenine, thymine, guanine, and cytosine. The two strands are at the outside of the structure and are hooked together across the space between them by the bases, which form pairs. The base pairs are thus the ribs of the structure. The bases are all-important, because their sequence along the strand is the only variable part of the structure. The crucial fact that Watson discovered is that the physical shapes of the bases limit them to just two kinds of pairs as they connect the strands: they pair with almost perfect fidelity adenine to thymine, guanine to cytosine. The bases are spaced along each strand so that exactly ten pairs occur in the length of a full turn of the helix. The pairs are stacked flat, with 3.4 angstrom units and a tenth of a revolution separating a pair from the one above or below. Three-point-four angstroms is about one-and-a-third hundred-millionths of an inch.

For the genetic instructions, the freedom afforded by the structure is almost unimaginably generous: along a strand, no physical or chemical rule determines the sequence of the bases—and though they are but 3.4 angstroms apart, the total DNA in a typical mammalian cell measures more than two meters. Yet at the same time the constraints are almost perfectly stringent: between the two strands, once the sequence on one is fixed, the base pairing—A to T, and G to C—altogether determines the complementary sequence on the opposite strand.

Form and function are one. The model illuminated at a glance the passing on of potential and its expression. The constraint imposed by the pairing rules makes it a unique consequence that if the strands part, each can assemble on itself a replica of its complementary previous partner, producing two identical double helices during cell division. That's gene transmission—in principle, at any rate, for it took several years to show that in life DNA does replicate this way and several decades to work out the biochemical details. The freedom of the sequence of bases along the strand allows the encipherment, as with a four-letter alphabet—A, T, C, G—of the entire specification for the substance and regulation of the organism. That's how the gene dictates the organism—again, in principle.

Like the oak in the acorn, the human genome project was implicit in the discovery of the structure of DNA. Until the discovery of the structure, biologists could do only as Mendel had done, inferring the presence and interactions of genes by their visible signs, the characters in the organisms as brought out in breeding trials. Mendelian genetics took this method to remarkable and subtle lengths. But the moment molecular biologists understood the gene as a specific sequence of bases in DNA, then they could begin to think of analyzing the growth and functioning of organisms from the inside out—as it were in parallel with the natural processes of the organism—by identifying functional base-sequences and finding out what they determine in the living creature. In 1953, this was a program inconceivably vast and difficult.

From Mendel to the first genetic map. Gregor Mendel published his experiments on plant hybridization in 1865, in a paper of fifty-five pages. Mendel, to be sure, did not think in terms of maps or sequences somewhere within the cell, but rather in terms of the statistics of inheritance of visible characters. His genius was to find in his pea plants certain discrete pairs of alternative characters—round seeds versus wrinkled ones, tall plants versus short, and so on, seven pairs in all—that were inherited in a remarkably unconfusing way. He traced these pairs through successive generations of hybrids. In the first place, of any pair each plant displayed either one trait or the other: the alternatives were nonblending. (For example, his pea plants were either tall or short, never of medium height.) Second, every pair was inherited independently of the other six pairs: the characters were free-assorting. (For example, a tall plant could have round seeds or wrinkled seeds, as could a short plant, and so on with the seven character pairs in any combination.) Third, a trait, one of a pair, could persist without being exhibited through one or many generations, yet eventually reappear: the latent trait, Mendel said, was *recessive* and the visible character *dominating*—terms that, of course, we still use. (For example, when he crossed a tall plant from a pure-bred line with a short plant from a pure-bred line, he got only tall plants in the first generation. Tall was dominant. But when he crossed those hybrids to each other, about a quarter of the next generation were short.) All this led him to see that each parent plant contributed one of a pair of factors, and no more than one, to determining each character in the makeup of each offspring plant. If an offspring were hybrid for a given character, one of those factors but not both being recessive, then in that character the plant would look like the dominating form but—an essential point—would pass the factors to its own offspring at random, about half dominant and half recessive.

These observations now seem simple and familiar. They were revolutionary. With them, Mendel established many of the fundamentals of genetics. We must analyze the heredity of an organism, at least in the first place, by its various unit traits. These are determined by factors that typically occur in pairs, with one of each pair received from each parent. The factors of a pair don't blend, and, further, one of a pair may be masked by the other without being lost. In determining the next generation, each pair is split up and distributed at random. Therefore, one must think about heredity statistically, looking at relatively large numbers of offspring in successive generations, to discern the patterns.[1]

Mendel's paper went unnoticed; his revolution hung fire for three decades. In that time, statistics developed enormously. So did cell biology, or cytology, a sister science that provides genetics its anatomical basis. Biologists made the first close observations, under the microscope, of the sperm cell entering and fusing with the ovum. They noted that cells have nuclei. In 1877 they saw that, as the cell is about to divide, thread-like structures appear in the nucleus; they termed these *chromosomes*. Microscopists teased apart the steps by which, in cell division, the chromosomes align themselves in matched pairs, then double and split up, so that complete and apparently identical sets go to each daughter cell. They distinguished a second kind of cell division, a dance of the chromosomes whereby the germ cells form themselves, each sperm or ovum containing a half-set of chromosomes, exactly one of each pair. The point was obvious: at the moment of fertilization the new individual receives a full complement of chromosomes with one of each pair from each parent.

In 1892, the noted German physiologist August Weismann put all this down in a book, *Das Keimplasma (The Germ-Plasm)*. He was able to say that at last it was settled that the male and the female parent contribute equally to the heredity of the offspring; that sexual reproduction thus generates new combinations of hereditary factors; and that the chromosomes must be the bearers of heredity. What the chromosomes bore were ultimate particles, determinants for each separate character of the organism. Weismann speculated about the number of these determinants in terms of the number of characters to be accounted for and of the sizes of molecules as chemists knew them. He needed perhaps millions of determinants and worried that they wouldn't fit in the chromosomes. Speculation, yes—but Weismann was first to think seriously about the hereditary material as a physical substance, and its necessary characteristics. Weismann did not go unnoticed. His books were translated promptly into French and English, *The Germ-Plasm* appearing in 1893. By 1896, the noted American cell biologist E. B. Wilson had elaborated a chromosome theory of heredity.[2]

Thus in 1900 when Mendel's rules and his paper were rediscovered, independently by several botanists, the community of biologists was prepared to consider heredity cytologically and statistically. Mendel was translated into English at once, by William Bateson, an English botanist. Developments came fast, and on both sides of that duality of transmission and expression. In 1902, an English physician, Archibald Garrod, reported in *The Lancet* his observations of a disorder, alkaptonuria—in which the patient's urine on exposure to air turns terrifyingly but harmlessly black—which he had traced through several generations in several families. It was inherited, he said, in a Mendelian fashion. Garrod was first to demonstrate Mendelian transmission of a character in man. (He thereby founded a distinguished line of research physicians whose work in the ensuing ninety years has been indispensable to the growth of human genetics.) Further, alkaptonuria appeared to be a blockage of one step—that is, of a specific biochemical reaction—in a metabolic pathway. He called the disorder an *inborn error of metabolism*. Garrod was thus first, also, to connect Mendelian heredity with the biochemical pathways of the individual organism. He identified several other inborn errors of metabolism, including cystinuria (which shows itself as a tendency to form bladder stones) and albinism.[3] By the fall of 1991, the on-line computer version of the established reference work *Mendelian Inheritance in Man*, edited by Victor McKusick, catalogued some 5,600 genes known or thought on good evidence to be inherited in Mendelian patterns.[4] A substantial majority of these are identified with mutations that produce some metabolic block—that is, the failure to produce a specific enzyme, as Garrod surmised. These genetic diseases are almost always recessive. In the past five years, as techniques have grown more sophisticated, an increasing number of dominant human genetic disorders have been traced to mutations of other kinds, in control elements or in genes for such things as structual proteins or cellular receptors. Garrod would have found these a comprehensible extension of his discoveries.

In 1903, Walter Sutton in the United States and Theodor Boveri in Germany independently made explicit the connection between the patterns by which pairs of Mendel's factors assort themselves and the precisely similar sorting and recombination of the chromosomes in formation of the germ cells and fertilization of the egg. Mendelian factors travel like chromosomes; they must be on the chromosomes. But Sutton also pointed out that the chromosomes in a cell are far fewer than the number of distinct characters to be accounted for. (Fruit flies have four pairs of chromosomes, humans 23.) Undoubtedly, many elements must travel and be inherited together, not sorting freely as do Mendel's seven pairs of traits in peas.[5]

Then between 1905 and 1908 analyses by William Bateson and his colleague Reginald Crundell Punnett, among others, showed that some genes modify the action of other genes. Working with the colors of cocks' combs and of sweet-pea blossoms, they demonstrated that more than one gene may be required to produce a normal character.[6] With Garrod as precursor, Bateson and Punnett and their followers established a line of research into gene expression that became known as biochemical genetics. By 1910, others showed that inheritance of a character that appears to vary continuously can be interpreted as Mendelian if one demonstrates that it is controlled by several pairs of genes whose effects can add up.[7]

Mendel's model had not been overturned, yet was distinctly altered: genes are on chromosomes, and may be linked in hereditary transmission or related in biochemical expression. But no particular gene had been associated with any particular chromosome. In 1910, Thomas Hunt Morgan, at Columbia University, was breeding fruit flies, *Drosophila melanogaster;* in this species a single mating will produce, in less than three weeks, hundreds of offspring themselves ready to breed. Morgan had begun in science as an embryologist, concerned, in other words, with how the fertilized ovum, one cell, develops into the adult organism with perhaps billions of cells of markedly specialized types—muscle, nerve, retina, blood, gland, and so on. But these problems seemed intractable, and after Mendel's rules were rediscovered Morgan turned to the transmission of those characters whose expression was so baffling. Among other things, he was interested in the question of what determines whether an individual is male or female. Gender, when you think about it, behaves much like a Mendelian character-pair. Cytologists had recently noticed, also, that higher organisms have a pair of chromosomes that is anomalous, in that in one sex the two don't look alike, one of them being conspicuously rudimentary. One sex (in flies and humans, the female) has two normal chromosomes of that pair, designated X and X; the other sex instead pairs the normal with the rudimentary chromosome, X with Y. In a line of Morgan's flies that had bred pure for twenty-odd generations, he noticed a male fly with white eyes; normal flies have red eyes. He tried the standard Mendelian crosses, and found that the gene for white eye was recessive to red—and, in an unheard-of result, was inseparable from the factor for sex. That is, it would show up in a male only when inherited from a female (and she could have red eyes if hybrid); yet a female could be white-eyed if her father was (and if her mother was at least hybrid).[8] In brief, the gene for white eye or its alternative, that for red, traveled with the X chromosome, the sex determinant. We now say the gene was *sex-linked* or, more clearly, *X-linked*.

Morgan's discovery of X linkage was soon seen to straighten out the bookkeeping for the inheritance of other characters. The first instance found in humans was red-green color blindness.[9] Another standard example is hemophilia, which entered most of the royal families of Europe from Queen Victoria by way of her many daughters. Hundreds of human characters are now known to be X-linked, including, among recently isolated genes, those for Duchenne muscular dystrophy and for chronic granulomatous disease, which is a blinding disorder of the eyes.[10]

The gene for white eye in fruit flies was the first in any creature to be assigned to a specific chromosome. It led directly to the first genetic map. Within a year or so, Morgan and his colleagues had identified a number of X-linked characters in their flies. They promptly did the obvious, carrying out matings to see whether these characters are inherited in sets, as predicted. The result: such genes do indeed travel together—but not always. For example, Morgan crossed a white-eyed fly that had normal wings with a red-eyed fly with a new X-linked character, rudimentary wings. After two generations, he was able to get a certain percentage of flies that had white eyes and rudimentary wings. Both mutant characters were now in the same flies, yet still showed the tell-tale X-linkage pattern of inheritance by males from female parents only.

The explanation of this result had to be that an interchange of materials sometimes takes place between two chromosomes of a pair—in this first analysis, the two X chromosomes of the female. The interchange they termed *crossing-over* (Figure 4). Imagine a pair of chromosomes as two strands of cooked spaghetti, one tomato-pink, the other spinach-green: if the two were to tangle, break at a point of overlap, and rejoin, the result would be one strand that was long pink then short green and the other short pink then long green. That's crossing-over, except that the occurrences were not then detectable by inspection. In the case of the white-eyed fly, of course, crossing-over could take place only in the female, since the male has but one X. But the idea of linkage that is sometimes broken by the interchange of genetic material obviously explained other recombinings of traits, not only those whose genes lie on the sex chromosomes.

In 1913, Alfred H. Sturtevant, a student of Morgan's, analyzed mating results for fruit flies with six different mutant factors each known to be recessive and X-linked. He traced each mutation and its normal alternate in relation to each of the other mutants, and thus calculated the exact percentage of crossing-over between the genes taken two by two. Some factors were more tightly associated than others. If crossing-over was indeed due to chromosomes exchanging portions, then two genes lying adjacent to each other would be less likely to be separated than two lying far apart. Sturtevant realized that the full table of crossover frequencies reveals the order in which genes are arranged on the chromosome; but more than just the order, it reveals their approximate relative distances from one another. These points established, he could extend the tables of data to take the genes three at a time. These calculations showed that sometimes two genes that lie far apart

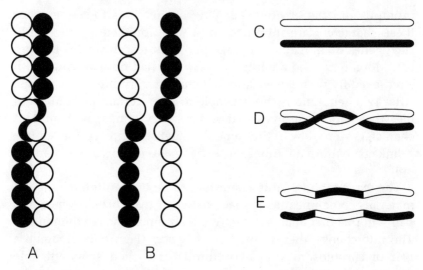

Figure 4 Crossing-over takes place as part of a complicated process in which germ cells—sperm and ova—form. Other cells of higher organisms have two full sets of chromosomes, making 8 in *Drosophila*, 46 in man, and normally carry genes in pairs on the matching chromosomes. A germ cell has only one full set—thus, 4 in *Drosophila*, 23 in man. At conception, these combine, one set from each parent, to form a new single cell with a full, dual complement of chromosomes (and genes), which then develops into the new individual. The precursor cells of the sperm or ova must multiply and at the same time reduce their number of chromosomes to one full set; this is called *meiosis*, or the reduction division, and how it works is still not fully understood. But at one step in meiosis, two chromosomes of a matching pair may exchange segments, in the manner shown schematically here *(A–B)*. A double crossing-over *(C–E)* may also occur. (Redrawn from T. H. Morgan, A. H. Sturtevant, H. J. Muller, and C. B. Bridges, *The Mechanism of Mendelian Heredity*, 1915.)

remain linked while a gene in between exchanges with the other chromosome: crossing-over sometimes happens more than once to the same chromosome pair. A double crossover interchanges segments from somewhere in the middle of the chromosomes— a strand of spinach pasta gets a tomato insert. Sturtevant also found triple crossovers. All these variations, however, confirmed the sequence of the genes and the assignment of their relative positions. His paper, "The Linear Arrangement of Six Sex-Linked Factors in Drosophila, as Shown by Their Mode of Association," includes tables of data, charts of crossover frequencies, and a single illustration—one long horizontal line with three genes shown clustered at the left end, two genes located slightly to the right center and quite close to each other, and the sixth at the extreme right. Sturtevant had drawn the first gene map.[11]

The map of defects. Geneticists soon realized that mutations of any particular gene occur rarely and at random; a mutation when it does occur is likely to be recessive to the trait the organism normally exhibits in the wild. Morgan's team soon found characters not X-linked that associate in heredity, constituting linkage groups. To define an organism's linkage groups is to solve a puzzle: as characters are added to one or another linked set, say A C B + E or L J G K + H, some sets turn out to overlap in several members, say D and I, and coalesce into a single linkage group. In such fashion, sure enough, the genes of fruit flies proved to fall into three linkage groups besides the X-linked set. Other geneticists found similar patterns in other organisms. Characters that lie in different linkage groups assort freely, just as in Mendel's peas, while the genes or their mutations within any one linkage group are indeed more or less tightly associated, crossing over with various frequencies. These discoveries of course generated maps—straight lines bristling like barbed wire with the marks showing relative positions of genes along them.[12]

In all creatures, almost all mutant characters are deleterious to some degree; many are lethal. Even in the shelter of a milk bottle with plentiful food in Morgan's fly room at Columbia University, those white-eyed flies reproduced less prolifically than their normal brood-mates, while flies with gross mutations of, say, wings or thorax obviously could never survive in the wild. Yet only a mutation—that is, an alternative to the wild type—could identify a unit character and open its heredity to analysis and its gene to mapping. Thus from the start the genetic map of any species—molds or flies, maize or humans—has been primarily the map of defects.[13]

Two questions arose inescapably: What causes mutations? and What exactly does a mutation change? Morgan's most original student was Hermann J. Muller; in 1921, before an audience of naturalists in Toronto, Muller read a paper about the nature of the gene—a paper of astonishing prescience. He conceived of the gene as a particle that despite its ultramicroscopic size must exhibit a complex structure of different parts. He drew geneticists' attention to the dual functions of the gene: its power to direct the organism to make a specific substance, be it enzyme, antibody, pigment, or whatever, that is different from the gene itself, and its power to dictate the formation of an exact duplicate of itself. That power of self-replication is an essential of the gene and was more remarkable than biologists realized, he said—for when a character mutates, the changed gene, too, is henceforth duplicated faithfully.

Understanding the gene as a substance with structure, Muller distinguished possible types of mutation in it—loss, gain, substitution, or rearrangement—and noted that other types of mutations, changes of the genetic material on a larger scale, must also take place. He said, "Some new and unusual method may at any time be found of directly producing mutations." He expressed simultaneously a commanding view of the central position of genetics in biology and the practical necessity to determine the physical nature of the gene—"to grind genes in a mortar and to cook them in a beaker, after all."[14] Then in 1927, Muller demonstrated that X rays cause mutations in fruit flies. The frequency of mutations increases with the dose of X radiation; but which genes will be affected is not predictable.[15] The next year, Lewis Stadler showed that ultraviolet radiation can cause mutations. By the 1940s, mutations were also known to be caused by a variety of chemicals, in a catalogue that's still growing.

In 1933, in a brief paper in *Science*, T. S. Painter announced that he had charted perceptible differences among chromosomes under the microscope—differences detailed enough to correlate crossing-over of genes as shown in the statistical tables with physical interchanges in the material of the chromosomes. The larvae of many species of flies, including *Drosophila*, have salivary glands that contain abnormally large chromosomes, elongated and segmented in a series of rings, or bands. When such a chromosome is stained with the appropriate dye, Painter found, the bands display a fine sequence—light and dark, thick or thin, bulging or pinched. He saw that the banding gives each pair of chromosomes its own characteristic pattern, which is the same from fly to fly. For the first time, chromosome pairs other than the X and Y could be distinguished unambiguously. Furthermore, rearrangements of portions of any chromosome—single or multiple crossovers, doublings or deletions, translocations or inversions—could be detected and correlated with genetic changes.[16] In a burst of papers in the next months, Painter and his students assigned the *Drosophila* linkage groups to particular chromosomes and located many individual genes on the chromosomes with precision. To be sure, the method appeared to be restricted to *Drosophila* and related species. Other organisms offer no giant chromosomes.

Genes and chemical pathways. Essential to understanding the expression of genes, from Garrod onwards, is the place of the gene in the biochemical processes of the organism. The central concern of biochemistry for a century and more has been the functioning of certain of the large molecules active in the chemistry of the organism, and preeminently the enzymes. The living cell contains an immense variety of different molecules, but most are either quite small or extremely large. The large molecules—as biochemists deduced by the 1890s and established beyond doubt by the end of the 1930s—are long strands built up of subunits linked together. These chains, called polymers, are rare in nature except as the products of living things, where they are common.

The large biological molecules are of four kinds: lipids, or fats; carbohydrates, or chains of sugar molecules; proteins; and nucleic acids. Lipids and carbohydrates, however long their chains, are chemically and conceptually boring—monotonous repetitions of one or a few subunits in entirely predictable order. The nucleic acids are deoxyribonucleic acid, now familiar to school-children as DNA, and its close chemical cousin, ribonucleic acid, or RNA. They were discovered late in the nineteenth century. Their chemical subunits, the nucleotides, with the four kinds of bases, were determined soon after. Until the late 1940s everyone supposed, though the evidence was meager, that the four nucleotides were repeated over and over again, in a structural unit called a tetranucleotide, making nucleic acids much like carbohydrates. Max Delbrück, one of the founders of molecular biology, observed that nucleic acids had long been taken to be "stupid molecules"—substances that could perform no interesting function, molecules that could not specify anything.[17]

The interesting molecules were the proteins. These are strung together from chemical subunits called amino acids; by the late

1940s, twenty-odd amino acids had been commonly found in proteins, and biochemists had isolated a number of others that were rare. Proteins were known to do virtually everything in the biochemistry of the organism. Many of the hormones are proteins. Immunoglobulins, or antibodies, are proteins—as, indeed, are many of the substances they react against. Hemoglobin is a protein, as are a whole set of other respiratory molecules in animals and plants. But the quintessential function of proteins is to act as enzymes, that is, as biological catalysts. The cell breaks down or builds up its component parts step by step, in a network of metabolic pathways of numbing intricacy. Each step in every pathway is made to occur quickly—thousands of times per second per cell for some reactions—by the action of an enzyme specific to that step.

In first approximation, then, the problem of gene expression was to determine the relationship of genes to enzymes and other proteins.

One gene, one enzyme. Perhaps the most influential early biochemical geneticist was an American, Sewall Wright, who trained at Harvard, in the period just before the first world war, where a group of geneticists were working not with flies but with small mammals. In a series of papers in 1917 and 1918, Wright analyzed the inheritance of coat colors in guinea pigs, mice, rats, rabbits, horses, and other mammals. He stated his program:

> It remains for genetics to assist embryology and biochemistry in filling in the links in the chain between germ cell and adult in specific cases. Variations of adult characters must be traced back through the contributing causes at each stage of development . . . and, on the other hand, the ramifying influences of unit variations in the germ cell must be traced forward through development.

In particular, Wright showed that production of the pigment making for coat color in mammals requires that several biochemical steps take place in fixed temporal order: he supposed that each step is mediated by a different, specific enzyme.[18]

In drawing attention to the specific enzyme related to the gene, and to the order in which enzymes act, Wright set up several presumptions. The first is a point suggested in earlier work which he now stated with power and generality: genes produce characters in combined action, meaning that genes can have secondary influences and that characters typically are modified by actions of more than a single gene. (We now say that many genes are *pleiotropic* and many characters *polygenic*.) The second presumption is that the gene is expressed through the action of an enzyme. The third, implicitly, is that some relationship ought to obtain between two different sorts of sequences, the map of genes on the chromosome and the temporal order of steps in the metabolic pathway. (This last turned out to apply well to micro-organisms, not well to higher organisms.) These were necessary approximations in the continuing development of the Mendelian model, and they illustrate the complexity and importance of the idea of genetic maps and sequences. Wright continued in this research—in a span of interests that included population genetics and evolutionary theory—for forty years and more.[19]

In the summer of 1935, in France, a young American from Morgan's lab named George Beadle collaborated with the leading French geneticist of that period, Boris Ephrussi, in a set of experiments using two microscopes and four hands. They transplanted tissue between larvae of fruit flies bred with various mutations of eye color and then watched the mature flies develop. They reversed biochemically the effects of the particular genetic crosses and determined the metabolic order of successive steps, catalyzed by whatever was produced by two specific genes, in the pathway that leads to the normal pigment of the wild-type red eye.

Back in the United States, Beadle began to seek out better organisms with which to do genetics—simpler, more easily manipulated creatures with a faster generation time than flies. He and a colleague, Edward Tatum, settled on a one-celled organism, *Neurospora*, a mold that grows on bread in the tropics. They found mutant strains of the mold, deficient in one or another factor necessary for growth, and were then able to carry out in a day or two experiments in biochemical genetics that with flies would take months. By 1941, their work convinced most geneticists of an idea long in the air, that what the gene does is to specify an enzyme. This they called the one-gene–one-enzyme hypothesis.[20]

The sequencing of biological polymers. A fundamental transformation took place in the late 1940s in the understanding of the chemical nature of large molecules of biological importance. Through the 1930s and virtually until the end of the forties, biochemists took it for granted that genes must be made of protein. But then in the mid-forties the possibility emerged that the material of the genes was not protein but a different class of macromolecule altogether—the nucleic acids, in particular DNA. Yet for some time this new identification of the genetic material explained little: proteins and nucleic acids alike were poorly characterized, their three-dimensional structures unknown, the physico-chemical basis for their biological specificity unguessed.[21]

The work of two biochemists, employing a new technical approach, proved decisive in changing the way people thought about specificity. The men were Frederick Sanger and Erwin Chargaff. The new method was *chromatography*. By chromatography and related methods, notably including electrophoresis, vanishingly small amounts of virtually identical complex substances can be persuaded to separate themselves for identification and measurement as they migrate, in solution, along a sheet of filter paper, down a column of inert granules, or across a slab of hard gelatin, some traveling faster and farther than others because of differences in weight or solubility or electrical charge among molecules.

In the course of nearly a decade, working at the University of Cambridge beginning in the mid-forties, Sanger used the new techniques to determine the amino-acid sequences of the bovine insulin molecule. When he began with insulin, not even its molecular weight was correctly known. Sanger showed that the molecule is of two chains, linked by cross-bonds at certain locations, and that the sequence of each chain is unique and always the same, meaning that (barring rare mutations) every molecule of cow insulin is exactly like every other. These results came slowly. He first worked with fairly rough chemical methods to knock off and identify amino acids at the tips of the chains. He then learned that natural digestive enzymes will break a protein chain at specific places, each type of enzyme cleaving only the bonds between two particular amino acids. With enzymes and some radioactive labels, he was able to break the molecules into several overlapping sets of fragments, separate the fragments chromatographically (and find each group on the filter paper by its radioactive label), and then analyze each fragment into its components. Finally, he related the fragments to each other by their overlaps, as though solving a word puzzle, to get the sequences of the chains and the entire molecule.

Even as Sanger worked, the news slowly spread and its implications sank in. The conclusion was forceful by mid-1949, when he went to the annual symposium on quantitative biology at Cold Spring Harbor, Long Island, summertime headquarters of Delbrück and his crew. In a paper published on the first of June of that year, he was already able to say, "There appears to be no principle that defines the nature of the [amino-acid] residue occupying the particular position in a protein."[22] The sequence of amino acids is entirely and uniquely specified. No general law, no physical or chemical rule, governs their assembly.

Sequence specificity in the protein demands a specific instruction from the gene. In 1950, Erwin Chargaff, who five years earlier had begun using chromatographic methods to analyze nucleic acids, published a paper reviewing his results. He and his colleagues had analyzed the DNAs of a variety of different organisms. The four kinds of nucleotides appeared in DNA in proportions that were constant in a given species but varied widely from one species to another. Only in a few species did they approach the equal representation required by the tetranucleotide hypothesis. Chargaff's paper demolished the notion that DNA was a repetitive molecule, a "stupid substance" that could not carry specificity. He established that DNA, after all, could be as specific in sequence as proteins.[23]

Molecular biology and the structure of the gene. The first concern of molecular biology has always been to understand the nature and action of genes and their products in terms of their physical structures as assemblages of atoms. From the early 1940s, scientists of that small band who later called themselves molecular biologists began by attempting to analyze the transmission genetics of the simplest creatures they could find. Where Beadle and Tatum were using *Neurospora*, the group around Delbrück worked with a strain of the common intestinal bacterium *Escherichia coli* and a set of the viruses, called *bacteriophage*, that prey upon *E. coli*. Bacteria are one-celled creatures without cell nuclei. Together with certain algae, they are called *prokaryotes* (from Greek roots meaning "pre" and "kernel" or "nucleus"). Higher organisms, whose cells have nuclei, are *eukaryotes* ("good kernels"). The total of different biochemical steps *E. coli* carries out—a different enzyme for each—is about two thousand, 3 or 4 percent of the number performed by the person in whose gut the bacterium resides. Bacteria can reproduce, by doubling, every half-hour. Viruses are still smaller and simpler, with a dozen genes or less. They are not able to reproduce by themselves. Instead, a virus invades a host cell and diverts it to reproducing more virus. A bacteriophage can multiply a hundred-fold in twenty minutes—and burst forth, killing the cell. These rates of increase mean that when one or a few bacteria are seeded onto a petri dish—a flat, low-sided glass plate—of nutrient jelly, within a few hours the colony grows to a spot visible to the naked eye; when a bacteriophage is introduced to the culture its progeny soon betray their presence as they clear a disc of burst cells in the bacterial lawn. Molecular genetics has been low in budget, high in ingenuity.

In the early 1940s, though, bacteriologists generally supposed that bacteria had no genes: they were not thought part of Mendel's realm. One of Delbrück's first discoveries, made with Salvador Luria, demonstrated that bacteria mutate: their genetic nature was proved. At about the same time, independently in German-occupied Paris, a French microbiologist (and Resistance leader) named Jacques Monod made the same discovery by a different method.[24]

The three-dimensional structures of nucleic acids and proteins had to be approached through a technology developed in the physical sciences—X-ray crystallography. The method was invented in 1912, when it yielded the arrangement of the atoms of sodium and chlorine in crystals of common salt. Physical chemists used it through the 1920s and 1930s to get out the precise spatial arrangements of the atoms composing molecules of ever-greater size and complexity. By 1940, when war interrupted, several laboratories had turned the methods of X-ray crystallography on crystalline forms of proteins and of nucleic acids.

In 1951, James Watson, an American postdoctoral biologist trained in bacteriophage genetics in the Delbrück group, came to England, to the Cavendish Laboratory, the experimental physics laboratory in the University of Cambridge, where he joined a small unit that was trying to determine by X-ray methods the structures of protein molecules. There he met Francis Crick, a physicist beginning to work on biological questions. (Crick had been acutely aware of the significance of Sanger's earliest results on the amino-acid sequence of insulin.) Eighteen months later, in the spring of 1953, Watson and Crick built and published their model of the three-dimensional structure of DNA, the double helix.[25]

———————

Sexual recombination in bacteria. Laymen and historians alike tend to misunderstand the importance of the discovery of the structure of DNA. The structure has elegant explanatory force. Its discovery was a marvelous drama. But in the development of molecular biology another discovery, made during the several preceding years, was equally instrumental. This was the discovery of bacterial sex.

Even when the genetic nature of bacteria (and viruses) was proved, bacteria were understood to reproduce asexually, the cell splitting into two identical offspring. But then in 1946 Edward Tatum and Joshua Lederberg showed that bacteria sometimes do exchange genetic material directly, in a process they named *conjugation*. By 1952, William Hayes, in London, showed that what happens in conjugation is that one cell pipes a copy of its genes into the other. This transfer takes some ninety minutes. Meanwhile, in 1951, Lederberg and Norton Zinder showed that bacteria sometimes exchange genes by another, indirect method, which they termed *transduction;* here, a virus mediates, by snaring bits of DNA while multiplying in one bacterial cell and transporting the bacterial genes into the next cells it infects. Then in 1955 Elie Wollman and François Jacob, at the Institut Pasteur in Paris, showed that bacterial conjugation could be stopped at any moment, when only some part of the genes had been piped across. They did this by agitating the culture violently in a Waring blender, and called the experiment, of course, "coitus interruptus."[26]

Conjugation, transduction, and similar practices provided the bacterial equivalent of matings in fruit fly, maize, mouse, or man. Indeed, experimentally they are more powerful and precise than matings in higher organisms, for they enable geneticists to move small, selected portions of genetical material into a bacterium and then observe the outcome—seeing just how the inserted genes express themselves and how they are switched on and off. These methods have led to most of the major discoveries of molecular biology.

The idea of a map or sequence of genes was rapidly becoming more precise yet more complex. At the Institut Pasteur, Jacob and Wollman were able to clock the transfer of particular genes from donors of one strain of *E. coli* into recipients of another by breaking off conjugation after different lengths of time and then observing the recipients for various biochemical mutations they had not previously possessed. They found that in any given strain the map of genes was always the same, and they displayed the map as a dial marked off in ninety minutes.[27] The map of *E. coli*'s genes is a direct descendant of the first maps of *Drosophila*'s X chromosome, for it depends on breakages and recombinations in the chromosome, detectable through comparing the action of gene mutations with the wild type.

Meantime, the discovery of the structure of DNA magnified the gene. It was no longer a point on the chromosome but rather a length of base pairs—and, therefore, it has an internal sequence. Thus, among other things, the structure of DNA also implies a theory of mutation. A gene can mutate in several ways: by deletion or addition of one or more base pairs, or by substitution of one base pair for another. Further, breaks and crossings-over or other recombination events should hit not merely between genes but within them. In the mid-1950s, Seymour Benzer, then at Purdue, a physicist recently recruited to the Delbrück group, devised an experimental setup to map mutations within a short genetic region of a particular bacterial virus. With many hundreds of mutants and many thousands of individual experiments, Benzer over a five-year period mapped recombinations of genetic material that often distinguished between mutational changes that had taken place at adjacent base pairs.[28] The gene had a fine structure that could be mapped by classical genetic techniques cleverly adapted.

A new problem for genes: Structure or control? Late in the 1940s, Barbara McClintock, an eminent American geneticist (one of the first women elected to the National Academy of Sciences; president of the Genetics Society of America in 1945) who had worked with maize for two decades, began to see in the pedigrees of her corn plants indications of genetic elements that were not ordinary genes yet that influenced ordinary genes—and not the character expressed but rather the rate of expression. Genes were being turned on or off, amplified or diminished, by other genetic elements. She found inhibitors, inducers, and others, which she soon grouped together as "controlling elements." Some of these, she divined, move unpredictably from place to place on the chromosomes. She began to publish papers about controlling elements in the early 1950s,[29] but they were largely ignored. Many of her new ideas appeared impossibly bizarre, especially those jumping genes.

At that time, in Paris, François Jacob and Jacques Monod began a collaboration, working on the expression of genes in E. coli that enable the bacteria to utilize galactose, a complex sugar, by breaking the molecules up for energy and raw materials. Over several years they found that the bacteria require three different proteins to digest galactose. But the curious fact was that the bacteria don't make the necessary enzymes unless galactose happens to be available. This brought them to a question of more general significance.

Even a bacterium, and still more a cell of a higher organism, has the genes for many more functions than it carries out at any given time. Indeed, if the cell did do all those things all the time it would rapidly exhaust its energy supply. But how does the cell turn functions on or off just as needed? Jacob and Monod found that E. coli's three genes for galactose metabolism lie next to each other in a fixed sequence on the bacterial chromosome, and that certain other elements are right next door. The first of these is a master turn-on for the whole series of genes, which they called the operator. They found other control functions as well. They located a gene for still another function, not an enzyme but a molecule that actually sits on the DNA at a specific place. There it locks the sequence of genes off. They called this the repressor. When a galactose molecule appears, it plugs into another spot on the repressor, where it acts as a key: the repressor changes shape and drops off the DNA and the sequence of genes is read out. As the available galactose is digested, the key unplugs and the repressor re-attaches to the DNA and turns the gene sequence off again. Jacob and Monod christened the entire coordinated sequence of genes and controls on the chromosome the operon.[30]

These discoveries were of highest importance. They established the existence of genetic regulation—of control functions located on the chromosome, mappable, in the DNA sequence. Henceforth, genes that read out as enzymes or other proteins have been distinguished, where necessary, as *structural* genes, as opposed to the newfound regulatory genes and other control elements. Further, Jacob and Monod established the existence of proteins that have dual specificities; these molecules respond to the presence or absence of one chemical substance (say, galactose) by changing their three-dimensional shape and consequently becoming reactive with another, chemically unrelated substance (say, a binding site on the bacterial DNA). Monod and Jacob recognized other repressors controlling other functions in bacteria and viruses. They saw that some enzymes, and even carrier molecules like hemoglobin, have similar double specificity depending on an induced change of shape; they christened these *allosteric* proteins (from Greek roots meaning "variable shape"). Allosteric proteins indeed have many functions in the organism. Still further, regulation of metabolic pathways in response to varying conditions requires biochemical versions of what information theorists and computer engineers had begun to call *feedback loops*—that is, systems whose outcome or end product influences the continuing earlier stages of the process. Monod thought that the negative-feedback model provided by the galactose operon and repressor was so elegant it would prove universal, but the feedback loops in living systems have turned out to be of every imaginable kind.

Regulatory genes and other control loci are elements in the genetic material that have different and sometimes subtle functions, compared to structural genes; they create new complications for the mapping of genes.

The messenger and the code. The work in Paris on gene regulation also threw off essential clues to the way the cell converts the sequence of base pairs on the DNA into the correct sequence of amino acids of the protein chain. In the 1950s, electron microscopy had shown the insides of cells to be filled with minute but well-formed anatomical structures, including vast numbers of a complex molecular organ now termed the *ribosome*. Biochemists then demonstrated that ribosomes are the sites where proteins are assembled. They are tiny knitting machines that select amino acids from the cellular soup in the correct sequence and link them together. Jacob, Crick, and others put the final missing component of this mechanism into place when they realized that the rates at which genes are turned on and off show that an intermediary must carry the genetic instructions from the DNA to the ribosomes. This intermediary is, in fact, a length of DNA's chemical cousin, ribonucleic acid. RNA is also a string of bases, but in single strands. In different forms it carries out a variety of functions in the synthesis of proteins by cells. To read out the gene, a new RNA molecule assembles along the DNA, the RNA bases lining up by the complementary-base-pairing rules. This step is now called *transcription*, and the assembled long strand is called *messenger RNA*, or mRNA for short. The messenger is then picked up at the correct end by a ribosome, which acts as a reading head. This step is called *translation*. The steps in protein biosynthesis were clear by the summer of 1959 and were published soon after.[31]

The specificity of genes thus turned out to be radically different from the specificity of proteins. The mystery was now solved. The specificity of the gene is linear, one-dimensional. A structural gene is an instruction, a message, and it is written as a unique sequence of bases on the DNA. In contrast, the specificity of the protein molecule is conformational, three-dimensional. Be the protein an enzyme, a hormone, hemoglobin, a repressor, or whatever, its specificity results from the contours, pockets, and local distribution of charges over the surface which result when the completed chains of amino acids fold up into their precise final shape. But the one-dimensional sequence dictates the three-dimensional form. The protein chain, once correctly strung together, folds itself automatically—by virtue of the shapes and charges of that particular sequence of subunits—into its correct specific functioning conformation.

The problem of the code remained: What particular sequences of the four bases in fact specify each amino acid? The summer of 1960 saw the first break. Marshall Nirenberg and Heinrich Matthaei, two young scientists at the National Institutes of Health, made an artificial strand of RNA in which all the bases were of just one kind. They put that as a messenger into a solution complete with ribosomes, amino acids, and biochemicals and other cellular components necessary for protein formation. They got out a protein chain in which all the amino acids were of just one kind.[32] Over the next six years, artificial messenger sequences proved the way to determine the full dictionary of the genetic code.

A triplet of bases in a particular order along the nucleic-acid strand specifies an amino acid. Such a triplet is called a *codon*. The bases cytosine-cytosine-guanine (CCG) on the messenger, for example, cause the ribosome to add the amino acid proline at the next position in the growing protein chain, while the same bases in reverse order, the codon GCC, translate as alanine. The genetic code has 64 codons (any of four possible bases in the first position, in the second, and in the third) and a lot of redundancy. Three of the 64 codons specify no amino acid but are, instead, stop signals, signifying the end of the chain.

The distinction between structural genes and regulatory loci, and the elucidation of the genetic code, of course redefined the ideas of the sequence and of the map of genes. The sequence becomes simply the linear run of bases, or base pairs, defining codons and making up genes. The map becomes more complex, however—no longer the line of genes on the chromosome but the line of structural genes and control elements, identified and located, the significant points of mutation within each portion also identified and explained, together with an account of the relationships of all these aspects to one another.

———

The emergence of recombinant DNA. At the end of the 1960s the founders of molecular biology were cocky. In the preceding twenty years, they had erected a coherent and satisfying framework of explanation for the genetically controlled processes of the cell. These functioned, they said, through the dual, complementary structure of DNA; through the information encrypted on the DNA and the mechanisms by which it is read out to direct the building of the organism; and the feedback loops that control the biochemistry of the cell and the functioning of the genes themselves. The machinery thus revealed was characteristic of all life. At its center was the universal genetic code and those regulatory loops. Structure and function, anatomy and physiology, had been unified at the molecular level.

Or, at least, molecular biologists had achieved all this in outline, for the simplest one-celled organisms, the bacteria and other prokaryotes. An immense amount of detail remained to be filled in, but the founders were confident that the rest of bacterial molecular biology would yield to intelligent drudgery, with few surprises. In this—including a few surprises—they were correct.

So by 1970 molecular biologists were looking for the next fields ready, as they liked to put it, "to go molecular." The first new task, Francis Crick said at the time, was to ask "the fundamental classical questions about genes and their products all over again for higher organisms."[33] But this was by no means a routine matter. The molecular biology that distinguishes eukaryotes from bacteria is, in effect, the molecular biology by which a single fertilized ovum becomes the multi-celled creature of differentiated tissues. This has always been the most intractable problem in biology.

Baffled by the problem of doing molecular biology all over again for eukaryotes, the younger generation of scientists turned to what had worked before, to see whether they could adapt the familiar, successful methods of genetic analysis in single-celled creatures. Eukaryotic cells are far larger than bacteria—proportionately as a horse to a bumblebee. They have hundreds of times more genes, and 500-fold more DNA. Yet since all the cells in the earliest stages of an embryo have exactly the same gene set, the problem of development and differentiation can be restated as one of finding the controls that switch whole blocks of genes on and off, in the correct order. Only by such regulation can one imagine that of two adjacent cells in an early embryo one is directed to multiply many times to become a liver while the other multiplies and differentiates further to become a brain.

As a prerequisite, the new molecular biologists taught themselves to grow animal cells not as coherent tissues but as cultures of loose, independent cells, like bacteria. They tried using the viruses that infect animal cells to bring out and transport bits of genetic material, the way bacterial viruses do with bacteria. Especially interesting here were animal-tumor viruses, because they do not kill the cell but take over and redirect its biochemical machinery.

The scientists learned to their delight that they could keep right on using bacteria, putting segments of DNA from eukaryotes into them. Bacteria have a single chromosome, but typically carry tiny extra rings of DNA, with a few genes on them, called *plasmids*. They pass plasmids among themselves. Biologists learned to add short stretches of DNA, a gene or two, to bacterial plasmids and reintroduce them into the bacteria; the bacteria then can multiply normally with the foreign DNA multiplying in step. Thus, starting with a few copies of a segment of genetic material from a higher organism, a biologist can grow bulk quantities of those genes. Molecular biologists took over from botanists the term *clone*, meaning the multiple identical descendants of a single cell, to describe such a culture of bacteria grown up to multiply transplanted genes. By still other tricks, the necessary control functions can be moved with the foreign DNA, or the bacteria's own control mechanisms can be taken over, so that the cloned genes can be got to function in the bacteria—transcribed, translated, and making the protein end-product.

Crucial to these methods was the discovery of certain enzymes that can manipulate nucleic acids. The first of these is the class of enzymes that copy nucleic acids, by stringing together nucleotides to form the new complementary strand when the double helix of DNA replicates or by transcribing DNA into messenger RNA. These are called *polymerases*, and cells of various organisms utilize a number of them, DNA polymerases and RNA polymerases. An important small subset of these comprises enzymes that transcribe in the reverse direction, making a DNA copy from an RNA message. Many viruses that infect higher organisms, especially animal-tumor viruses, carry their genes as a strand of RNA rather than DNA; this they inject into a cell, infecting it. Biologists had thought the RNA was copied directly into RNA as the virus forced the cell's machinery to make more virus. But then in 1970 Howard Temin and David Baltimore, publishing independently, showed that many such viruses, once their RNA is in the host cell, transcribe it into DNA that is then integrated into the host's DNA. An enzyme does this, called *reverse transcriptase*.[34] Using such an enzyme, a biologist can start not with a DNA gene but with a strand of messenger RNA, use an RNA virus to put it into the cell, read the message back into DNA, grow up the clone, and then get the gene expressed to see what it does.

Others, in many different laboratories, in a series of small discoveries of massive cumulative import, found enzymes that break or repair nucleic acids. Bacteria turn out to have protective mechanisms that disable alien DNA, as when a virus invades. These include *restriction enzymes* (called that originally because they restrict the range of hosts in which a virus can survive), which act like submicroscopic wire cutters to snip DNA between specific sequences of bases. Cells have other self-protective enzymes with which they repair their own DNA when it breaks or is miscopied.

In the early years, at least, the aim was always to find ways to move small, identified segments of genetic material from one cell to another, and then to see what difference the inserted genes made. Often, for analytical convenience, the move was from a species that was poorly understood into a different species, typically bacterial, that was well known. Yet rather suddenly, by the end of 1973, molecular biologists realized that they had assembled a remarkable tool kit. They gave this set of techniques the name *recombinant DNA*. Journalists also call it *genetic engineering*.

By such means, for example, although the practice is still tricky, large quantities of rare biochemical substances can be grown. Among early successes were production, in bacterial cultures, of human growth hormone and human insulin, both potentially valuable in medicine and lucrative to their makers. Starting with such methods, also, several teams of researchers are now attempting to find ways to insert particular genes into the cells of infants suffering an inborn error of metabolism, actually to cure the genetic defect. Nonetheless, the technology of recombinant DNA was invented and is today being elaborated not primarily to make money, nor even to cure disease, but to open up that profound and heretofore intractable problem of development and differentiation—to find genes in higher organisms and learn what they do.

The genetics of eukaryotes is full of surprises. For example, the genome of a rat pup or a child is unexpectedly fluid. Genes interact with one another, affecting characters all across the field of development, more intensely and intricately than was ever surmised. One most unsettling discovery was that in higher organisms many genes are interrupted, in the sequence on the DNA, by stretches of bases that have no relation to the gene, that have no known function. These are termed *intervening sequences,* or *introns;* the stretches of DNA that are part of the structural gene are called *exons* (Figure 5). The phenomenon complicates the behavior of messenger RNA. After the gene is transcribed, the mRNA must be edited, the introns snipped out, to form the messenger. Only then can the mRNA strand pass through the nuclear membrane

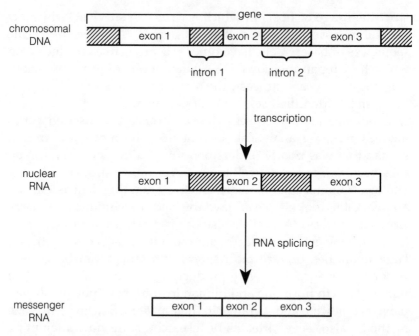

Figure 5 The genetic material of higher organisms, the DNA of the chromosomes, occurs in very long strings, in which some regions carry instructions for assembling protein chains while other regions have other functions—or no function yet known. In many genes the DNA sequence that codes for proteins may be interrupted by stretches of noncoding DNA. The interrupting stretches are called *introns*, while the coding regions are called *exons*. In the cell nucleus, the stretch of DNA that includes all the exons and introns of the gene is initially transcribed into a complementary RNA copy (RNA is a close chemical relative of DNA). This transcript is called *nuclear RNA*, or *nRNA* for short. In a second step, introns are removed from nRNA by a process called RNA splicing; the resulting edited sequence is called *messenger RNA*, or *mRNA*. The mRNA is able to leave the nucleus for the cytoplasm or outer portion of the cell, where it encounters cellular bodies called *ribosomes*, which are in effect microscopic machines for producing protein chains. Pulled along a ribosome, the mRNA, which now carries the gene's instructions, dictates the specific sequence in which the subunits of the protein chain are to be strung together.

to the cytoplasm, where ribosomes and the rest of the translation machinery reside. The controls on eukaryotic transcription and editing are by no means fully understood. Indeed, control mechanisms in higher organisms are turning out to be unexpectedly different from those in bacteria, and far more diverse. A disappointment was that the elegance of bacterial operons and repressors has no obvious analogue in higher organisms. More surprises are surely in store.

The sequencing of nucleic acids. After solving the amino-acid sequences of the insulin molecule, Fred Sanger turned to the problem of sequencing nucleic acids. This was technically more difficult by a couple of orders of magnitude. When Sanger began with nucleic acids, he had no way to secure a supply of molecules that he could be confident were all alike. Again, the amino acids in proteins are twenty, which means that sets of overlapping sequences are easy to distinguish and to puzzle together into longer stretches; but the bases in nucleic acids are only four. Sanger first tried single-stranded RNA, from viruses. As technology for manipulating DNA became available, he switched over.

He chose to work with the DNA of the bacteriophage known as ΦX174("phi-X 174")—a shrewd choice, for the virus is one of the tiniest, with few genes and little DNA. Its DNA is not a double helix but a single strand, in the form of a circle. It has a total, we now know, of 5,375 nucleotides. In contrast, the two chains of the insulin molecule total 51 amino acids. Nonetheless, ΦX174 gave Sanger a plentiful source of molecules all bearing parts of the one genome.

Sanger had sequenced protein by breaking the chains in a few fixed places with specific digestive enzymes, segregating the fragments chromatographically into homogeneous sets, breaking each fragment into subsets, and so on. Repeated degradation and separation of fragments and subfragments of the ΦX174 DNA, though, demanded immense labor yet often produced materials too scant to work with.

His solution was elegant. Instead of breaking DNA down, he grew it. A supply of identical single-stranded DNA with a known starting point can be replicated in a solution that contains the necessary components including free nucleotides. When one adds a DNA polymerase, the system will string together DNA that is complementary, by the base-pairing rules, to the original. Sanger set up such systems four at a time, manipulating each by supplying DNA strands of all different lengths, together with radioactive nucleotides—but in each case leaving out one of the four sorts of nucleotides. In each solution, the polymerase then continues every strand; when the strand reaches a place where the ΦX sequence calls for the missing nucleotide, growth stops. If the missing nucleotide is A, the result is a mixture where every chain, whatever its length, requires A as its next base. Similar incubations minus T, minus G, and minus C produce, in all, a set of four samples, each a mixture of lengths where the identity of the next base, beyond the tip of the strand, is known.

These polymers of different lengths can then be separated by electrophoresis, a variant of the basic idea of chromatography, in which a weak electric current causes the molecules in solution to migrate across a slab of gelatin, smaller chains moving faster and so farther. Gel electrophoresis is sensitive enough to discriminate pretty accurately between small groups of molecules differing in length by as little as one nucleotide. The result is a series of bands, one for each strand length, detectable by their radioactivity when a sheet of photographic paper is laid on the slab. When the four samples are fractionated by electrophoresis in parallel on the same gel, the result is "a pattern of bands" in not one but four tracks, from which, Sanger said in 1975, "under ideal conditions, the nucleotide sequence can be read off." A baffling mystery was reduced to a graphic chart.[35]

Two years later, Sanger made a neat simplification. Normal nucleotides have a growth point, a site on the molecule where the polymerase attaches the next of the chain. Artificial nucleotides can be made where that growth point is blocked. These are called *dideoxy* nucleotides. When the polymerase attaches such a nucleotide to a growing chain, growth of that chain then ceases. Sanger made dideoxy nucleotides his radioactive labels. The DNA to be sequenced he then incubated in a solution of the polymerase and all four nucleotides in their natural forms, adding one in the radioactive, dideoxy form. In the solution containing dideoxyadenine, after a while every polymer came to a stop bearing a radioactive adenine at its tip, and similarly with the other three incubations. Electrophoresis then proceeded as before. (See Figure 6.)[36]

During the same period, Walter Gilbert and Allan M. Maxam, at Harvard, devised a method for sequencing DNA using chemicals rather than enzymes, but with comparable simplicity and directness. Since then, the two methods have been standardized, speeded up, in large part automated.[37]

The anatomy of the human genome. The genetics of man presents obstacles both biological and ethical that have required develop-

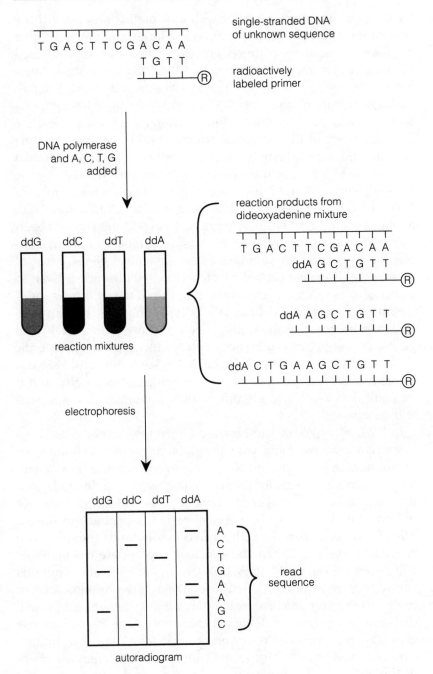

Figure 6 Enzymatic DNA sequencing using dideoxy nucleotides as radioactive labels: Frederick Sanger's technique.

ment of an armamentarium of technical methods beyond those needed for most other studies. Even in a large family, the number of children is small for pedigree analysis. To be sure, sophisticated statistical techniques were developed to wring more significance out of the meager data, yet the practical results were limited. Humans are among the most long-lived mammals, with a generation time exceeded by only a few creatures, like sea turtles; the life expectancy of the geneticist embraces no more than a couple of generations of human subjects. Geneticists may not conduct human breeding experiments—yet suppose you found the gene for longevity and tried transplanting it into a human embryo. How could you expect to publish the results? Until about 1970, human genetics had hardly progressed beyond the sort of observation available to Garrod. An increasing number of human characters were known to be inherited as Mendelian genes. Most of these were disorders caused by biochemical deficiencies. Even to associate these in linkage groups, other than the group on the X chromosome, was difficult: as late as 1967, only nine human linkages had been established, and seven of those were merely two genes. To assign even such rudimentary linkage groups to specific human chromosomes was impossible. Indeed, not until 1956 did cytologists get the number of human chromosomes right, and it was still not possible to identify particular human chromosomes with assurance.

In 1967, Mary Weiss and Howard Green took a crucial first step in human gene mapping with the publication of a technique for fusing human cells and mouse cells growing together in one culture. This is *somatic-cell hybridization*. The technique depends on a mammalian-tumor virus called Sendai virus. Added to a culture of human and mouse cells, a Sendai particle will attach to several cells; far smaller than the cells, it necessarily holds them in close contact. Weiss and Green showed that appropriate biochemicals will soften the cell walls; some cells then fuse; some of the hybrids survive and multiply in the culture. Though the hybrids tend to throw off the human chromosomes as they go through successive divisions, lines can be stabilized that reproduce the full mouse genome—on twenty chromosomes—and one or a few human chromosomes as well. Mouse and human chromosomes are easily distinguished under the microscope; the protein products of an increasing number of human genes are known, of course, and can be detected in the biochemical production of hybrid cells. In a further refinement, geneticists found that by using X rays on a culture of hybrid cells carrying a single human chromosome, they can break off pieces from the two arms of that chromosome, creating a dozen or more different cell lines each carrying a fragmentary human chromosome of a different length than the others. Powerful mapping deductions became possible. The technique was obviously of immense promise. The geneticist Guido Pontecorvo called it "parasexual." J. B. S. Haldane called it "an alternative to sex."[38]

In 1970, Torbjörn Caspersson, L. Zech, and colleagues, in Sweden, published the first method for staining human or other mammalian chromosomes in such a way that banding patterns appear, like those Painter found in the giant chromosomes of fruit flies nearly forty years earlier. Caspersson used fluorescent dyes, which require an ultraviolet microscope. Within a year, several other laboratories came up with staining methods that do not require special instrumentation. The banding technique most widespread today uses a dye called giemsa.[39]

These methods revealed a physical anatomy of the human chromosome set. They brought out patterns of bands, thick and thin, that make each chromosome unambiguously identifiable. One can now pair them up by matching stripes and distinguish them from others in the cell—the two of chromosome 6, say, obviously different from the two of chromosome 7, and so on. One can pick out any particular chromosome—11, say—in cells from different individuals. One can spot changes in individual chromosomes—inversions, translocations, deletions, duplications—that alter the banding pattern. One can tell exactly which human chromosomes remain in a stable line of hybrid mouse-human cells.

Somatic-cell hybridization took off when banding became available; indeed, the first human-gene mapping conference, in 1973, was based for the most part on the rapid development of mapping by somatic-cell hybridization. Geneticists began to assign genes to specific chromosomes and even to subregions of the chromosomes—sometimes to specific bands. To do this, they used as the human component in a cell-fusion experiment a cell line in which the usual chromosomes were altered in some way, say by reciprocal translocations. The techniques, though complex in detail, were in concept the direct elaboration of the classical gene mapping of

Morgan, Sturtevant, and Painter. Banding and somatic-cell hybridization made linkage and crossover studies fruitful with human genes.[40]

The anatomy of genetic material is exploited at a finer scale, at the level of the molecular sequences of DNA itself, by a method developed in the late 1970s that combines close analysis of pedigrees with basic tools of genetic engineering. Restriction enzymes of high specificity became available, enzymes that cut DNA precisely at positions defined by longer sequences of bases—for example, at the downstream end of every sequence ATTGTCA and nowhere else. Such an enzyme, as we've noted, breaks a given sample of DNA into fragments of varying length. By 1978 and 1979, studies by David Botstein, then at the Massachusetts Institute of Technology, now at Stanford University, and others had begun to find that when a restriction enzyme is applied to DNA from different individuals, the resulting sets of fragments sometimes differ markedly from one person to the next. The reason had to be that the location of restriction sites—the specific sequences the particular enzyme recognizes—varies from one person to another, because of mutations in base sequences. This offered the chance to get at variation between individuals not at the level of the character or the enzyme, as in classical Mendelian genetics, but at the level of the genetic material. The odds would be against the variations hitting within a particular gene of interest; but several geneticists saw at once that variations anywhere in the neighborhood of a gene could usefully serve as markers.[41]

Such variations in DNA are called *restriction fragment length polymorphisms*, or RFLPs (pronounced "riflips"), and their study is proving immensely productive. Several instances of great clinical promise have gained press attention recently, beginning in 1983 with publication of a large investigation of an extended family in Venezuela that is scourged by Huntington's chorea. A paper later in this volume discusses the Venezuelan study in detail: briefly, the geneticists demonstrated that family members with the disease showed a distinct and characteristic pattern of restriction fragment lengths. Some other relations in whom the disease had not begun its course also showed the characteristic pattern—and these eventually came down with it.[42] The same methods have yielded results with cystic fibrosis, adult polycystic kidney disease, Duchenne muscular dystrophy, and others. In many instances the method has located the gene on a specific chromosome. The next step, geneticists hope, will be to isolate the gene itself for such a disorder, sequence it, and thus identify the gene's product—and so arrive at its mode of action and thus at the possibility of a therapy. The approach has been called *reverse genetics*. Beyond the clinical applications, RFLPs are playing a major part in mapping the human genome.

Another big technical step was taken in 1980, with what is called *in situ hybridization* of genetic material—and this technique is altogether new, with no obvious precedent in classical genetics. The approach is direct, though it requires high technical precision. The trick begins with a length of DNA that contains a gene or even a short stretch of a gene. Often, for example, the geneticist will start with a piece of messenger RNA and use a system containing a reverse transcriptase to read that back into DNA. This is then cloned in bacteria, to grow up a suitable quantity.

For *in situ* hybridization, the geneticist grows the cloned DNA with a radioactive label. The resulting supply of labeled genetic material is called a probe. He then prepares a spread of the chromosomes of a human cell, treated so that the DNA of those chromosomes is opened up and can form base pairs with complementary sequences of DNA. He applies the probe to the chromosome spread. The probe will form hybrids with the cellular DNA but only at the place where the base sequences pair up. The geneticist then makes an autoradiograph of the spread, producing a sheet of film, a picture of the chromosomes with black dots due to the radioactivity at the points where the genes are located. The idea had been tried by many. It worked well with genes that are present on the chromosome in multiple copies. However, genes of greatest interest are likely to offer just one copy in the genome, and with them the amount of radioactivity was too low to show up. Then in 1980 Mary Harper added one ingredient to the recipe. She found that dextran, which is a carbohydrate, when put in with the probe forms a mesh or tangle that makes the molecules of the probe cluster; as a result, the hybridization site has enough radioactivity to show up clearly. In this way, in 1981, Harper and two colleagues mapped the gene for insulin. She prepared an autoradiograph that showed a black blob from the radioactivity at the tip of the short arm of chromosome 11.[43] That year, mapping by *in situ* hybridization became a standard method.

Thus today, whenever a gene gets cloned the geneticist has not completed its characterization until he determines which chromosome is carrying it and where. The gene is first run through a panel of somatic-cell hybrids. After it is found to be on a particular chromosome, the geneticist confirms the assignment and narrows it down to a specific part of that chromosome by *in situ* hybridization. One great advantage is that the gene does not have to be actively expressed in the cell. The geneticist finds the gene directly, not by its products.

The third wave of gene technology. In the mid- to late 1980s, molecular biologists developed a third wave of gene technology—tools essential for manipulating DNA in bulk, even as the idea of mapping and sequencing entire genomes came to be taken seriously. The first generation of methods for genetic engineering, those discovered in the early 1970s that gave recombinant DNA its name, included mammalian-cell culture; restriction enzymes and DNA-repair enzymes to cut and paste segments of DNA; reverse transcriptases to read RNAs back into DNA; bacterial plasmids and viral vectors to carry bits of DNA into cells; and cloning to grow genes and their products in bulk. DNA sequencing, somatic-cell hybridization, chromosome banding, restriction-fragment length polymorphisms, *in situ* hybridization—these were inventions of the late 1970s. Chief among the latest tools, to 1991, are DNA synthesis, pulsed-field gel electrophoresis, yeast artificial chromosomes, automated DNA fluorescence sequencing, and the polymerase chain reaction. These are not discoveries, exactly, but rather inventions providing the technological infrastructure for making discoveries.

A primary requirement for almost everything later was some way to synthesize strands of DNA of any desired base sequence. Early in the 1980s, Marvin Carruthers, at the University of Colorado, devised a method that begins with a single, known base, fixed to a minute solid bead, and then adds new bases, by chemical means, one by one, as desired (Figure 7). This technique will construct fragments of DNA of predetermined sequence from five to about seventy-five base pairs long. Carruthers and Leroy Hood, who was then at the California Institute of Technology, then invented instruments that make such fragments automatically.

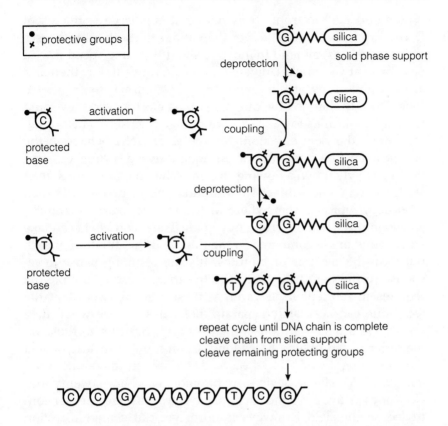

Figure 7 Chemical synthesis of DNA of any desired sequence is carried out in a small, glass column filled with minute beads of silica, which provide a solid, inert anchor for the growing DNA chains. The technician begins by affixing to the beads one molecule each of the first base of the sequence—say, cytosine, C. The bases have been treated so that the site where the next base attaches is temporarily blocked by a small chemical protective group (indicated here by small dots); thus, no sequences of CC or CCC can form. (Protective groups also block reactive sites other than the extension site on the bases, as symbolized here by the small crosses attached to the bases.) The technician then pours chemicals through the column in an alternating sequence. The first step is a solution that removes and flushes away the protective groups on the extension sites, leaving those sites open. The second step is a solution of the next desired base—say, thymine, T—that has been chemically activated (shown here as the attachment of a triangle to the protected base) so that it will couple to the first base. These bases, too, have their extension sites blocked; they bond to the Cs on the beads, giving the sequence CT throughout the column but no CTT, CTTT, etc. The first step repeats, flushing away protective groups; then the second step, a solution of the next predetermined base with the extension site protected; and so on. Today the process is largely automated (see Figure 14). A computer directs the order in which the bases are added and ends the cycle when the DNA sequence is complete. Then the chains are chemically separated from the silica beads and the last set of protective groups is removed.

Standard gel electrophoresis does not sort large segments of DNA very well. Strands longer than about thirty thousand bases (thirty kilobases) all tend to move at about the same speed. David Schwartz, at Columbia University, and Charles Cantor, then also at Columbia, now at the University of California, found that if they abruptly changed the direction of the electric field they could force the molecules to pause and rearrange themselves before setting off in the new direction. Molecules of DNA behave in the electrical field like strands of an aquatic weed strung out and floating slowly down a flowing stream. When the flow stops, then starts again in a new direction, a strand bunches up on itself; then a portion, typically somewhere in the middle, starts migrating, and soon the strand is strung out again. The second field can take any desired angle compared to the first, and the duration of each pulse can be seconds or minutes. It's the resulting pause in response to the new field that does the trick: the time a molecule requires to reorient itself varies with size so precisely that the technique can accurately separate molecules by size when they are on the order of a thousand bases long, right up to those on the order of ten million bases long. Cantor and Schwartz named the technique *pulsed-field gel electrophoresis* when they announced it in 1984.[44] The latest devices are hexagonal and have twenty-four electrodes arrayed around them; the electrodes are computer controlled so the field is always uniform yet can change direction instantaneously.

To sequence such great lengths of DNA efficiently—significant portions of an entire mammalian chromosome or genome—requires technical means to clone up quantities of any such portion. For several decades, some biologists have found that they can do genetics with yeasts in a way that is transitional between bacteria and multi-celled creatures. Yeasts are eukaryotes that have seventeen chromosomes, more or less, depending on the species. But they are single cells and are simpler than most eukaryotes in other ways, as well. In 1987, Maynard Olson, at Washington University in St. Louis, and colleagues introduced a method whereby DNA from other sources can be united to a stripped-down nub of yeast chromosome, creating an artificial chromosome to be reintroduced into a yeast cell (Figure 8). There it divides in step with the cell's usual chromosomes, doubling

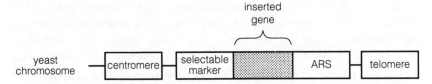

Figure 8 DNA of 100–1,000 kb in length may be cloned in yeast cells by inserting the fragments into a specially constructed yeast artificial chromosome (YAC). In order to replicate correctly, a YAC must have the following features of a natural yeast chromosome: the centromere, which controls the movement of the chromosome during cell division; the autonomously replicating sequence (ARS), which initiates DNA replication; and the telomere, the region that denotes the end of the chromosome. In addition, a selectable marker must be present so the YAC may be maintained in the yeast cells. The artificial chromosomes so constructed, including the sequence to be cloned, may then be inserted into a yeast strain for reproduction.

every couple of hours. The method is broadly analogous to the use of bacterial plasmids to multiply selected segments of DNA in bacteria. But where a genetic segment inserted in a bacterial plasmid is on the order of one hundred bases in length, yeast chromosomes are three or four hundred kilobases long. The cell's machinery can handle artificial chromosomes from one hundred kilobases up to ten times that. Olson called his invention, awkwardly enough, *yeast artificial chromosomes*, often abbreviated YACs.[45] He provided biologists a way to clone long chosen segments of DNA—a thousand to ten thousand times longer than the fragments that could be cloned in bacteria.

At that same time, Kary Mullis and others at Cetus Corporation, in Berkeley, California, invented a method, the *polymerase chain reaction*, for multiplying DNA sequences *in vitro*, that is, without needing to put the sequence into a cell. The method depends on the fact that when a solution of DNA is heated, the two strands of the double helix, held together only by the hydrogen bonds between complementary bases on opposite chains, come apart. When the solution is cooled (the term is *re-annealed*) the DNA forms double helices again—complementary sequences finding each other with very high precision. The polymerase chain reaction begins with a sample of DNA, any length, that is known to contain somewhere within it the desired genetic segment. It requires that two short DNA strands be synthesized, one complementary to a sequence known to be on one chain somewhere to the left of the target segment, the other matched to a sequence on the other chain and to the right. These probes (or primers), the sample of DNA, and plenty of nucleotides are put in solution. The method also requires a DNA polymerase—but a polymerase with the unusual quality that it resists heat. The solution and the polymerase at hand, the procedure is astoundingly simple in concept. The biologist first heats the solution to 95 degrees Celsius and keeps it there for two minutes; he then cools it to 30 degrees and adds the polymerase. After another two minutes, he repeats the cycle of heating and cooling. Each time the solution is hot, the double helices come apart. During each cool-down, the helices re-anneal—and the DNA probes attach to their complementary strands. The polymerase then goes to work, replicating only the portion of the DNA in between the two probes, including the target segment. In the next heating and cooling, everything comes apart again—but this time, on re-annealing, the solution contains twice as many copies of the DNA to which probes attach. After, say, twenty cycles, the portion of DNA including the target has been multiplied a million-fold (Figure 9).[46] The procedure is direct and fast—cloning without using cells. The polymerase chain reaction has been called the most revolutionary new technique in molecular biology in the 1980s. Cetus patented the process, and in the summer of 1991 sold the patent to Hoffman–La Roche, Inc. for $300 million.

Sequencing itself had to be speeded up enormously—without losing accuracy. This came with the invention of the automated DNA fluorescence sequencer, by Hood and a regiment of scientists and technicians at Caltech and Applied Biosystems, Inc. Hood and colleagues began where Sanger's sequencing had left off—with the use of those dideoxy nucleotides, which halt further elongation of a DNA strand. They found that they could label the primers, with which dideoxy sequencing begins, not radioactively but by attaching to each a small molecule of a fluorescent dye. They used four reaction mixtures, each with its primers labeled with one of four different fluorescent colors; they then supplied each reaction mixture with a different dideoxy nucleotide, and incubated the DNA to be sequenced, up to 450 or 500 base pairs. Thus, in one mixture every chain ended in dideoxyguanine and was labeled red, in the next every chain ended in dideoxyadenine

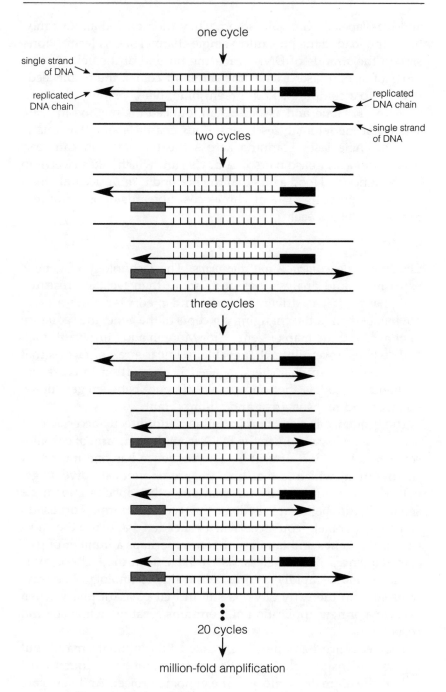

Figure 9 DNA replication by the polymerase chain reaction.

and was labeled blue, and so on. They then pooled all four mixtures and loaded the pool into a single channel of gel electrophoresis. As the strands of DNA reach the far end of the gel, in strict
order of size, a laser excites the different colors; the result feeds
directly to a computer. Red, green, blue, blue, yellow—that's G,
C, A, A, T. Hood and associates announced the method in 1986.
The latest model analyzes twenty-four channels simultaneously,
with a single laser scanning across the bottom, and can read
twelve thousand base pairs in a day's run, which takes twelve to
fourteen hours. Hood avers that the method is more accurate than
manual sequencing. The machines cost upwards of one hundred
thousand dollars each.[47]

Conclusion: the sequences and the maps. The technology of genetic
experiment and analysis has done more than facilitate research
and theory. It has driven research and theory—nowhere more
obviously than in the changing concepts of the gene, the sequence
of genes, and the map of genes. For Morgan and Sturtevant, map
and sequence were interchangeable, a linear array of factors that
determine known characters, spaced in proportion to crossover
frequencies. Today, the conceptual difference between genetic sequences and maps is great and all-important.

The sequence is now simply defined and directly accessible. It
is the order of base pairs in the DNA from the long arm of chromosome 1 to the short arm of Y. The sequence has one important
complication, which is that humans are enormously diverse genetically, more diverse by far than any other species. No single
sequence can be established as the type or norm. For useful
analyses, parallel sequences will be needed—and for the most
interesting genes and families of genes, perhaps a number of parallel sequences. Nonetheless, as we have seen the technology is
available and is rapidly getting better: this technology, like, say,
transistors in the early days, is still in that pleasant phase when
every major new application of it entrains great reductions in unit
costs.

The sequence has often been called "the ultimate map," but
this is a considerable semantic mistake—for the sequence is,
rather, the ultimate territory of the genome project. And maps are
never the territory mapped. They are at once less than the territory

and more. They are abstractions from the territory, leaving much out; and they are labeled, the features of the terrain identified. Every natural territory generates many different maps, and the differences spring from the purposes to which the maps are to be put. The political map differs from the road map which differs from the weather map or the map of natural resources as seen from mapping satellites, and so on. Yet all maps share one general characteristic: the abstracting and labeling establish a set of relationships. In short, by omitting much of the territory maps gain explanatory power. By being less than the sequences, the maps of the genome will be far more. They will locate and distinguish all the different components of the genome, establishing their functions and interactions, and thereby move us toward the eventual goal of understanding in the fullest useful detail the relationship between the sequence—the genome—and the living cell and organism.

The sequence by itself is dry and uninformative—an all-but-endless string of A, T, G, C, in no predictable order. The map is going to be more complex than we can yet imagine. The contrast is rooted in a single cause: the human genome was not designed, it evolved. The relationship of the map of genes to the living creature will have to be analyzed almost instance by instance.

Even at the simplest level, the locations of structural genes, mapping has progressed far enough to establish certain daunting negatives. For reasons rooted in the evolutionary history of mammals, genes for one organ or one system or one function may not confine themselves to some one chromosome. Genes for the enzymes for the several steps in a biochemical pathway are not to be found lying together in an operon as in the bacterial model; they may be on one chromosome—or on different chromosomes. Even in the simplest case where a protein is made up of several different chains, all too often the genes for the different chains are not on the same chromosome. Thus, the gene for the alpha chain of hemoglobin is located on chromosome 16, that for the beta chain on chromosome 11.

At the next level, the maps must account for regulation, for genetic sequences that are not structural but control elements. This will necessarily include whole classes of controls we have hardly glimpsed. The map must also sort out the problem of the immense amount of DNA that has no function we know, or that may be functionless. In doing so, it will have to relate the various segments of structural genes broken up by intervening sequences, and show how the genes get put back together again.

The penultimate level of the maps is the developmental: what we want the map to show us is what gets turned on and off in temporal order during the human creature's lifetime, to answer the intractable questions of embryology, of differentiation. Special versions of the developmental map will be clinical, dealing, say, with inborn errors of metabolism and their cure by gene therapy, or with the cancers and their prevention.

The ultimate map will extract from the sequences and from all the preceding levels of mapping, like the layers of an archeological dig, whatever suggestions about the evolution of man the steps of that evolution have not obliterated.

GENETICS, TECHNOLOGY, AND MEDICINE

II

WALTER GILBERT

A Vision of the Grail

3

The genome project is not just an isolated effort on the part of molecular biologists. It is a natural development of the current themes of biology as a whole. In the simplest sense, the idea of determining the sequence of the human genome is an attempt to define all of the genes that make up a human being. The information carried on the DNA, that genetic information passed down from our parents, is the most fundamental property of the body. To work out our DNA sequence is to achieve a historic step forward in knowledge. Even after we have made that step, we will still need to refer back to the sequence, to try to unravel its secrets more and more completely. But there is no more basic or more fundamental information that could be available.

The DNA sequence has a simple numerical expression: it is composed of three billion base pairs. That is enough information to code for about 100,000 to 300,000 genes, each gene being a region of DNA that can specify a protein or some other structure that carries out a function in the organism. Nobody knows how many genes are really involved, because we do not know the average size of a gene in the human body. Our estimate of 100,000 assumes that there are about 30,000 base pairs per gene, which is a reasonably good guess. But many genes are only 10,000 base pairs long, so perhaps there are as many as 300,000. Many of the most interesting of those genes have multiple RNA splicing patterns; that is, the messenger RNAs transcribed from a single gene may splice together different parts of the DNA sequence of the gene. The functions of these patterns must be understood in order to study an individual human gene. So saying that a human is made up of 100,000 genes underestimates the complexity of the human being, because many of those genes may encode ten or twenty different functions in different tissues.

The three billion base pairs of the human genome contain the amount of information included in a thousand thousand-page telephone books. What we hope to learn by working out the sequence of these base pairs is a list of all the genes that make up a human being. This information raises three striking questions about the nature of humans. The first is the question asked by developmental biology: how does a human being develop from an egg? The best way of studying the beginnings of animal development is to study model systems, such as the worm or the fruit fly; modeling of this sort is a dramatic and central part of modern biology. The second question is: what actually specifies the human organism? What makes us human? This is what medical science is about—the *specific* ways in which we are different from animals. The third question one might ask is: how do we differ from one another? This is the question of population biology—the *variation* of humankind across the species. These three questions are posed in order of increasing complexity.

The human genome project answers the second question—not the third. It is directed toward a molecular biologist's view of a species rather than a population biologist's view. The latter views a species as the envelope of all possible variants that can breed together; the importance of that envelope is that different aspects of a species population will be drawn forth if you change the environment. Molecular biologists generally view the species as a single entity, sharply defined by a set of genes and a set of functions that makes up that entity. Both of these ways of viewing humans are correct and not wholly antithetical. One emphasizes the variation that evolution works upon, while the other emphasizes the essential underlying features that define a species. The population geneticist, or the classical biologist, in defining the species, can point to a type specimen, an organism, and say that it exemplifies the species. The molecular biologist's view is that this organism is defined by its DNA. That DNA molecule can be sequenced to reveal the essential information that defines the type organism and hence the species.

Can we understand all the genes that make up a human being? Could we understand all their interactions and their differences across our species? These questions concern the totality of biology, and they are far beyond the human genome project. The human genome project cannot answer all of those questions, but it develops the human sequence as a research tool. The genome project also will determine the sequences of the genomes of simple model organisms. Together, the human genome sequence and those of the model organisms will be powerful tools for all biologists to use to approach a variety of fundamental questions.

The problem of working out the human genome can be broken up into three phases requiring inputs that differ by orders of magnitude. First the DNA itself—two meters in length—must be broken into ordered smaller fragments, a process called *physical mapping*. The best estimates of how long the mapping process should take are on the order of a hundred person-years. The second phase—actually determining the sequence of all the base pairs of all the chromosomes—will take three thousand to ten thousand person-years. The third phase—understanding all the genes— will be the problem of biology throughout the next century: about a million person-years, a hundred years of research for the world.

In addition to determining base-pair sequences, two types of maps of the genome will be made by the project (see Figure 10). Genetic maps will trace the inheritance of DNA regions in human populations and connect specific DNA regions to disease. Physical maps will provide DNA research material. The first effort to produce a complete genetic map of the human chromosomes was published a few years ago. It consists of approximately 150 polymorphic markers distributed along each of the human chromosomes; the markers are separated by about 20 megabases (20 million base pairs), about 20 centimorgans apart. As this map becomes finer in resolution over the next few years, more and more diseases will be pinpointed on it.

Physical maps also come in two types. One is created by measuring the distances along each chromosome in terms of the sequences at which restriction enzymes cut. That provides an abstract distance map for the size of the chromosomes and some points within them. The second type of physical map is called a

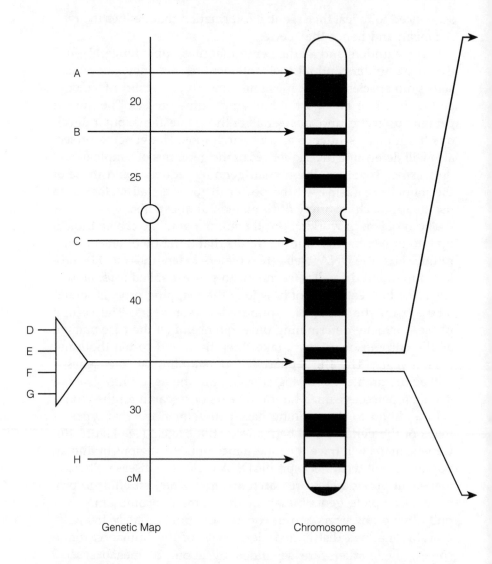

Genetic Map   Chromosome

Figure 10 The ultimate goal of the human genome project is a determination of the sequence of base pairs that make up every human chromosome. The *sequence* will be the most detailed map of the human genome. Less detailed maps have been constructed since the beginning of this century; the maps being made today are essential not only to the sequencing program but also to other research goals—such as finding the genes responsible for hereditary diseases like Huntington's chorea.

Genetic maps are representations of disease traits, physiological traits, or random polymorphisms that can be assigned to particular chromosomes and mapped relative to one another by following the transfer of alternative forms of these traits in families. This model of a genetic map shows the location of 8 markers (called here *A–H*), such

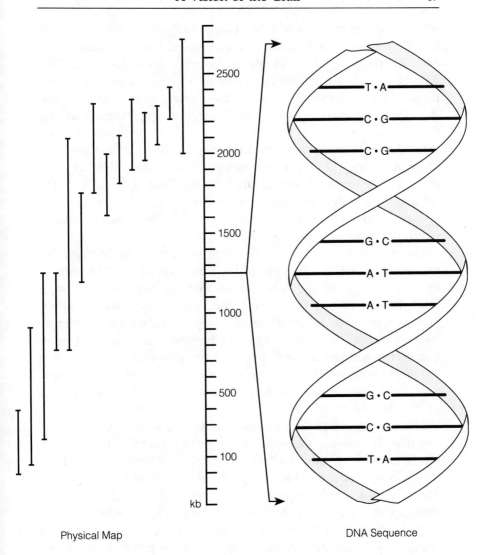

Physical Map   DNA Sequence

as genes or polymorphisms, along the chromosome. The triangle maps a small part of
the chromosome in greater detail, as may be needed for an intensively studied part of
a chromosome (perhaps a region suspected of being the location of a disease-causing
gene). Distances in genetic maps are measured in centimorgans (cM, about 1 million
base pairs).

Physical maps are not representations but overlapping collections of DNA fragments.
DNA is snipped into fragments by the action of restriction enzymes, and then cloned
and stored in a variety of forms—such as cosmids in bacteria or YACs in yeast. These
tiny fragments (measured in kilobases, kb) may then be analyzed by various means to
discover the base-by-base sequence of DNA.

cosmid map—it consists of DNA pieces each about fifty thousand bases in length, each cloned in a separate bacterial strain, and each overlapping other identified cosmids on either side. This map is actually a physical collection of bacteria, about a hundred thousand in number, containing the clones, known as cosmids, that span the entire genome. It is roughly equivalent to having the physical material in one's hand for each of the hundred thousand different human genes. Mapping by this method involves "fingerprinting" DNA pieces—recognizing sequence features that demonstrate that two cosmid DNAs share DNA sequences (or fail to) and hence overlap one another (or fail to). Assembled in a known overlap pattern, the cosmids provide the physical material to study the genome further.

The genetic map can be correlated with the cosmid map, because the genetic (or polymorphism) map specifies DNA regions of a known genetic distance apart that are detectable through DNA hybridization. Comparing the genetic and physical maps makes it possible to detect immediately whether any particular fragment or polymorphism lies on a particular cosmid. The correlation between the cosmid map and the genetic map provides a physical structure that makes it possible, if a polymorphism is near some disease gene, to determine where that disease gene lies on the cosmid map.

The simplest approach to sequencing the whole genome is to begin with a map—a cosmid map of the human being—and to sequence each of the cosmids separately. So, one strategy is to take a chromosome, which might be represented by a thousand cosmids, and simply sequence one cosmid after another to determine the entire structure of the chromosome. In the whole human sequence, the first 1 percent of the sequence—that is, the first thirty megabases (thirty million bases)—will probably be determined by sequencing regions near genes that are biologically or medically interesting. The next 10 percent (three hundred megabases) to be discovered will be the sequences of individual small chromosomes. The last 90 percent (2,700 megabases) will be everything else. As a result of further development of sequencing technology, these three tasks probably will each take a roughly equal amount of time.

Before 1976 it was essentially impossible to sequence DNA. When Allan M. Maxam and I worked out one of the first DNA regions in 1971, it took us two years to do twenty base pairs. A throughput of this magnitude is impractical in terms of working out a whole gene. In 1976 Fred Sanger, in England, and Maxam and I discovered two rapid ways of analyzing DNA sequence, which made it possible for a single individual to decipher about five thousand base pairs a year—equivalent to the structure of a small gene. Fifteen years later that rate is somewhere between ten thousand and a hundred thousand base pairs of DNA a year, making it relatively easy to work out the structure of single genes.

Most of the time is spent not in sequencing the gene but in preparing DNA fragments suitable for sequencing. Currently, these procedures involve converting the genome into smaller fragments of DNA that are then cloned into suitable recombinant DNA vectors. Typical clones contain inserts of DNA ranging in size from 15,000 base pairs to 50,000 base pairs. These must be broken down, or subcloned, into smaller DNA fragments about 300–1,000 base pairs long, fragments that are suitable for DNA sequencing. How much work is done in the DNA sequencing depends on the strategy. Two alternative strategies can be employed—ordered and shotgun. In the ordered strategy, the DNA is sequenced in a linear and consecutive manner. In the shotgun strategy, a large piece of DNA is randomly sheared into smaller fragments, the smaller fragments are randomly sequenced, and then these small strings of sequence are assembled by computers into the final sequence. The shotgun process requires that each stretch of DNA be sequenced on average five or six times. If the sequencing process was focused and the preparation of the clones and the DNA made simple and routine, probably even today sequencing speeds of the order of a million bases per person-year would be possible. By directed effort, speeds of about 10 million bases per person-year could ultimately be achieved.

In dollar terms, the cost to sequence DNA at a million bases per year per person would run about ten cents per base. At that rate, a working group of about 300 people would take ten years to do the entire genome. This would be using the best of today's technology or the technology that is immediately on the threshold. The rate of DNA sequencing has gone from virtually zero ten years ago to about 20 million bases of DNA sequence in 1990. In December 1990, 50 million bases of DNA were collected in data bases. The world rate of DNA sequencing is increasing very rapidly—accelerating at 60 percent per year. However, if one were to continue sequencing only at the current rate with no large, directed efforts, it would take several hundred years to sequence the human genome; only a few percent of all the genes have been sequenced at this point.

A million bases per year amounts to about five thousand base pairs per day. Two techniques can sequence at this rate. There are machines that can produce about ten thousand bases of rough sequence per day. A process called genomic sequencing can identify about thirty thousand base pairs per day of rough sequence. Because they both take a shotgun approach, these processes produce only about one-fifth as much final sequence.

The human genome project can be viewed as a purely technological effort to obtain the DNA sequence, put it into a computer data base, and study it. Today we know a variety of techniques for analyzing that DNA sequence. In fact, if one is presented with an arbitrary DNA sequence, there are techniques that will, broadly speaking, identify the gene structure with approximately 90 percent accuracy. What does that mean to biology? For example, if we were given a sequence data base of the human genome today, could we understand anything from it? The answer, I think, is yes; we could understand a tremendous amount. Today we learn a great deal about the functioning of genes by looking at the sequence of proteins they produce. For example, there is a set of about a hundred genes called oncogenes: each one identified as a DNA fragment, isolated from a tumor line or tumor cell, that will endow a normal cell with the ability to grow indefinitely. The normal form of each of these genes, a proto-oncogene, influences some aspect of cell growth. At first these genes were just a random collection of names with only an ability to control cell growth in common. But by looking at the protein sequences predicted from the DNA sequences, a lot of information about their functions can be inferred.

For example, we can recognize, by sequence comparison, that one oncogene is related to a receptor for a hormone. We can see that other oncogene products are slightly altered growth factors. We can recognize that some of the oncogene products bind to DNA and probably influence genes by determining the way the DNA is transcribed. Thus oncogenes may be divided into distinct categories by sequence, from which we infer their function. All of these insights, which at one level suggest experiments and biological relationships, appear immediately from sequence comparisons.

Another example is a characteristic amino-acid sequence motif called a "zinc finger," which was first recognized in a protein regulating the transcription of a particular gene. This motif is a string of amino acids with two cysteines and two histidines in a specific relationship that enables them to bind zinc. Scientists studying this short sequence of amino acids recognized that it denotes the ability of proteins containing this motif to bind to RNA or DNA. Then biologists realized that this sequence appears in a number of transcription factors. Many genes that control pattern formation in *Drosophila* specify proteins that have this same amino-acid motif, and one could therefore infer that these gene products function by producing proteins that bind to DNA. When a novel gene is sequenced, the presence of the amino-acid motif predicts that this new, and otherwise uncharacterized, gene product is a transcription factor. So, we *can* learn a great deal just by looking at sequence of genes or proteins.

Many of the genes in our bodies are members of large gene families. The ability to recognize gene-family relationships comes from an analysis of the gene sequences. There is now about a 50 percent probability that when we isolate a new gene we will see that it is related to something that has been previously identified. As the human genome project progresses, cross-relationships among gene products are going to become more evident, opening up the possibility of postulating functions for new genes and then proposing biological experiments to test these ideas.

We would like to know the three-dimensional structure of these gene products, the proteins, but that is not a problem for the human genome project. The structure-function problem is a very well defined problem in biology, and it is the crucial problem that underlies our understanding of proteins. We can go from the DNA sequence to the amino acid sequence by computation, but can we go from an amino-acid sequence to a three-dimensional structure and a function? So far the answer is no. But this is a well-posed theoretical problem with two approaches for its solution. One is to try to devise better computer programs to fold amino-acid sequences by energy calculations that seek stable configurations (the lowest-free-energy forms). With this approach protein structure could be obtained from first principles. The second approach is simply to have enough three-dimensional structures from known proteins to be able to recognize in a novel protein the small motifs that serve as building blocks and then to predict its structure as a melange of motifs whose structures are known. This very powerful approach is just now beginning to work and will probably lead to the solution of the protein-folding problem in the next five years.

If that happens, and if we also have a data base of DNA sequence, we can expect that we will be able to predict not only the protein sequences encoded by the genes but also the three-dimensional structures of the proteins. It is here that a theoretical biology will emerge. It will be a science of pattern recognition—extracting from the genome sequence the identity of human genes, their interrelationships, and their control elements. This information will be used to predict how the genes and their proteins function. A scientist will then use laboratory procedures to test these conjectures. Thus the future data base will enable us to approach human biology in an entirely new manner. Today we cannot identify which genes are expressed in the brain or in the heart. We know that the body has mechanisms to express one set of genes in the heart and a second set in the brain. These mechanisms define molecular addresses, the control elements on the DNA that direct the tissue-specific expression of the genes. In the future we will be able to use these molecular addresses to classify genes and organs. One can immediately think of global questions that can be asked if one has an appreciable fraction of the total information on human genes. But one cannot ask these questions today. Our present technology only allows us to go after genes one at a time, and to work out the relationships of the new genes to previously discovered genes.

The genome project is an application of scientific technology to produce a certain end—the information content of the genome. Science is going to change quite drastically over the next ten years in ways that we have not even begun to realize. A major change is occurring in the relationship of molecular biology to the rest of biology. Over the last decade it has become clear that molecular techniques are a powerful way of studying almost any question in biology, ranging from questions of development to those of evolution and population biology. All sorts of questions are now being studied by finding a gene and seeing what it does to the organism, or by deducing a feature about the pattern of inheritance.

As a successful science, molecular biology has become a set of cookbook techniques. Its very success is producing an odd sort of reaction: all of these wonderful techniques can simply be looked up in a handbook, and biologists seem to spend their time reading techniques and then cloning genes, or reading techniques and then sequencing DNA. Where is the biology? We are witnessing the last stage of the development of a technology and, as has happened a number of times before, many of the techniques of molecular biology will very soon leave the research laboratories entirely. We will purchase them externally as services; they will not be performed by research scientists.

Thirty or forty years ago students were taught how to blow glassware because they were expected to make their own apparatus. Today we buy plastics and throw them away, and we would not expect a student to know how to make a condenser out of glass. About fifteen to twenty years ago, when restriction enzymes had just been discovered, every graduate student in my laboratory made restriction enzymes. We wanted to work with DNA, we had to have those proteins, so every student made one or more restriction enzymes, knew how to purify a protein, and had to maintain the supply of associated materials. Now we buy the enzymes and even the bacterial strains needed for making clone libraries. Research scientists no longer have to grow bacteria themselves in order to generate competent hosts. Now they are beginning to buy ready-made clone libraries; soon they will supply a probe and order a specific clone from a commercial source. Over this next decade, DNA sequencing and many recombinant DNA techniques will move out of the laboratories. Five years from now, instead of struggling with cloning particular genes, we will simply buy the clone or the sequence. DNA sequencing will become centralized in very large-scale service organizations that will sequence DNA on demand. The science will have moved on to the problem of what a sequence *means*, what the gene actually does. Biologists who recognize and adapt to these changes will find that biology will continue to be active and stimulating, although it will be different.

Sometime in the 1990s, the world sequencing rate will reach about a billion bases a year, and we will have completed the human sequence as well as a variety of model sequences. At the end of the genome project, we will want to be able to identify all the genes that make up a human being. For example, we will compare the sequences of the human and the mouse and be able to determine the genes that define a mammal by this comparison, because the regions of DNA that code for protein are very well conserved over evolutionary time whereas the regions that do not have important functions are not well conserved. So by comparing a human to a primate, we will be able to identify the genes that encode the features of primates and distinguish them from other mammals. Then, by tweaking our computer programs, we will finally identify the regions of DNA that differ between the primate and the human—and understand those genes that make us uniquely human.

The possession of a genetic map and the DNA sequence of a human being will transform medicine. One immediate change, which will emerge over the next decade, will be a knowledge of genes that cause rare genetic disease. More important, however, will be the identification of genes for common diseases. When we have a detailed genetic map, we will be able to identify whole sets of genes that influence general aspects of how the body grows or how the body fails to function. We will find sets of genes for such conditions as heart disease, susceptibility to cancer, or high blood pressure. Along with many other common afflictions, these will turn out to have multiple genetic origins in populations, as will such mental conditions as schizophrenia, manic-depressive illness, and susceptibility to Alzheimer's disease. A whole variety of human susceptibilities will be recognized as having genetic origins.

One of the benefits of genetic mapping will be the ability to develop a medicine tailored to the individual: drugs without side effects. The side effects of drugs are often due to truly different responses of an individual to the chemical agent, because the variation between individuals is great enough to produce a different biochemistry. For example, there is a recessive gene in European populations that controls a sensitivity to medications for high blood pressure. The 5 percent of the population that shows this gene can use blood-pressure medications only in amounts one-hundredth or so of the normal dose. Genetic typing for such differences will produce novel medicines suited to specific patients.

DNA-based tests for all diseases, including the neurological ones, will result in subdivisions of those diseases into many categories, which will require different treatment programs. Consider badly defined entities such as mental diseases—for instance, schizophrenia. We will probably be able to identify a set of genes that lead to similar mental states. The ability to test for these genes will mean both a sharper diagnosis of the condition and a sharper prognosis of what happens to people who have it. Our knowledge of specific genes with brain-related functions—for example, a dozen genes that specify receptors for a single neurotransmitter—will affect treatment as well. The ability to isolate the multiplicity of receptors for a given neurotransmitter means that we can try to discover specific drugs that identify and affect in turn each of those receptors separately. These drugs will target specific receptors on specific cells. In general, knowledge of common features will lead to replacement medicines that will supply natural components of the body to enhance a natural function of the body.

Gene typing and genetic mapping could also have very strong social effects. However, the problems posed by the knowledge are not insurmountable and can be dealt with in a democratic society. First, we are going to have the ability to recognize defective genes in the embryo by simple techniques. That will mean a constant improvement and extension of prenatal diagnosis, which will lead to the elimination of much human misery. But better and more intrusive prenatal diagnosis will exacerbate the abortion discussion, with which society is wrestling at the moment.

What about genes in the workplace? What will happen if individuals who are susceptible to toxic chemicals can be identified as well as individuals who are resistant? Should society permit or resist genetic analysis of workers for environmental or work-associated susceptibilities? Medical insurance is already creating problems in this regard. We have examples of insurance companies arguing that a congenital illness is a previous condition and is therefore not covered by medical insurance. Should we permit that use of genetic analysis? Both of these examples suggest that our society adopt attitudes or pass laws that preserve the privacy of the individual. But will society do that? The problem of testing is already with us today in the matter of infection with the AIDS virus, HIV. Should one test for it? And what, if anything, should one do with that knowledge? These questions will be raised again and again as we gain more and deeper genetic knowledge.

Racism is a danger. Will the ability to analyze the genetic structure of individuals be used to try to define improved individuals and thus fan the flames of racism? Or will society recognize in a healthy fashion the worth of each individual? All human beings share an underlying DNA structure necessary for functioning. One of the ways of ensuring that that structure is not misconceived is implicit in the way in which the human genome project will be done. As we understand human DNA by sequencing the chromosomes of different individuals from around the world, we will put together a sequence that represents an amalgam of the underlying human structure. It will reflect our common humanity.

I think there will also be a change in our philosophical understanding of ourselves. Even though the human sequence is as long as a thousand thousand-page telephone books, which sounds like a great deal of information, in computer terms it is actually very little. Three billion bases of sequence can be put on a single compact disk (CD), and one will be able to pull a CD out of one's pocket and say, "Here is a human being; it's me!" But this will be difficult for humans. Not only do we look upon the human race as having tremendous variation; we look upon ourselves as having an infinite potential. To recognize that we are determined, in a certain sense, by a finite collection of information that is knowable will change our view of ourselves. It is the closing of an intellectual frontier, with which we will have to come to terms.

Over the next ten years, as a consequence of the advance of our biological knowledge, we will arrive at new understandings. We will understand *deeply* how we are assembled, dictated by our genetic information. Part of that understanding is, of course, to realize that genetic information does *not* dictate everything about us. We are not slaves of that information. We must see beyond a first reaction that we are the consequences of our genes; that we are guilty of a crime because our genes made us do it; or that we are noble because our genes made us so. This shallow genetic determinism is unwise and untrue. But society will have to wrestle with the question of how much of our makeup is dictated by the environment, how much is dictated by our genetics, and how much is dictated by our own will and determination.

One consequence of the human genome project is that we will see more and more clearly how connected all life really is. Research in early development tells us that the genes that form our bodies are similar to the genes that produce worms and fruit flies and every complicated organism. Those genes were created before the branching off of any of the higher organisms that are on the earth today. The data base of the human genome, coupled with our knowledge of the genetic makeup of model organisms, promises to reveal patterns of genes and to show us how we ourselves are embedded in the sweep of evolution that created our world.

CHARLES CANTOR

The Challenges to Technology and Informatics

4

Three numbers characterize the goals of the human genome project. The first number, 24, is an accurate count of the different chromosome types in a normal human. The second number, 3 billion, is an estimate of the total base pairs in the human DNA sequence; it was originally a guess, but it is turning out to be a remarkably good guess. The cost of this project scales up roughly in proportion to the square of that number, so if it were 6 billion base pairs we would be in trouble; but 3 billion is probably right to within 5 or 10 percent. The third figure, 100,000, is said to be the total number of human genes, but, so far as I can tell, it has basically been pulled from thin air.

The entire human genome can be seen at the level of the light microscope, in the form of the 46 human chromosomes. Procedures used in clinical cytogenetics typically show a banding pattern on the chromosomes that contains roughly 600 bits of information, which is comparable to the level at which we currently can analyze a whole genome. With cytogenetic methods we can immediately spot any change in the genome that alters a few million base pairs. We can also tell, for example, that a particular individual is a male because there is one X and one Y chromosome and that, unfortunately, he has Down's syndrome because there are three copies of chromosome 21. Analysis at this level is terribly coarse. The human DNA sequence contains ten million times as much information as a chromosome banding pattern—a very large improvement. Thus, a way of summarizing the goal of the human genome project, which aims to specify that sequence, is to say that it will provide a view that is more detailed than a chromosomal one by ten-million-fold.

The human genome project must be approached in stages. We will make a series of maps to describe the genome in different ways and also at increasing levels of resolution. The first of these is the genetic map, which is essential to *all* studies of the genome, because it is the only map on which traits can be located that are defined as a phenotype, or characteristic feature, like schizophrenia, Alzheimer's disease, or cystic fibrosis. In today's human genetic map the average spacing between markers is roughly the size of a chromosome band, or ten million base pairs. A current goal of the human genome project is a genetic map with markers every two million base pairs. A second type of map is a restriction map, a map illustrating a set of ordered DNA fragments, which provides a much finer description. Today a number of such maps have been constructed that have an average spacing across significant regions of the genome of a million base pairs or better. It is relatively easy to make these maps, but it is relatively difficult to work with the resulting information.

Ultimately, before sequencing begins, one needs to have a still finer map. With today's technology, this map would be an ordered library of clones, stretches of human DNA measurable in kilobases ("kb," for short, meaning thousands of bases) of length, that can be multiplied by insertion into another organism. Whether the library consists of cosmids (inserts roughly 50 kb in length), or yeast artificial chromosomes (YACs—roughly 300 kb inserts), or bacteriophage clones (roughly 15 kb inserts), one *needs* to have the DNA in a form that can be sequenced directly. The most important of the maps, ultimately, is the DNA sequence, because it is the only map that affords a *ghost* of a chance of interpreting genes directly in terms of function.

The goals of the project thus span six or seven orders of magnitude in size, from the two-million-base-pair resolution of the genetic map to the one-base-pair resolution of the complete sequence. We know how to do all of these maps today—which is to say that the project could be done with current technology, but the cost would be unacceptably high. Thus, one of the major thrusts of the whole human genome project is to improve the mapping technology and, especially, the sequencing technology tremendously, a point I will return to again. Then, too, this set of objectives, to take just the mapmaking, is incredibly boring. Maps are just tools that are needed to find all the genes and to facilitate novel biological and medical studies.

Even if we had the complete sequence of the three billion base pairs in a human genome, however, we are not yet in a very good position to exploit it. We would be paralyzed for two reasons. First, today's ability actually to interpret DNA sequences in terms of structure or biological function is very limited (another point that I will return to later). Second, biologists are ultimately experimental scientists, and it is very difficult to do experiments on human beings. Since we humans are long-lived, one cannot study many generations; one cannot control matings; and ethical considerations severely restrict any type of human experimentation. All of these facts make human beings an extremely poor experimental animal, especially for genetic experimentation. Suppose you have identified a gene with a suspected function: if you wanted to prove its function, your experimental procedure of choice would be to destroy the gene in some organism and determine the resulting phenotype. This type of experiment cannot be done on a human being. It is absolutely necessary that the human genome project encompass a series of genome projects on experimental animals, like the mouse, fruit fly, and nematode (roundworm). Politicians and the public really do not want to hear very much about these experimental animals, so the titular description of the project stresses the *human*, but the genome project actually includes the genomes of these fellow travelers.

The largest continuous DNA sequence yet compiled is roughly a quarter of a million base ("mb," for megabase) pairs—about the size of the smallest known chromosomes. It is accordingly correct to say that entire chromosomes can be sequenced, but only the very small ones of certain yeasts. The largest object for which a complete, ordered library of DNA fragments exists is of the bacterium *Escherichia coli*. The *E. coli* library, which consists of 400 fragments, each preserved in a clonable form by inserting it into a different cloning vehicle (the bacterial virus lambda phage), was done by a single Japanese graduate student, Yuji Kohara (admittedly a rather special graduate student). The largest robust and complete restriction maps cover DNA sequences not much larger than the size of an entire yeast genome, about fifteen megabases. The smallest human chromosome is three times longer, yet the difficulty of mapping does not increase linearly with size but as the square of the size, which means that the range of difficulty in going from yeast to human DNA and from the smallest human chromosome to the biggest human chromosome is quite substantial. The increase in difficulty does not scale linearly because one ultimately spends most of one's time finding mistakes and correcting them, and mistakes multiply as the square of the number of pieces that need to be assembled.

Sequencing the human genome with its three billion base pairs will be an enormous task. Early in the genome project, the prospect of dealing with such very large numbers led its policymakers to decide to rely on a continually evolving technology. Taking this approach, the project during its first five years plans to invest very heavily in improving the technology of mapping and sequencing, which will ultimately reduce costs, and invest very little in large-scale DNA sequencing, which is currently very expensive. It is clear that if we do not improve the sequencing technology by an order of magnitude, the project will fizzle. In the second five years, the technical advance of the first five will pretty much have to be repeated; the methods have got to become more efficient and reliable by at least another factor of ten. By then, the genome will have been mapped and the sequencing can begin. In the last five years of the project, somewhere, somehow, the rest of the genome will be sequenced and all of the genes will likely be found.

At the moment, the genome project resembles many other large-scale efforts in their first phase: it is rather chaotic, marked by significant redundancy and disorganization. A few groups are doing very well; others, very poorly. Many different methods are being used for sequencing and for mapping. Although it is not now clear which are the best, during the next five years—the second phase of the project—it will be necessary to settle on a more limited set of approaches. In the last phase, when we will seek to complete the sequence, resources will probably have to be concentrated at a relatively small number of locations.

There are two general approaches to mapping. In a top-down approach, one takes a purified chromosome, cuts it into pieces, puts them in order, and analyzes each piece to obtain a finer map.

The process can be repeated with each piece *ad nauseum* until a sequence is obtained. In a bottom-up approach, one starts with an intentionally randomized set of DNA fragments or clones of the fragments, all deriving from an original long stretch of genome. The fragments are then fingerprinted—that is, identified by determining marks, such as patterns of base pairs—and pasted together into contiguous lengths at points of overlapping fingerprints. The result is a physical map of the original stretch of genome. The advantage of the bottom-up approach is that one actually possesses clones—DNA in a form that is replicatable and, hence, immortal. In the bottom-up approach, however, the maps are difficult to complete. With the top-down approach, one can make a finished map, at least in principle, but the DNA is not maintained in a form that is as convenient for further analysis.

At Berkeley, Cassandra Smith and I have been attempting to make a restriction map of chromosome 21, using mainly one restriction enzyme—it is called *Not* I—which cuts that chromosome less frequently than any other known restriction enzyme. Our results so far have been essentially confined to the long arm of chromosome 21, because it has been extremely difficult to find DNA probes that correspond to the short arm. As of May 1990 our map, a physical map, encompassed some 43 DNA pieces and totaled about 47 million DNA base pairs. It is far more detailed than the pattern bands seen on the chromosome in the light microscope, but it is still not continuous; it is a broken map, and we do not yet know whether the breaks indicate pieces of missing DNA. We do know that chromosome 21 contains at least five million additional base pairs of DNA that are not on this map. We suspect that most of the missing DNA belongs to the short arm of the chromosome but we cannot yet prove the suspicion.

Even though after several years of effort we have mapped most of chromosome 21, much hard work lies ahead. The truth about mapping projects is that it is very easy to start them and very difficult to finish them. Perhaps 90 percent of the effort is expended in obtaining the last 10 percent of the map. The last 10 percent further challenges the scientist's tenacity because it yields progressively less information. The temptation is to publish a map that is 90 percent done and let it go at that. But it would be unfair to claim—as some have done—that such a partial map is "the map" of the chromosome. A complete map is one that is genuinely finished.

In typical mapping and sequencing projects, the strategy evolves from one designed to finish most of the map to another—we call it the end-game strategy—that will permit the completion of the map in a finite period of time. We have begun following an end-game strategy with chromosome 21. We know that we are missing just a small number of the chromosome's pieces. To find those pieces, we employ two powerful new strategies. The first, called the *polymerase chain reaction* (PCR), multiplies any DNA sequence a million-fold with the use of short, complementary DNA sequences *(primers)*, one for each DNA strand, at either end of the region to be multiplied. The second involves choosing as the PCR primers a highly repetitive human DNA sequence—it is present on average once every 5,000 bases—called *Alu*. Applying *Alu* PCR amplification to chromosome 21 yields a series of human DNA fragments, each of a unique length and bounded by *Alu* primers.

As a source of chromosome 21 we use a human-mouse hybrid cell that has been manipulated to contain this chromosome as its only human component. First the DNA of the cell is cut into large fragments with the enzymes *Not* I. Then these are fractionated by size. In each size fraction where we suspect there may be an unidentified human DNA fragment, the fragments are replicated by PCR using *Alu* primers. Any human DNA that is thus amplified will be located on that fragment, and it can be used to determine the location of the fragment by the standard methods used in making physical maps. The power of this approach is that instead of picking probes at random, hoping one will land on the putative fragment of interest, we can start with the fragment itself and generate a probe from it.

At the finish, we will have only a restriction map, which is not useful to the broad community of investigators who want access to the DNA of chromosome 21. However, that DNA is likely to be efficiently obtainable. Our PCR amplifications of chromosome 21 DNA fragments could be used to identify clones of corresponding fragments from the same chromosome that have been inserted into yeast artificial chromosomes. Since the ordering of the *Not* I fragments is known, the identifications would largely determine the order of the random fragments that have been PCR-amplified. This type of ordering is a hybrid of the top-down and bottom-up approaches because although we use a random library we order it from the top down—by comparison with the YAC fragments—instead of from the bottom up.

This approach has in principle been made both practical and attractive by a dramatic change in the technology of YAC production that occurred in the early months of 1990. At that time, scientists in Paris managed to construct YAC libraries of the human genome that averaged 400 to 500 kilobases rather than 100 to 200 kilobases in size. Since the average *Not* I DNA fragment is a million base pairs, it could be spanned by just three or four YACs and the problem of ordering these YACs would become very simple.

The sudden advent of that prospect spotlights what has been an ongoing problem, and a challenge, in the human genome project—namely, the methodological opportunities change every six months. The fast pace frustrates attempts to fix on any particular technology, but it also excites anticipation that some marvelous new techniques may lie just around the corner.

Let me peer through a very cloudy crystal ball and look ahead fifteen years to delineate the possible changes that the completion of the human genome project will bring about in medicine, technology, instrumentation, biology, informatics, and biotechnology. I am optimistic that we will ultimately find all one hundred thousand or so human genes, most before the completion of the project, some afterward. Long before completion, we will have very accurate diagnostics for most inherited diseases. Indeed, the diagnostics will be accurate enough ultimately to understand the functions of each gene involved even in polygenic diseases. Genetic diagnostics will be followed by therapeutic and preventive measures. The design of both will be aided by the systematic construction of animal models for human diseases, either by modifying the animals' own genes to produce the disease or by inserting the human disease genes into their germ lines in place of their own normal genes. But while diagnostics based on newly discovered genes will come very quickly—it is already here in the case of cystic fibrosis—therapeutic improvements will come more slowly. It is sobering to remember that, although the molecular defect at the base of sickle cell anemia has been known for years, that knowledge has not yet brought about significant therapeutic benefits. Therapeutic and preventive benefits arising from the discovery of the genes for a disease could lag twenty to fifty years behind the diagnostics.

Both the public at large and much of the scientific community tends to think that inherited diseases are rare, that "only other people are at risk for it." The belief is a mistaken one. True enough, most *single-gene* diseases are relatively rare. Huntington's disease, which gets a lot of publicity, is very rare; so is familial Alzheimer's disease. Even cystic fibrosis, the most common genetically recessive disease, is not very frequent. Yet some single-gene diseases—for example, familial hypercholesterolemia—are common enough so that most people know someone who suffers from one. Anyone who has died of a heart attack at age forty-five is likely to have been a victim of familial hypercholesterolemia. *Polygenic* disorders are much more common, however—probably because they represent many genes, a much larger target in the genome. Genes figure in familial polyposis, a common cause of colon cancer; schizophrenia; bipolar disease, manic depression; hypertension; coronary disease; alcoholism; diabetes; and obesity. These diseases are not purely genetic, but probably every reader of this chapter is at risk for at least one of them because of his or her genes.

In fifteen years, we will probably be able to apply a single multiplex test to fetuses in utero, babies at birth, or, in many cases, parental carriers, a test that will detect somewhere between 100 and 1,000 of the most common genetic diseases, disease predispositions, and genetic risk factors for environmental insults, drug dose responsiveness, and the like. We will be able to do this extensive fingerprint for any individual, but we will, at least initially, be unable to offer any help based on this information. Such impotence in the face of information exposes one of the serious social issues raised by the genome project. So long as no effective therapy can be offered, an individual's right to reject diagnosis must be protected. Already, today, even though a test for the presence of the Hungtington's gene is available, most people at risk do not want to be tested for it. As a rule, people prefer optimism to pessimism. To be told that you will come down with a disease for which there is no therapy is virtually to be robbed of hope. There must be no social stigma attached to anyone's choosing to reject genetic testing.

With appropriately polymorphic DNA probes, it is now possible to distinguish every individual on the earth, to identify individuals with high precision from a single hair or sperm cell. Once we understand the function of many normal genes—that is, how they generate physical features such as eye color, hair color, and details of facial physiognomy—it may well prove possible to extrapolate enough from a single hair to be able to draw a picture of the person. Leaving a hair anywhere could tempt others to invade the data bases and track a person into every room he or she ever enters. The example may be somewhat exaggerated, but it is no exaggeration to say that the storage of human genetic information in data bases will pose new threats to privacy. One of the agencies most interested in the genome project is the FBI, a technologically very capable organization.

———————

The enterprise that the genome project will change most is ordinary biological research, particularly how biologists deal with information. The project will yield quantities of information that will dwarf by orders of magnitude anything encountered before, so modes of dealing with data will have to change. Current methods of information handling will simply not work.

Unlike virtually any previous biological venture, the human genome project should, ideally, spend perhaps half its funds—one hundred million dollars annually, if the recommended funding level is achieved—on the development of techniques and technologies. In biology, the monies available for such activities have previously been small. Typically, grants to develop new analytical methods have been difficult to come by. But the genome project will foster dramatic changes in the technologies of biological research and emancipate much of that research from its routine, error-prone, and boring procedures. Change will be brought about by, for example, robotics and automation; robots are already being taught to perform tasks that lab workers find frustrating.

At Berkeley, we have deployed robotics to convert the growth patterns of bacteria, yeasts, or viruses on a plate from an essentially random pattern to an orderly one. The point is to keep better track of all the clones—almost a million clones in the case of a complete human clone library. The orderly pattern is also necessary to allow subsequent automated manipulation of the clones. Today, ordered clone sets are made by hand; someone picks and distributes the colonies one at a time. We have been trying to teach a programmed Hewlett-Packard robot to perform this task.

It has not yet mastered the job, but it has learned to pick colonies and arrange them. We hope eventually to have a fully automated system that will make an ordered library of any cloned DNA—an accomplishment that will relieve investigators of a great deal of monotony.

The genome project will also bring biology into the age of nanotechnology—the era in which we have the ability to detect and work with single molecules. Although it is currently unclear exactly what forms this technology will take, it will change biology dramatically. Finally, the project will force the direct coupling of experimentation and data-base development. Inefficiency and errors arise inevitably when someone must manually enter experimental data into a computer. I was surprised to learn recently that 20 percent of the data entered into the European Molecular Biology Laboratory data base is still entered by somebody who reads the relevant journals and types in the information. The process could and should be entirely automated, given that most journals are computer-typeset.

The human genome project already has available a variety of fairly sophisticated technologies that make it possible to map and sequence on a moderate scale. We can purify chromosomes by flow cytometry and large DNA fragments by pulsed-field gel electrophoresis; we can do automatic four-color DNA sequencing by fluorescence; we have robots that can perform sequencing reactions and handle clones; we can do automated nucleotide synthesis; and we can employ PCR, of course, to identify clones and, eventually, to amplify small DNA fragments for sequencing. None of these methods existed ten years ago, and it would be extremely presumptuous to assume that any of them will necessarily be in common use ten years from now. It is likely that, by the end of the genome project, fifteen years hence, the methods that will be widely employed will be different from those in use today.

For all their merits, current methods suffer handicaps, the most important of which is limited sample throughput, which slows mapping and sequencing. Major advances in sequencing could come from any one of six technological areas. (1) Scanning-tip microscopes move an extremely fine point across a molecule and feel its presence in a number of different ways. The process is quite analogous to someone reading Braille. In principle, their resolution is high enough to see individual atoms. (2) Single-molecule manipulations may be done by suspending a DNA molecule by one end, successively cutting single bases off the other, and detecting which is which as they are swept away from the DNA. (3) Sequencing by hybridization, an idea that seemed ludicrous until quite recently, involves reading the order of the DNA bases many at a time instead of one at a time. One analogy is reading by words instead of by letters. The difficulty is that there are many different words and one needs a specific reagent for each. (4) Mass spectrometry can measure the weight of DNA very accurately because the four different DNA bases have different weights. In principle, mass spectrometry could be used in a way analogous to fluorescent DNA sequencing, except that it is potentially much faster. (5) X-ray scattering provides information about the distances between atoms. If a sample of aligned DNA molecules was exposed to an X-ray source, one could in principle use the information about distances to reconstruct the sequence of the bases. (6) Single-molecule groove scanners would take advantage of the fact that enzymes like DNA polymerase or RNA polymerase can read the sequence of DNA very rapidly when they make a DNA or RNA copy from a DNA template. If we could figure out physical or chemical tricks to make these enzymes call out what they are reading as they move along the template, they would tell us the DNA sequence.

Even without these promising technologies, however, most laboratories are already generating map and sequence data faster than they can analyze the information. The problem is not that the data are so very difficult to analyze. It is that our current informatics capability is the weakest link in the mapping and sequencing efforts.

The genome project must be concerned with cost. In principle, large-scale sequencing could be done today by hiring 2,000 people, buying 2,000 automated fluorescent sequencers, and then proceeding to sequence the entire genome. But the best sequencing in the world today is estimated to cost a dollar a base, for finished DNA sequence, including the generation of the DNA samples, the sequencing process, and the assembly of the resulting data into finished DNA strings. This estimate is based on the facts that the total annual cost for a skilled sequencing person is $100,000 and that the best people in the world sequence about 100,000 bases a year—which works out to one dollar per base. A typical graduate student or postdoctoral fellow does about ten times worse, a cost of about ten dollars per base.

In ten years, we should be able to sequence at 10 percent of the current least cost—that is, at ten cents per base, or one million bases per person per year, for any sequence. In my judgment, this target is too pessimistic. It would allow us to grind out the human sequence, but it would not allow us to do any of the parallel studies in humans and other model systems that would enhance the interest of the sequence data. I believe that we should aim for a cost of a penny a base. To meet that cost objective for raw data, a skilled sequencer would have to sequence DNA at one base per second, every second, with no time off for lunch. To meet it for finished data, the rate would have to be roughly ten bases every second. Clearly, reducing costs to the penny-a-base level will require incredibly fast or multiply parallel equipment or both.

It should be obvious that in the early stages, when the cost per base is high, the genome project ought not to be sponsoring the sequencing of DNA that may, like junk DNA, be biologically uninteresting. Any pilot projects to sequence hundreds of thousands or millions of base pairs of DNA should be chosen so that the targets are biologically or medically rewarding and the data may be interpreted. Only when the cost per base drops significantly should sequencing reach to the presumptively barren regions of the genome.

Whatever the details of the scenario may be, in the next fifteen years we are going to generate an enormous amount of sequence data. The amount of data will be gigantic even if in 2005 the project is only half-completed. Today's biologists are rightly accused of being less computer literate than most other scientists. With so much data to hand, virtually all biologists will depend heavily on computers; certainly, any biologist who wishes to exploit the human genome data base will have to use them expertly. Laboratory notebooks will disappear because mapping and sequencing data cannot be conveniently stored in a hard-copy archive. Traditional data records will have to be replaced by something in electronic form, including images. For example, the effort at Berkeley has generated so many photographs of gels that we can no longer keep track of them in that form. All are being put into an image data base for access.

An analogy shows why genome data will be unmanageable manually. The complete human sequence, written out in type typically found in a telephone book, would occupy 200 volumes of the thousand-page Manhattan directory. Clearly, no one could scan by hand a data base of three billion entries. The ultimate human data base, one that would incorporate all human diversity, would be at least four orders of magnitude (roughly ten thousand times) bigger. It should be obvious that our approach to dealing with biological data must change.

Contrary to original expectations in the project, establishing a centralized genome data base will not be the optimal way to handle the new data. A data base is only as good as the people who maintain it. Sequencing is a tremendously error-prone technique; the accuracy of current sequencing may be no better than 99 percent; the sequences are constantly updated, the data constantly changing. As scientists are learning about that data, they are constantly annotating the data base. And no one will care about keeping track of the day-to-day changes in a data base of three billion items. As a result, I expect a trend to constructive balkanization, the establishment of hundreds or thousands of data bases. The caretakers will be individuals who want to understand and deal with a small region of a genome. Those data bases will be perfectly up-to-date; they will be accurate; they will be well annotated. The key challenge will be to integrate them so that they look and respond like a single virtual data base. To that end, we will need to create a new tool—a human genome workstation—a terminal at which one can address *all* data bases without having to know anything about their internal structure or the hardware on which they reside. Such a scheme may be a bit beyond the current state of the art in computer science, but it is both necessary and achievable.

The human genome project will inevitably transform opportunities in biological research. It has been said that understanding the function of a single gene is, on average, the work of a scientific lifetime. In fifteen years, we will know one hundred thousand genes—which can potentially occupy one hundred thousand scientific lifetimes. The number of molecular biologists is far smaller than that number. Thus, either we must become highly selective in choosing the most interesting genes to analyze, or the field of biology must undergo rapid expansion to take advantage of the identification of all human genes. Our society may not wish to support this expansion, but the medical arguments in its favor will be compelling. Then, too, we will want to do parallel studies on the genomes of other organisms. And once we have vastly more efficient methods for obtaining and manipulating very large quantities of DNA sequence data, it will be feasible to design projects to explore biodiversity at the molecular level. Technical developments will thus open up the possibility of other large-scale biology projects, especially in the areas of ecology and evolutionary biology.

C. THOMAS CASKEY

DNA-Based Medicine: Prevention and Therapy

5

The rapid advances in DNA-based technology seen in recent years provide a powerful set of tools for studying biological events and also promise a dramatic change in our practice of medicine. We have already attained an ability to examine the human body, from individual cells to nuclear DNA and patterns of gene expression, that would have been unthinkable only twenty years ago.

DNA-based technology has been rapidly applied to the study of the mechanisms of disease and to new drug production. Our ability to diagnose inherited disorders, such as sickle cell anemia, or acquired genetic diseases, such as neoplasia, would not be possible today without this technology, nor would we be able to produce therapeutic agents like insulin. Figure 11 illustrates how quickly basic scientific results have been taken up in the practice of medicine; the expansion of technology has accelerated throughout the 1970s and 1980s (see also Table 1). Today, DNA-based methods are routinely applied in the areas of surgery (transplantation), medicine (cancer), pediatrics (genetic diagnosis), and obstetrics/gynecology (prenatal diagnosis). The newest techniques are presently available only in academic centers, however, since relatively few DNA-based methods have been transferred to the medi-

The assistance of Belinda Rossiter and Elsa Perez in the preparation of this chapter is appreciated.

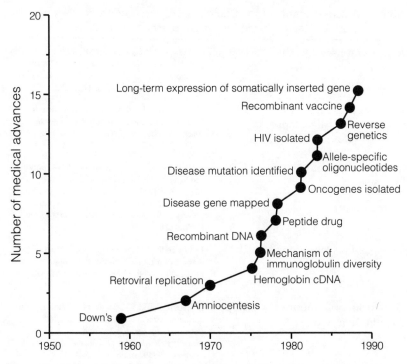

Figure 11   The number of biomedical advances made possible by DNA-based technology has increased rapidly over the last thirty years. The dates plotted on this graph were assigned according to the year of the initial report of success with a new procedure. See Table 1 for a brief statement of the nature of the advances noted in this figure and for references.

cal practitioner. If transfer is to occur, practitioners will need to be better educated and trained in genetics and biotechnology. Not only practitioners, but also consumers of medical care need to be made aware of the new directions in medicine that molecular biological methods are making possible, and we all need to think about the new medical management issues that will almost surely be generated in the near future.

Many leading scientists believe that the genome project will achieve a description of the human genetic map and DNA sequence in the next fifteen to twenty years.[1] This optimism is shared by scientists in the USA, Japan, France, the UK, the USSR, Italy, and other countries. The project will require cooperation and interactive research alliances that are without precedent in biology and medicine. Unlike many large physics projects, where

Table 1   A select sample of medical advances resulting from DNA-based biotechnologies, listed by year of relevant publication

Down's chromosome abnormality recognized	1959	J. Lejeune, M. Gautier, and R. Turpin, "Etudes des chromosomes somatique de neuf enfants mongoliens," *Comptes Rendus Academie des Sciences* (Paris), D, 248 (1959), 1721–1722
Amniocentesis and cytogenetic analysis developed	1967	C. B. Jacobson and R. H. Barter, "Intrauterine diagnosis and management of genetic defects," *American Journal of Obstetrics and Gynecology*, 99 (1967), 796–807
Retroviral replication cycle discovered	1970	D. Baltimore, "RNA-dependent DNA polymerase in virions of RNA tumour viruses," *Nature*, 226 (1970), 1209–1211; H. M. Temin and S. Mizutani, "RNA-dependent DNA polymerase in virions of Rous sarcoma virus," *Nature*, 226 (1970), 1211–1213
Hemoglobin cDNA cloned	1975	F. Rougeon, P. Kourilsky, and B. Mach, "Insertion of a rabbit β-globin gene sequence into an *E. coli* plasmid," *Nucleic Acids Research*, 2 (1975), 2365–2378
Mechanism of immunoglobulin diversity elucidated	1976	N. Hozumi and S. Tonegawa, "Evidence for somatic rearrangement of immunoglobulin genes coding for variable and constant regions," *Proceedings of the National Academy of Sciences*, 73 (1976), 3628–3632
First clinical use of recombinant DNA technology	1976	Y. W. Kan, M. S. Golbus, and A. M. Dozy, "Prenatal diagnosis of alpha-thalassemia: Clinical application of molecular hybridization," *New England Journal of Medicine*, 295 (1976), 1165–1167
Peptide drug synthesized by recombinant method	1978	L. Villa-Komaroff et al., "A bacterial clone synthesizing proinsulin," *Proceedings of the National Academy of Sciences*, 75 (1978), 3727–3731
Disease gene mapped by RFLP linkage	1978	Y. W. Kan and A. M. Dozy, "Polymorphism of DNA sequence adjacent to human β-globin structural gene: Relationship to sickle cell mutation," *Proceedings of the National Academy of Sciences*, 75 (1978), 5631–5635

scientific results may be uncertain, this endeavor will be inherently productive in understanding man's genetically determined biology.

The project, if successful, will identify the 50,000 to 100,000 genes present in the human genome. These in turn will be available as diagnostic markers and, in some cases, as therapeutic agents for many genetic diseases. Some 5,000 heritable disorders

Table 1   (continued)

Oncogenes isolated	1981	R. C. Parker, H. E. Varmus, and J. M. Bishop, "Cellular homologue (c-src) of the transforming gene of Rous sarcoma virus: Isolation, mapping, and transcriptional analysis of c-src and flanking regions," *Proceedings of the National Academy of Sciences*, 78 (1981), 5842–5846
Disease mutation identified by molecular methods	1981	R. F. Geever et al., "Direct identification of sickle cell anemia by clot hybridization," *Proceedings of the National Academy of Sciences*, 78 (1981), 5081–5085
Allele-specific oligonucleotide detected	1983	B. J. Conner et al., "Detection of sickle cell β^s-globin allele by hybridization with synthetic oligonucleotides," *Proceedings of the National Academy of Sciences*, 80 (1983), 278–282
HIV isolated as causative agent of AIDS	1983	R. C. Gallo et al., "Isolation of human T-cell leukemia virus in acquired immune deficiency syndrome (AIDS)," *Science*, 220 (1983), 865–867; F. Barré-Sinoussi et al., "Isolation of a T-lymphotropic retrovirus from a patient at risk for acquired immune deficiency syndrome (AIDS)," *Science*, 220 (1983), 868–871
Reverse genetics: disease gene characterized from map position	1986	B. Royer-Pokora et al., "Cloning the gene for an inherited human disorder—chronic granulomatous disease—on the basis of its chromosomal location," *Nature*, 322 (1986), 32–38
Recombinant vaccine produced	1987	M. R. Hilleman, "Yeast recombinant hepatitis B vaccine," *Infection*, 15 (1987), 3–7
Long-term expression of somatically inserted gene	1988	E. A. Dzierzak, T. Papayannopoulou, and R. C. Mulligan, "Lineage-specific expression of a human β-globin gene in murine bone marrow transplant recipients reconstituted with retrovirus-transduced stem cells," *Nature*, 331 (1988), 35–41; J. W. Belmont et al., "Expression of human adenosine deaminase in murine hematopoietic cells," *Molecular Cell Biology*, 8 (1988), 5116–5125

have been clinically characterized; of these, approximately 800 have been delineated at a biochemical level and the responsible gene sequenced.[2] A deeper understanding of our genes—how they are involved in development, immune responsiveness, central nervous system development, disease susceptibility, germline and somatic mutations, evolution, and reproduction—will require enormous additional research effort. The technologies to resolve these important biological questions are highly diverse; researchers may choose among cell culture techniques, gene expression studies, transgenic animal methods, and many other new developments that have been spurred on by recent successes in molecular biology.

How the "genome approach" has provided an efficient means of discerning the genetic basis of an inherited disease is illustrated by the fragile X syndrome. This X-linked mental retardation disorder is associated by linkage with a rare, fragile site located at the chromosomal address Xq27.3. The chromosome is readily broken at this point, resulting in inactivation of the gene present there and, in consequence, the syndrome. The culprit gene has recently been cloned in our laboratory. Rather than attempt to generate recombinant DNA clones derived specifically from the region of interest, our approach was to generate yeast artificial chromosome (YAC) libraries from the whole X chromosome, to map the location of randomly chosen clones, and then to analyze any clones found in the area of Xq27.3. Twelve such clones were isolated, four pairs of which contain overlapping regions, as determined by a rapid PCR-based fingerprinting method.[3] It was from one of these clones that the fragile X gene was identified.

The benefit of this approach is that cloned material is generated from a fairly large portion of the genome (the X chromosome) without discrimination and only later is this material scanned for clones that seem interesting. This is actually more efficient than preparing clones from a specific region. The method is of particular importance when many genes are sought in one region because cloned material remains available for future study.

The genome project will significantly affect our ability to screen for diseases at birth, during pregnancy, and in all stages of adult life. Currently, physicians screen for heritable disease risk at birth and in adulthood during the age of reproduction, usually prior to conception, for people at risk of transmitting a heritable disease. In 1961, Robert Guthrie introduced a simple and inexpensive metabolite inhibition assay with the capacity to detect severe, treatable inborn errors of metabolism.[4] This soon led to successful prevention of mental retardation from phenylketonuria (PKU) and galactosemia. It later became possible to prevent neonatal deaths associated with sickle cell disease and retardation related to thyroid deficiency. Today, newborns are regularly screened for a vari-

Table 2   Diseases commonly tested for in newborns by detection of protein factors

Disease	Incidence	Therapy options
Phenylketonuria	1:15,000	Dietary
Galactosemia	1:70,000	Dietary
Homocystinuria	1:100,000	Dietary
Maple syrup urine disease	1:250,000	Dietary
Biotinidase deficiency	1:70,000	Pharmacologic
Sickle cell disease	1:400[a]	Hematologic and antiseptic
Cystic fibrosis	1:2,500[b]	Pulmonary and dietary
Congenital hypothyroidism	1:4,000	Thyroid replacement
Congenital adrenohyperplasia	1:12,000	Steroid replacement
Duchenne muscular dystrophy	1:3,500	Physical therapy

a. Pertains to black population in United States.
b. Pertains to white population in United States.

ety of genetic diseases, as listed in Table 2. Successful screening programs are those that target diseases of high incidence, that are administered at low cost, and that offer therapeutic options and parental education about the implications of the screening results, including the possibilities of therapeutic abortion or postnatal treatments of the child.

Newborn screening procedures will be expanded to test for additional diseases as a consequence of improvements being made in techniques for isolating disease genes; we now have the ability to detect genetic mutations by simple DNA-based methods. No longer will newborn screening be limited to detection of a circulating metabolite or blood component. Now, for example, hemoglobinopathies are more accurately detected by DNA methods than by protein methods. We would need to search for only four alleles to detect 50 percent of the phenylketonuria cases in the United States. Tests for both PKU and the hemoglobinopathies, which are already a part of a national newborn screening program, could be carried out by a single, automated DNA-based method in place of current tests, which require a variety of technical skills and interpretive ability. To be effective, the DNA-based methods would require detection of only six hemoglobinopathy alleles and the four PKU alleles. Multiple mutant alleles are involved, however, in a number of common disorders—such as galactosemia and Duchenne muscular dystrophy. Hence, general DNA methods must be devised to detect different disease alleles in virtually every family examined.

Newborn screening methods that utilize the traditional methods (the Guthrie test, hemoglobin electrophoresis, and radioimmune assay) are highly accurate. Failures to detect affected patients result from technical or procedural errors. Screening for disease by allele identification must offer a similar high standard of accuracy as well as cost-effectiveness. Taking into account the factors of disease incidence, severity, and therapeutic availability, I have forecast a number of diseases that will be amenable to DNA-based screening procedures in newborns (Table 3).

Although a more ambitious list of treatable newborn disorders could be compiled (one that would include deficiencies of adenosine deaminase, factor VIII, factor IX, and growth hormone, for instance), Table 3 illustrates the complexity involved in newborn screening by DNA methods. I have estimated that between 4 and 100 alleles are responsible for each disease. DNA methods currently exist for studying nine independent genetic loci, and these methods are amenable to full automation. Recently a detection method for multiple human leukocyte antigen (HLA) alleles was reported by scientists at the Cetus Corporation.[5] Considering that we are now able to study independent loci and multiple alleles at each loci, screening of newborns by DNA methods is sure to come.

Given this forecast, it is important that we consider the medical and ethical issues that screening raises. One issue is immediately obvious: DNA methods identify heterozygotes (carriers). Many of the selected diseases shown in Table 3 are autosomally recessive diseases or X-linked disorders; for these diseases the heterozygote is free of the disease but at increased risk for bearing an affected offspring. A heterozygote generally enjoys normal health, but the information that the person carries a defective allele would be part of his or her medical record for the purpose of reproductive decisions. This information would have to be kept private, and the individual would have to be protected against negative repercussions such as loss of insurance coverage and job opportunities, should this information become public. Furthermore, experience with adult screening programs for sickle cell disease, β-thalassemia, and Tay-Sachs disease have taught us the importance of

Table 3 Diseases that could be tested for in newborns by DNA-based methods

Disease	Incidence	Number of alleles to be tested for	Effect of disease	Therapy options
Phenylketonuria	1:15,000	≈4	Mental retardation	Dietary
Galactosemia	1:70,000	≈4	Mental retardation	Dietary
Hemoglobinopathies	1:1,000	≈6–8	Anemia, sepsis	Hematologic and antiseptic
α₁-Antitrypsin deficiency	1:8,000	≈2	Emphysema, hepatic disease	Pulmonary, smoking avoidance
Gaucher's disease	1:2,500	≈4	Anemia and/or spleno-megaly	Hematologic and antiseptic
Urea cycle defects	1:10,000	≈8	Mental retardation, sepsis, hyperammonemia	Dietary
Glucose-6-phosphate dehy-drogenase deficiency	1:100	≈4	Anemia	Drug avoidance
Type II hyperlipidemia	1:2,500	≈10	Coronary artery disease	Dietary, lipid-lowering med-ications
Duchenne muscular dys-trophy	1:3,500	hundreds	Muscle wasting	Supportive
Cystic fibrosis	1:2,500	≈50	Pulmonary failure	Supportive
Neurofibromatosis	1:3,500	hundreds	Neurological failure	Supportive
Adult polycystic kidney disease	1:5,000	≈5	Renal failure	Dietary, antihypertensive medications
Huntington's chorea	1:100,000	1	Dementia	Supportive

educating those screened regarding the significance of heterozygote (carrier) status. Screening programs for β-thalassemia and Tay-Sachs disease have reduced tenfold the incidence of these diseases, but there has been no reduction in sickle cell disease.

A second issue is also obvious: DNA predicts the onset of a disease before symptoms are experienced. In some cases the disease occurs early in life (for example, Tay-Sachs disease, Gaucher's disease, Duchenne muscular dystrophy), but other diseases, such as adult polycystic kidney disease or Huntington disease, appear much later. Presymptomatic diagnosis is possible for both types of disease, but ethicists and physicians would contend that there are significant differences between the prediction of diseases of childhood and the prediction of adult-onset diseases. One clearly needs to set objectives for the identification of presymptomatic diseases and to determine if the benefit-to-risk ratio argues for use of the new technology. Overzealous use of diagnostic technologies will thwart their long-term acceptance in society. Still, the general principle of applying neonatal screening procedures when it is possible to provide care to the affected individual would appear prudent.

It can be important to screen people of child-bearing age for risk of heritable disease. Carrier screening procedures for hemoglobinopathies and Tay-Sachs disease are well established in the United States. Sardinia has reported a remarkable (tenfold) decrease in the incidence of β-thalassemia through an integrated program of carrier screening, genetic counseling, and prenatal diagnosis.[6] Carrier screening for β-thalassemia in Sardinia is logistically simpler than hemoglobinopathy screening in the United States, where a wider number of disease alleles exist within the Italian, Greek, and black populations. Carrier screening for Tay-Sachs disease has also been highly successful: incidence of the disease has been reduced tenfold in the United States. The current test for carriers of Tay-Sachs disease is the presence in blood of an enzyme, hexosaminidase A: if the enzyme is present, the individual does *not* carry the Tay-Sachs gene. Recently, the first gene defects that cause hexosaminidase deficiency in Tay-Sachs disease were identified.[7] Given the high incidence of Tay-Sachs disease among Ashkenazi Jews, who constitute the majority of U.S. Jews, one expects—and finds—the disease to arise from a limited number of mutant alleles. DNA-based methods can therefore be anticipated to replace the enzyme test for detecting Gaucher's disease, another common disorder of Ashkenazi Jews, thus making DNA-based carrier detection possible for this common disorder. The method used today, testing for the presence of the enzyme gluco-cerebrosidase, is not sufficiently accurate. In the cases of β-thalas-semia, sickle cell disease, Tay-Sachs disease, and Gaucher's disease, more than 95 percent of the mutations in the at-risk population could be determined by DNA methods.

The most common disease allele in the Caucasian population, which is responsible for cystic fibrosis (CF), has recently been identified and is amenable to carrier detection by DNA methods.[8] This allele is present on 70–75 percent of chromosomes carrying CF mutations; it and an additional four alleles account collectively for 83 percent of all CF cases in the U.S. population. Controversy surrounds the question whether population-based CF screening should be initiated in the United States. The debate does not concern those families clearly at risk of carrying the genetic defect but, instead, those couples who would be left with ambiguous DNA results. Consider, for example, a couple in which one partner is a carrier of a cystic fibrosis allele (the alleles are recessive), and the other partner tests negative for *known* alleles (but *may* carry alleles not yet identified that cause CF). This outcome would occur in 7.5 percent of couples tested. Before testing, the couple has a one in 2,500 risk (the average risk, being the frequency with which CF appears in the population) of producing an affected child; if one partner carries a CF allele (as the test has shown), the couple has a one in 396 risk. When both partners lack *known* CF alleles (92.3 percent of couples), the risk would be one in 39,200, and when both have CF alleles (0.2 percent of couples) the risk is one in four. The human genetics community is divided over whether to proceed with CF carrier detection or to wait until the tests achieve an accuracy of 95 percent. Is the cup half-filled or half-empty?

Table 4 provides a list of diseases compatible with carrier screening. Undoubtedly, the list will expand as other autosomally recessive disorders are better characterized. Couples at risk of producing affected offspring have the opportunity to undertake pregnancies with the knowledge that accurate prenatal diagnosis is available and that, in accord with legal doctrine in most Western countries, elective termination of pregnancy is an option. To date, no distinction has been made in the law between termination of pregnancies associated with severe heritable disease and those not associated with fetal disease. In the United States, as long as freedom of family choice is supported by the Supreme Court at least for severe heritable disease, genetic screening for heterozygotes will no doubt continue to increase. Should pregnancy termination for severe heritable disease *not* be an option, the utilization of genetic screening would be reduced. This is a time of uncertainty regarding legal decisions that could affect the prevention of severe, untreatable heritable disease.

In the United States, not all pregnancies receive genetic attention because patients and physicians inadequately understand carrier screening. Approximately 100,000 prenatal diagnoses are done annually in the United States, predominantly to test for chromosomal abnormalities (in offspring of mothers over age 33) and neural tube defects, but also to test for a few other diseases that have been identified in the family histories of the parents. The "high-risk" population currently receives genetic evaluation, counseling, and carrier screening, but only 5 percent of current pregnancies in the United States are considered high risks. Only in the case of that segment of the population known to be at risk for Tay-Sachs disease does a couple receive genetic screening on a routine basis. Effective evaluation for the carriers of other diseases is lacking in the United States. Wider implementation of carrier screening will require considerable public and professional education.

Table 4   Diseases for which carrier screening by DNA-based methods is possible

Disease	Number of alleles involved	Disease frequency	Carrier frequency	High-risk population
Phenylketonuria	≈4	1:15,000	1:100	Broad
Hemoglobinopathy				
β^S and β^C alleles	2	1:400	1:12	Black
β^{39} and β^{112} alleles	2	1:2,000	1:20	Mediterranean
α_1-Antitrypsin deficiency	2	1:8,000	1:50	Broad
Cystic fibrosis	≈50	1:2,000	1:20	Caucasian
Gaucher's disease	3	1:2,500	1:25	Ashkenazi Jewish
Tay-Sachs disease	2	1:4,000	1:30	Ashkenazi Jewish

Table 3 includes two common autosomally dominant diseases (Type II hyperlipidemia and adult polycystic kidney disease [APKD]) and one uncommon autosomally dominant disorder (Huntington's disease) that can be predicted at birth but that does not affect the individual until adulthood. It is hoped that the coronary artery disease resulting from type II hyperlipidemia may be avoided by changes in diet and by drug therapy. For APKD and Huntington's disease, however, the outlook for intervention is less optimistic. For these diseases, as for CF and neurofibromatosis, therapy is presently focused on symptomatic relief. Collectively, these disorders represent both a considerable challenge to medical science and a new responsibility: once we can predict disease risk at birth, how should we use that information to improve the care provided to the individual?

Genetic screening raises a variety of other questions, too. First there is the matter of the consent of informed adults for participation in screening and state-mandated screening of minors. The cost-effectiveness of adult screening is yet to be estimated, given that some disorders have variable therapeutic options and that dietary and medication compliance among adults has not been evaluated. The confidentiality of screening information will need to be ensured and its potential adverse effect on insurability and employment considered. For example, will a patient with type II hyperlipidemia be an unacceptable employee? What is the self-image of an individual at risk for an adult-onset disease? Will the asymptomatic heterozygote for Huntington's disease or APKD be at jeopardy for health care coverage? Is genetic disease considered a "prior existing medical condition"?

These questions are presently unresolved at the national level, and if they remain undecided they could negatively affect the acceptance of adult screening programs. Public discussions are needed, as are policy decisions by insurance carriers. DNA-based data could reverse some current exclusions on health and life insurance. Some, but not all, insurance carriers will not qualify patients with a familial history of APKD, Huntington's disease, factor VIII deficiency, or other disorders. Since DNA testing can very accurately identify an individual as a heterozygote or normal, uncertainty over who is and who is not likely to develop a disease can be clarified. Fifty percent of those previously denied coverage would be shown to be normal and therefore could be assigned to the normal pool, but anyone identified as abnormal will truly be at high risk. It would appear reasonable to me that those with high risk could be allocated to an insurance pool much as noninsurable drivers are now assigned. One alternative is universal health care coverage by government as exists in the United Kingdom and Canada, where the issue of coverage is irrelevant.

DNA-based newborn screening raises the volatile issue of genetic data banks. Newborn screening, with the exception of sickle cell disease and β-thalassemia, is not now used for carrier identification. Furthermore, although parents of children with autosomally recessive disease are considered obligate carriers, no means of maintaining a record of carriers exists for the parents or their relatives. DNA-based newborn screening identifies the carriers unequivocally and thus demands debate about the propriety of genotype data banks. I lack sufficient confidence in the security of data banks to encourage their establishment, and I think that a good deal more public discussion of the subject is required.

The establishment of a national genetic data repository needs to be considered. The issues concerning an adult carrier data repository are different from those concerning a data repository for newborns. Those participating in the carrier data repository would be consenting adults rather than newborns. Carrier data constitute private and confidential medical information. Genetic screening is fundamentally a private study of family genetics. This form of carrier screening does not address national public health objectives or population screening. For the future, one could envision cost savings by use of a genetic data repository. Data repositories may inform physicians which individuals have an established genetic risk for common disorders (offspring of carriers) and thus deserve study. It is likely that simple, cost-effective testing for family study will exceed in efficiency a national, computer-based data repository. Furthermore, I expect that keeping genetic test results private will be more acceptable to the public and that the public will want genetic information related to disease risk to remain a private medical record.

Despite the success of prenatal diagnosis, it has resulted in a reduction of less than 5 percent in the incidence of genetic disease. Wider application of prenatal diagnosis is advocated for prevention of severe, untreatable genetic disease. For example, the fetuses of women over a certain age are commonly tested for aneuploidy—that is, an abnormal number of chromosomes—but more than 90 percent of aneuploid newborns (those with trisomy 21, 18, or 13) are born to mothers younger than those in the high-risk age group. There are two choices for improving care: increasing the use of prenatal diagnosis for all age categories, or making cheaper and more accurate the detection methods for chromosomally aneuploid pregnancies. Cost-benefit analysis argues against the first approach.

Technical innovation is emerging in the detection of aneuploidy. First, there have been discoveries of defects in single genes whose functions are known in yeast to lead to aneuploidy. Second, chromosome 21, one of those most frequently involved in aneuploidy, has been taken up for study by the human genome project. Third, recombinant DNA methods now allow a simple rapid method for aneuploidy diagnosis. It appears likely that effective methods to diagnose and understand the mechanism of aneuploidy will be improved.

New mutations in single-gene disorders represent an additional challenge. Presently, index-case diagnosis—that is, the first case of a disease in a particular family that indicates other members or subsequent offspring may have the defective gene—alerts family and physician to the possibility of recurrent risk in the family. This is particularly true of X-linked recessive and autosomally dominant diseases. Diseases encoded on the X chromosome include Duchenne muscular dystrophy (DMD), ornithine transcarbamylase deficiency, factor VIII and factor IX deficiencies, Lesch-Nyhan syndrome, and fragile X syndrome, among others. Autosomally dominant disorders include Marfan syndrome, osteogenesis imperfecta, neurofibromatosis, and retinoblastoma.

The problem of detecting the heterogenous mutations of Duchenne muscular dystrophy (DMD), Lesch-Nyhan syndrome, and ornithine transcarbamylase deficiency has been overcome by recombinant methods. In particular, the polymerase chain reaction (PCR) has enabled the efficient detection of mutations, and it has the potential for automation. At Baylor College of Medicine, our policy is to screen samples for deletions in the DMD gene by means of a multiplex PCR, which detects 81 percent of deletions (46 percent of all mutations) in this gene.[9] The multiplex reaction takes only a few hours to perform, whereas Southern blot analysis, the test used previously, takes several days. The search for mutations in the HPRT gene, which cause Lesch-Nyhan syndrome, is initially similar to the test for DMD in that a multiplex PCR is performed. For Lesch-Nyhan, however, the assay provides diagnosis of only 15 percent of the mutations; multiplex PCR serves more to provide material for the next stage of the analysis, which is automated sequencing of the reaction products.[10] This combination of procedures offers the potential to identify almost all mutations in the HPRT gene leading to Lesch-Nyhan syndrome.

Deficiency of the enzyme ornithine transcarbamylase (OTC) is caused by mutations too large to consider sequencing as the routine method of detection. PCR is used to generate single-stranded DNA from both normal and patient samples. These species are then hybridized together and the positions at which mismatches (the mutations) occur are subject to chemical cleavage, whereas two normal strands of the OTC gene hybridized together are not cleaved.[11] With this method family members can be identified as carriers or not once the mutation in the original patient (index case) has been identified. It remains to be determined if this technology can be applied to all pregnancies in an effort to detect a new mutational event. Two high-frequency disorders appear likely candidates for detection of germ-line mutations. New mutations that cause DMD and neurofibromatosis occur at a frequency of one in 3,500. At this time it appears that methods for prenatal detection of new mutations are too costly and too inaccurate, but it is my prediction that major technical revisions will bring about rapid and inexpensive DNA-based methods capable of detecting new mutation events *in utero*.

Inherited disorders often have an organ- or tissue-specific pathophysiology. In some cases physicians may use organ or tissue transplantation between a compatible donor and a recipient to correct the pathology of inherited diseases. An abbreviated list of successful transplantation efforts is given in Table 5.

Whether it will become possible to transplant other tissues to cure other diseases will depend on improvements in our ability to regulate graft-versus-host disease. Presently, immune suppression with cyclosporin has been the mainstay of transplantation success. Undoubtedly, molecular understanding of immune responsiveness will further enhance success with transplantation. We may be able to draw upon the research experience in the field

Table 5 Correction of genetic disease by organ or tissue transplant

Transplanted tissue	Disease	Pathophysiology
Bone marrow	Combined immune deficiency associated with adenosine deaminase or purine nucleoside phosphorylase enzyme deficiency	Excessive deoxyadenine or deoxyguanine
	Sickle cell anemia	Sickle cell hemoglobin
	β-Thalassemia	Insufficient hemoglobin
Liver	Ornithine transcarbamylase deficiency	Hyperammonemia
	α_1-Antitrypsin deficiency	Cirrhosis and emphysema
	Type II hyperlipidemia	Coronary artery disease
	Hemophilia (VIII [A])	Factor VIII deficiency
Kidney	Adult polycystic kidney disease	Cystic renal failure
	Fabry's disease	Glomerular trihexyceramide lipid accumulation, renal failure
Heart	Congenital defects and/or myopathy	Cardiac failure
Heart and lung	Cystic fibrosis	Pulmonary failure and cor pulmonale

of autoimmune disease, since common features mark the rejection of transplanted tissue. If dominant surface epitopes (antigenic markers) and corresponding T cell response genes can be identified, immune manipulation can be considered. It remains to be determined if selective regulation or elimination of killer T cells can be employed to alter the ability of these cells to reject foreign tissue. Recently, monoclonal elimination of T cell clones responsible for a disease in mice similar to multiple sclerosis has been achieved.[12] In this disorder, which causes the immune system to attack myelin, clinical amelioration was achieved. Bone marrow and tissue transplantation will improve substantially as a safe treatment for both inherited and acquired diseases. The addition of immune blocking peptides or monoclonal regulation of T cell rejection could substantially alter the success of transplantation therapy.

Substantial progress has already been made toward genetic correction of heritable disease. The genetic approach avoids the potential complications of transplantation, since the normal gene is introduced into the patient's own somatic tissues. The normal gene (cDNA) is cloned into an expression vector—an agent that carries the cDNA to the target tissue, where, under the regulation of a promoter (part of the DNA sequence that activates genes), it becomes active. These expression elements are built into a defective virus capable of reproduction in a helper cell line. The defective virus is highly efficient at gene delivery (bringing the gene to the target site) but cannot replicate. Retroviruses are the current vector of choice, but recently adeno-associated virus and herpes virus have been reported as viral gene delivery systems. Ideally, the correcting gene is delivered into a self-regenerating stem cell, which replicates the transferred gene as it replicates itself, thus eliminating the need for repetitive therapy. Long-term expression of human adenosine deaminase (ADA) in mice has now been achieved by several groups.[13] ADA genes have also been delivered to cultures of human bone-marrow cells, where they have been able to express the ADA enzyme. The first experiment approved in the United States to follow the viral introduction of a marker into cells rendered to cancer patients has been reported recently, as has a clinical trial for genetic therapy of ADA deficiency.[14] An overview of gene-therapy processes is outlined in Figure 12.

An alternative strategy is emerging that avoids the biohazard of viral vectors while achieving replacement of the defective gene with a normal gene. Legitimate recombination has been accomplished in mouse and human cells in culture using large DNA segments introduced by either microinjection or electroporation (in which target cells are induced by electrical currents to take up the foreign DNA). The direct injection of DNA into mouse muscle has been reported to result in expression of the transferred genes.[15] It has been possible in mice to target normal replacement sequences to the mutant gene, thus correcting the genetic defect.[16] Although the method is not as yet very efficient, it has the advantage, when it succeeds, of *replacing* defective genes with normal genes, rather than integrating the correct genes (via a retroviral vector) among the cells that carry the defective genes. The method is being used extensively as a means of developing transgenic mice. Table 6 lists a number of diseases under investigation for treatment by gene replacement or gene transfer into somatic tissue. This list will expand significantly if successful expressions of transferred genes can be demonstrated in human subjects.

Gene transfer therapy for somatic tissues raises only limited ethical questions, since the success or failure of the effort will affect only the patient with disease. The issue brings out the concerns typical of human experimentation, notably the ratio of risks and benefits to the individual. The risks that accompany the use of viral vectors, including their ability to infect progenitor cell lines and the potential of harmful insertion, will need to be examined carefully.

The transfer of genes into human embryos has little practical appeal, and it generates considerable ethical apprehension. It is likely that embryo diagnosis will become a reality of medical care, as it has been realized in mouse studies. Given this option, for, say, a couple wishing to avoid passing along a recessively inherited disease, it would appear more logical to permit the implantation of normal embryos (three out of four) rather than to attempt correction of an affected embryo (one in four). Current gene transfer and replacement technologies have low success rates (between one in a thousand and one in a hundred thousand) and high rates (between one in ten thousand and one in a million) of illegitimate recombinations, in which the gene inserts itself into the wrong place, sometimes in the middle of another gene. Illegitimate insertion of transgene sequences has produced disease in mouse embryos.[17] In other words, germ-line correction not only raises considerable controversy but also offers little of practical value for man.

I would reserve one area for consideration of germ-line manipulation—genetic advantage. In veterinary research, intensive investigation of disease resistance is ongoing. Should human disease resistance also be considered? I remind the reader of man's species-wide genetic loss of uricase (the disease consequence being gout), vitamin C synthesis (the disease consequence being scurvy), and the influenza resistance gene (the disease consequence being influenza). It is conceivable that at some point in the future genetic manipulation of an individual's germ-line may be undertaken to introduce or reintroduce disease resistance. If it

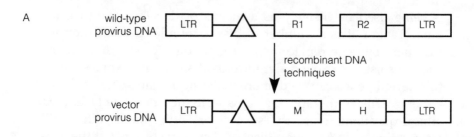

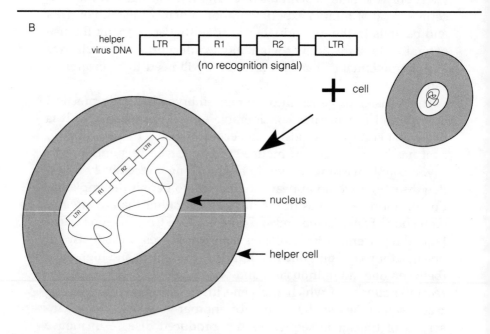

Figure 12   Gene therapy is the insertion of genetic material directly into cells for the purpose of altering the functioning of the cells—to correct a genetic defect, for example. The most promising agents for "delivering" the genes are the *retroviruses*, whose genetic material is not DNA but RNA, which comes packaged in a protein envelope. A retrovirus, once it infects a host cell, makes a DNA copy of its RNA, and this DNA is integrated into the cell's chromosomal DNA. At this stage it is called a *provirus.*

The first step *(A)* in gene therapy is to alter a wild-type provirus so that its DNA carries with it the human DNA to be delivered to the patient, as well as a selectable marker to identify the cells that contain the altered provirus. The recombinant provirus, known as the *vector virus,* has the regulatory elements required for inserting its DNA into another cell's genome (the long terminal repeats, *LTRs,* and a recognition signal), but it cannot produce the proteins needed to make a virus particle. Another cell is needed to provide this "packaging." A *helper cell* is constructed (step *B*) by adding to a cell's own DNA a *helper virus,* which is a provirus that has the retroviral genes needed to make the viral package but from which the recognition signal that initiates the packaging has been removed.

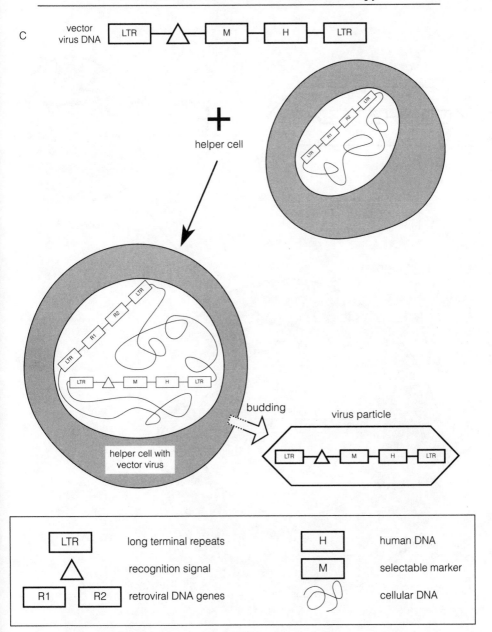

Figure 12 (continued) Now the vector itself may be produced (step C). The vector virus DNA is inserted in the helper cell (with the helper virus already integrated in its DNA), where it is copied into RNA. Then the helper virus DNA makes the viral package for the vector virus RNA; the product is a virus particle that can "infect" the target cell with the human DNA fragment needed to correct the genetic defect.

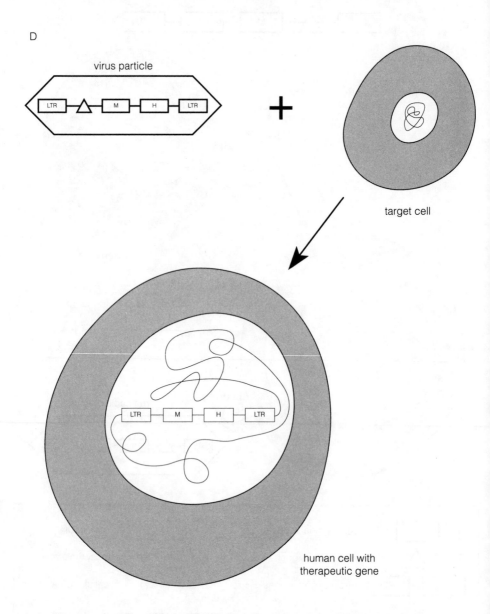

Figure 12 (continued)   Finally (step *D*), the viral "package" is delivered: target cells are mixed with helper cells or with the fluid from a culture of helper cells. When the target cells (for example, bone marrow cells) are thus infected with the vector virus, the correcting human gene is integrated into the cells' own DNA, and the cells now have the instructions they need to function properly.

Table 6 Diseases that are candidates for gene transfer or gene
replacement therapy

Disease	Target tissue
Adenosine deaminase deficiency	Bone marrow
Thalassemia, sickle cell disease	Bone marrow
Ornithine transcarbamylase deficiency	Liver or small bowel
Phenylketonuria	Liver
Duchenne muscular dystrophy	Muscle

is, the risk-to-benefit considerations will have to change significantly from those of the current Institutional Review Board guidelines: a major ethical issue will be risk of injury now in comparison
with benefit to the health of future generations.

The role of somatic mutations in acquired disease is already
evident in T and B cell neoplasias. Since DNA-based methods
are precise and highly sensitive, we can be confident that this
technology will have an expanding role in early diagnosis of malignancy. What is not clear is whether a substantial number of
disorders have a gene for susceptibility that can be detected. I
place in a separate category disorders such as xeroderma pig-
mentosum, Bloom's syndrome, and Fanconi's anemia, where
DNA repair defects result in an increased susceptibility to mutations at many loci. The models of retinoblastoma, neurofibromatosis, von Hippel–Lindau disease, Gardner's disease, and others provide our first insight into susceptibility to malignancy. In
these cases, a heterozygotic allele renders an individual susceptible to developing malignancy. It appears likely that our ability to
identify genetically susceptible individuals will improve, and the
technique will become part of genetic disease surveillance. We
presently use survey methods for early detection of breast cancer
(mammography), large bowel cancer (X ray), and leukemia (blood
study). It is virtually certain that DNA-based genetic surveillance
will add to the accuracy of early diagnosis and to therapeutic
efficacy. Table 7 lists some of the neoplasias resulting from the
disruption of known oncogenes; these are diseases for which it
might be possible, given the ability to search for mutations within
oncogenes, to predict susceptibility.

Genetic analyses and diagnostics raise significant implications

Table 7   Neoplasias resulting from genetic mutations or from oncogenes

Cause	Neoplasia	Genes involved
Translocations	B and T cell leukemias	*myc* and heavy-chain immunoglobulin gene
	Chronic myeloid leukemia	*abl* and *bcr*
Point mutations	Bladder, lung neoplasms	*ras*
Recessive oncogenes	Retinoblastoma Large bowel carcinoma Wilms' tumor, aniridia	

for the education and training of physicians. Little modification in present training is needed for physicians to use pharmacologic agents emerging from biotechnological advances. However, considerable modification of present training is needed if physicians are to understand the cell biology related to new drugs. Physicians can now increase erythropoiesis in patients with renal failure with exogenous erythropoietin; they can accelerate recovery of bone marrow with interleukins following chemotherapy, and they can stimulate growth in Turner syndrome patients with exogenous growth hormone. Each of these peptide growth factors is made in abundance by recombinant methods. We have achieved clonal T cell inhibition in mice with monoclonal antibodies, and synthetic peptides have been used to inhibit immune response; thus the potential to influence autoimmune disease is appearing. Physicians of the future will need to be well educated in cell biology to grasp the concepts and opportunities for cellular manipulation in their patients. The rapid expansion of medical applications, illustrated in Figure 11 indicates our need to examine the medical education process. An understanding of genetic principles will be expected of all physicians, since they will be responsible for patient care, disease prevention, and surveillance of genetic high-risk patients.

The human genome project, it is estimated, will be completed in fifteen years. We presently require twelve years to train a physician specialist (from high school to specialist certification). The new era of molecular medicine demands a revision of undergraduate schooling, medical school curricula, and postgraduate education. The clear message is that a high proportion of practicing physicians will need "retrofitting" to acquire the requisite understanding of DNA-based medicine.

Scientific expertise among specialists must be accompanied by public understanding or problems will surely arise. If DNA analysis is to be more widely used in the future, then the general population must be provided with a basic genetic understanding—not that everybody should be a molecular biologist, but people should comprehend the implications of the information that becomes available. In particular, the issue of genetic carrier status must be fully explained. Some of these educational issues include the significance of carrier status to personal health, job selection, insurability, and informed options for childbearing. Moreover, the importance of improving science education, from kindergarten through twelfth grade, cannot be overemphasized.

The technology associated with the human genome project has much to offer the medical field in terms of cloned genes, genetic markers, and improved procedures with which to perform DNA analysis. Although there is the potential for genetic information to be misused, this is no excuse for not pursuing the project. We must anticipate the problems and be prepared to deal with them.

LEROY HOOD

Biology and Medicine in the Twenty-First Century

6

During the past twenty years, brilliant advances in technology and fundamental new insights have led to a striking revolution in biology, which is slowly beginning to change medicine. This revolution will be accelerated as we move into the twenty-first century by even more far-reaching developments, especially the deciphering of the human genome, our blueprint for life. The human genome project is on its way to creating an encyclopedia of life, giving biologists and physicians direct computer access to the secrets of our chromosomes. The project is daunting in scope and scale, and accomplishing it will require still more advances in chemistries, techniques, instrumentation, and sophisticated computational hardware and software. If we succeed, the infrastructure of biology will be enriched, and the revolution that has begun in the practice of biology and clinical medicine will accelerate.

The human genome project is the first major biological initiative that takes the development of technologies as one of its major objectives. Some of these techniques are necessary to create and analyze the three types of maps essential to the genome project. We already know how to draw the genetic and physical maps, but improved technologies will increase enormously the rate at which they can be generated. We also must develop DNA sequencing techniques that are a hundred to a thousand times more rapid than what is currently available before we can seriously

Table 8 Genome sizes of model organisms

Organism	Million bases
E. coli	5
Yeast	15
Nematode (worm)	100
Drosophila (fly)	180
Mouse	3,000
Human	3,000

embark upon the task of sequencing the entire human genome. The development of hardware and software is required to organize the data from the three maps—genetic, physical, and sequence—of the human genome so that physicians and biologists can have computer access to this information for attacking fundamental problems in biology. Genomic analyses of model organisms—such as bacteria, yeast, worms, fruit flies, and mice—are also a part of the genome project (Table 8). These model organisms will provide valuable insights into how the genes shared with humans function, and the mouse genome (the only other mammalian genome in the list) will aid in defining human genes and regulatory regions through the identification of sequence regions conserved between the two species.

The timetable of the human genome project (see Table 9) divides into three five-year periods. During the first two five-year periods, the major focus will be technology development and creation of the genetic and physical maps. It is likely that only after the first ten years will the technology for large-scale sequencing have been developed to the point (faster than the current rate by a hundred-fold) where it will be feasible to do large-scale sequence analysis of the human and of many of the model organisms. Thus, the genome program proposes to carry out the bulk of the DNA sequencing only after appropriate rates of DNA sequencing are reached.

Just as the complex road system of the United States has transformed transportation in the country by permitting ready access to virtually any city, street, or house, so will the creation of genetic, physical, and sequence maps greatly facilitate our ability to access interesting genes. Currently, each time a new disease gene is to be isolated, a road must be built to that specific gene through

Table 9   Timetable for the human genome initiative

Time	Objectives
1–5 years	Technology: 5–10-fold improvement Informatics Crude genetic map Physical map for 5–10 chromosomes Sequence some biologically interesting regions (<1 percent) Model organisms: map and start sequence
5–10 years	Technology: 5–10-fold improvement More informatics Refined genetic map Physical map finished Sequence more biologically interesting regions (<5 percent) Model organisms: finish sequence
10–15 years	Technology: more Informatics: more Sequence: finished (95 percent) Additional model organisms sequenced

the techniques of recombinant DNA technology. Indeed, multiple roads are often independently built to interesting genes by competing groups. Once the three genome maps are available, the task of finding disease genes will be enormously simplified and the cost greatly reduced. Thus the maps of the human genome can be viewed as powerful tools that will significantly enrich the infrastructures of biology and medicine.

The benefits that will arise from having a complete sequence map of the human genome, sometime early in the twenty-first century, fall into four categories. First, the development of the requisite technologies necessary to accomplish the human genome project will revolutionize many other aspects of biology and medicine. Second, computer access to the genome maps will dramatically alter the practice of biology. Third, access to the genetic and sequence maps will fundamentally change the practice of clinical medicine. Finally, the information generated by the human genome project, as well as the new technologies that emerge from this endeavor, will ensure the United States a highly competitive position in the worldwide biotechnology industry.

The genome program will necessitate the development of more powerful technologies for DNA handling, mapping, sequencing, and analysis. There is potential for significant improvement in the physical and genetic mapping techniques and, indeed, the success in sequencing the human genome will require at least a hundred-fold increase in throughput for DNA sequencing. There are also challenging computational problems associated with the genome project. The improvement of technology impinges on four areas—the development of new techniques, automation, increased throughput, and increased sensitivity of analysis. In general, the key to technology development will be a multidisciplinary approach combining the powerful tools of applied mathematics and physics, chemistry, engineering, and computer science as well as biology.

Let me illustrate the power of this approach by describing the Science and Technology Center for Molecular Biotechnology that I head, in which we are developing an interdisciplinary group committed to the development of new technologies for biology. These interdisciplinary areas include expertise in protein chemistry, mass spectrometry, nucleic acids chemistry, large-scale DNA sequencing, genetic mapping, DNA diagnostics, and computational techniques (Figure 13). Cross-fertilization among these groups has led to the development of techniques and instrumentation that have had or will have a significant impact on the genome effort.

For example, in the early 1980s it became obvious that we needed to acquire the capacity to synthesize automatically small fragments of DNA (oligonucleotides), ten to fifty bases in length. These oligonucleotides or probes were useful for cloning genes and sequencing DNA, and later they served as primers for the polymerase chain reaction (PCR), a technique for amplifying any particular small region of DNA a million-fold or more. We automated a manual technique that attached the first DNA base in the oligonucleotide to a small, inert bead (solid support) and then carried out successive chemistries on this base-substituted solid support, adding one base at a time to the growing DNA chain (Figure 14).[1] The automation of this technique enormously increased the throughput for DNA synthesis, both by decreasing the cycle time (approximately five minutes) and by allowing multiple chains to be synthesized simultaneously (four-column machines).

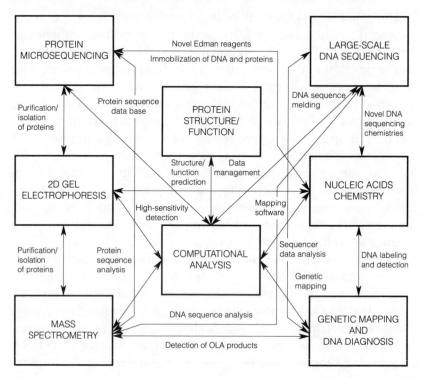

Figure 13 The major groups within the Science and Technology Center for Molecular Biotechnology collaborate on a variety of related projects.

To put this increased throughput in perspective, in the early 1970s it took H. Ghobind Khorana and twenty-five postdoctoral fellows almost five years to accomplish the first synthesis of a small gene, a task that now can be done in a single day by one technician with several four-column machines. With the scaling-up of the genome project, literally hundreds of thousands of DNA probes will be needed for PCR, DNA sequencing, and cloning procedures. Accordingly, we need to develop DNA machines that can synthesize fifty to a hundred fragments of DNA simultaneously, quickly and inexpensively.

The concept of sequence tagged sites (STSs) has fundamentally altered the approach to physical mapping.[2] An STS is a stretch of genomic sequence, generally 100 to 1,000 base pairs, that is "uniquely" defined by a pair of polymerase chain reaction primers. The fragment is "unique" because the PCR primer pair amplifies only a single sequence in the presence of the entire comple-

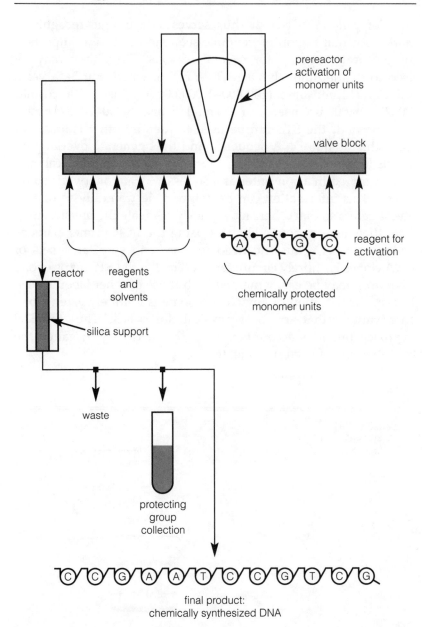

Figure 14 The steps in the chemical synthesis of DNA outlined in Figure 7 are now carried out automatically by a machine, as diagrammed here. One current form of this machine can simultaneously synthesize four DNA fragments, adding bases at the rate of one per 5 minutes for each chain. DNA fragments up to 100 bases can easily be synthesized.

ment of genomic DNA; it thus serves as a unique recognition marker for that region of genomic sequence. STSs are important for physical mapping for several reasons. First, they may be used to define uniquely each DNA clone, whether it be a yeast artificial chromosome (100,000–1,000,000 bp insert), a cosmid (30,000–45,000 bp insert), or a lambda clone (5,000–20,000 bp insert). Second, the STS can be used to identify other clones that share this unique DNA sequence and thus generate overlapping inserts for a physical map (see Figure 15). Third, the physical map may be stored in a computer as a series of STSs from overlapping clones. This information can be shipped electronically to distant investigators so that others may quickly re-create the physical map from their own genomic libraries using the PCR primer pairs as screening tools. Thus the need to store and ship large sets of DNA clones is entirely circumvented. Finally, the STS maps of one laboratory may be easily merged with those of other laboratories. Hence, the STS map of any chromosome is infinitely extendable, an advantage that arrays of physical clones lack. Thus the STS approach provides accountability in that it is easy to gauge the contributions of each investigator.

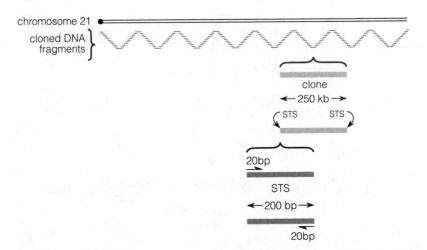

Figure 15 Sequence tagged sites (STSs) are short stretches of unique DNA sequence that will be used to create a physical map, each representing at least one clone in an overlapping set of clones for chromosome 21. In the illustration, the clones are an average of 250 kb in length, and each one begins and ends with an STS. The STSs are defined by unique PCR primers (these are 20 base pairs in length) and can be used to identify other clones that share the same DNA or STS sequence.

Genetic maps are created by following the segregation (passage from parents to children) of DNA polymorphisms in families. The genomes of humans are highly polymorphic: one base in five hundred will differ between any two individuals. If we are to develop a two-centimorgan map as a part of the genome project, more than 1,600 evenly spaced genetic markers have to be identified (the human genome is approximately 3,300 centimorgans in length). We've developed a technique that automates the analysis of DNA polymorphisms by using a robotic workstation that can handle plates with 96 small wells—thus 96 genetic markers can be analyzed simultaneously and automatically. This procedure enables us to: (1) amplify the segment of DNA that is to be examined for a polymorphism by PCR; (2) analyze the polymorphisms to determine which forms are present; and (3) automatically read and store the results directly in a computer.[3] It has the capacity to increase enormously the throughput analysis of genetic markers—indeed, 1,200 assays can be carried out per day by a single technician using a robotic workstation (Figure 16). With it we will be able to analyze the markers necessary to create the genetic map and to determine rapidly the location of interesting new genetic markers without employing the time-consuming techniques of conventional genetic mapping, such as RFLP mapping, which are difficult to automate. Indeed, it uses polymorphic STS markers to create a genetic map, and these can in turn be employed to generate a physical map (Figure 15). Hence, this technique leads to a merging of the genetic and physical maps.

The heart of the genome program is the sequence analysis of the twenty-four different human chromosomes. The development of fully automated techniques for DNA sequencing is an imperative for the genome project. We have begun this process by developing an automated DNA sequencing machine that uses four different fluorescent dyes to color-code the four DNA bases.[4] The sequence of the bases may then be read as colored bands migrating on an electrophoretic gel (Figure 17). This machine can analyze more than 12,000 base pairs of DNA sequence per day—the approximate amount of sequence a scientist in the early 1980s could determine in an entire year.

It is important to point out that large-scale DNA sequencing is a multi-step process.[5] It is necessary to purify the DNA, fragment and map it, electrophorese the fragments, assemble each string of fragments into larger strings (ultimately the length of each chro-

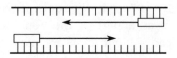

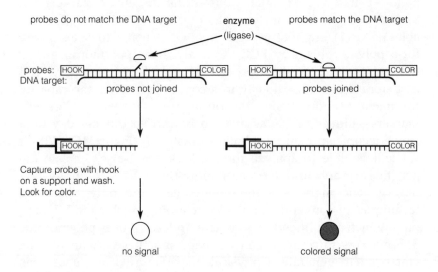

3. Determine outcome: computer reads color and
   makes diagnosis

Figure 16 Three steps of one automated technique for genetic mapping of identified DNA polymorphisms: (1) The oligonucleotide site of the polymorphism is amplified by PCR. (2) The oligonucleotide ligase assay is carried out. Two adjacent DNA probes are synthesized. The left-hand probe has a hook (biotin) attached to its end, whereas the right-hand probe has a colored reporter group attached to its end. The base at the right-hand end of the hook probe is located at the polymorphic site. The two probes are hybridized to the target DNA. If the polymorphic base of the hook probe is complementary to the target base, then the enzyme DNA ligase can join the two probes, and when the hook is used to remove the left-hand probe from the reaction mixture, it also brings the right-hand probe (and color). Conversely, if the base of the hook probe is not complementary, then DNA ligase fails to join the two probes and removal of the hook brings with it only the left-hand probe (no color). In other words, only those samples that show a colored signal contain DNA with the target polymorphism. (3) A computer reads the presence or absence of color in each of the 96 wells and performs the genetic mapping calculations.

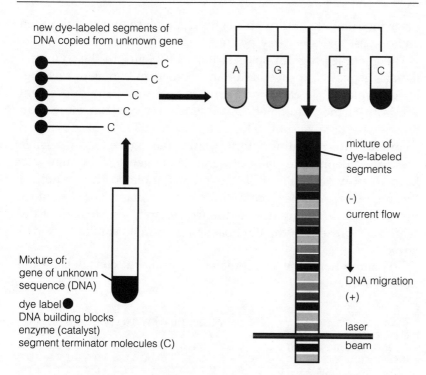

new dye-labeled segments of
DNA copied from unknown gene

C
C
C
C
C

A   G   T   C

mixture of
dye-labeled
segments

(-)
current flow

DNA migration
(+)

laser
beam

Mixture of:
gene of unknown
sequence (DNA)

dye label ●
DNA building blocks
enzyme (catalyst)
segment terminator molecules (C)

Figure 17   An illustration of the automated fluorescent DNA sequencing technology. The Sanger or enzymatic procedure for DNA sequencing (see Figure 6) is employed to synthesize four nested sets of DNA fragments all ending, respectively, in C, T, G, or A. The primer used to initiate synthesis for the C fragments is labeled with a red fluorescent dye, for the T fragments with a gold dye, for the G fragments with an orange dye, and for the A fragments with a green dye (here shown as varying shades of gray). These four DNA fragment mixtures are pooled and then fractionated by electrophoresis on a gel that has the capacity to resolve DNA fragments differing by one base. The laser beam activates the fluorescence of the dyes in each lane as the bands migrate past and this signal is picked up by a detector, which sends the information to a computer. Thus the color of the band identifies the end base on the DNA fragment, and the order of the colored bands as they migrate past the detector translates into the DNA sequence. The commercial form of this machine runs 24 sequences simultaneously, each reading out 450–500 bases.

mosome), and analyze the sequences. We need to automate virtually every step in this production line to eliminate the many potential bottlenecks to high-throughput DNA sequencing.

The chance is perhaps 50 percent that in ten years some entirely new approach to DNA sequencing will be employed—scanning-tip electron microscopy, mass spectrometry, or something else. However, the current approach for DNA sequencing has the potential to be improved a hundred-fold or more. I envision in ten years instruments and/or strategies that will sequence one to ten million base pairs per day per technician.

The genome program poses striking problems for the computational sciences. Improvements are needed in signal processing, for example; if we can speed up our analysis of the fluorescent bands from the automated DNA sequencer, we can more than double the data output. The data bases will require advanced techniques for inputting, storing, and making readily accessible the three billion base pairs of genome sequence; they may also have to provide a hundred-fold more annotated description of this sequence. Another computational problem is string matching, the comparison of any new sequence generated against all of the sequences present in the data base to determine similarity of patterns.

To comprehend the string-matching problem, consider the following sequence:

```
TGCCTGGACTTCGCGCGACTATAGAGCGCGAGCGGCGTGAGC
GAGACCAGTTCGCAATGACTACGGTGCCACGCAAGGGTCGTG
CCTGGCTCACGAAGGGTAGTCCTTAGTGAAGTGGCGGCTTAT
GCCTGGACTTCGCGCGACTATAGAGCGCGAGCGGCGTGAGCG
AGACCAGTTCGCAATGACTACGGTGCCACGCAAGGGTCGTGC
CTGGCTCACGAAGGGTAGTCCTTAGTGAAGTGGCGGCTTATG
CCTGGACTTCGCGCGACTATAGAGCGCGAGCGGCGTGAGCGA
GACCAGTTCGCAATGACTACGGTGCCACGCAAGGGTCGTGCC
TGGCTCACGAAGGGTAGTCCTTAGTGAAGTGGCGGCTTATGC
CTGGACTTCGCGCGACTATAGAGCGCGAGCGGCGTGAGCGAG
ACCAGTTCGCAATGACTACGGTGCCACGCAAGGGTCGTGCCT
GGCTCACGAAGGGTAGTCCTTAGTGAAGTGGCGGCTTATGCC
TGGACTTCGCGCGACTATAGAGCGCGAGCGGCGTGAGCGAGA
CCAGTTCGCAATGACTACGGTGCCACGCAAGGGTCGTGCCTG
GCTCACGAAGGGTAGTCCTTAGTGAAGTGGCGGCTTATGCCT
GGACTTCGCGCGACTATAGAGCGCGAGCGGCGTGAGCGAGAC
CAGTTCGCAATGACTACGGTGCCACGCAAGGGTCGTGCCTGG
CTCACGAAGGGTAGTCCTTAGTGAAGTGGCGGCTTATGCCTG
GACTTCGCGCGACTATAGAGCGCGAGCGGCGTGAGCGAGACC
AGTTCGCAATGACTACGGTGCCACGCAAGGGTCGTGCCTGGC
TCACGAAGGGTAGTCCTTAGTGAAGTGGCGGCTTATGCCTGG
ACTTCGCGCGACTATAGAGCGCGAGCGGCGTGAGCGAGACCA
GTTCGCAATGACTACGGTGCCACGCAAGGGTCGTGCCTGGCT
CACGAAGGGTAGTCCTTAGTGAAGTGGCGGCTTATGCCTGGA
CTTCGCGCGACTATAGAGCGCGAGCGGCGTGAGCGAGACCAG
TTCGCAATGACTACGGTGCCACGCAAGGGTCGTGCCTGGCTC
ACGAAGGGTAGTCCTTAGTGAAGTGGCGGCTTATGCCTGGAC
TTCGCGCGACTATAGAGCGCGAGCGGCGTGAGCGAGACCAGT
TCGCAATGACTACGGTGCCACGCAAGGGTCGTGCCTGGCTCA
CGAAGGGTAGTCCTTAGTGAAGTGGCGGCTTATGCCTGGACT
TCGCGCGACTATAGAGCGCGAGCGGCGTGAGCGAGACCAGTT
CGCAATGACTACGGTGCCACGCAAGGGTCGTGCCTGGCTCAC
GAAGGGTAGTCCTTAGTGAAGTGGCGGCTTATGCCTGGACTT
CGCGCGACTATAGAGCGCGAGCGGCGTGAGCGAGACCAGTTC
GCAATGACTACGGTGCCACGCAAGGGTCGTGCCTGGCTCACG
```

This stretch of sequence represents about one-millionth of the human genome. We must be able to extract from a sequence like this a variety of information, including the boundaries of genes, the presence of regulatory elements, and the presence of sequences that may relate to specialized chromosomal functions such as replication, compaction, and segregation. The key to extracting this information is the ability to compare this sequence against all preexisting sequences to test for similarities. We have approached the string-matching problem by the development of a specialized coprocessor, the Biological Information Signal Processor (BISP), which converts the Waterman-Smith algorithm, the most general approach for sequence similarity analysis, into a silicon chip (Figure 18). The BISP is about one centimeter square and contains 400,000 transistors; it is the most complex chip that the Jet Propulsion Laboratory at Caltech has ever designed. Its performance, measured against that of far more expensive computers, is strikingly rapid (Table 10). Clearly, close cooperation between biologists and computer scientists will be not merely advantageous but necessary to solving the complex and difficult problems inherent in the human genome project.

Interactive environments like the Science and Technology Center for Molecular Biotechnology, where many different disciplines can be focused on the development of the wide spectrum of techniques needed, are key to the success of the genome project. The human genome project needs to attract talented scientists from computer science, applied physics, applied mathematics, engineering, and chemistry, as well as many disciplines within biology itself. Scientists in these disciplines may be momentarily interested in biological problems such as the human genome project, but it is difficult to persuade them to make a long-term commitment. A critical question is, How can more scientists from other disciplines be brought into these efforts?

One approach to the problem is to create a new kind of biologist—mainly by establishing Ph.D. programs in biotechnology that build bridges to other disciplines. Such programs would select students who wish to major in one area of biology, such as molecular biology, and in another discipline, such as computer science. The student would have a mentor in each area and take appropriate qualifying examinations in each. The objective would be to choose, for example, a fundamental problem in molecular

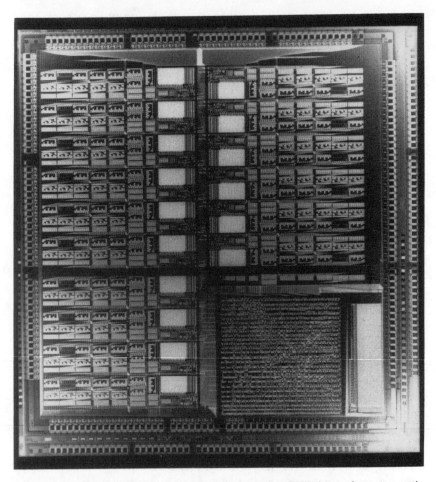

Figure 18   The Biological Information Signal Processor: a BISP chip is about one centimeter square and contains 400,000 transistors.

Table 10   The superior performance of the Biological Information Signal Processor: completion times for four systems comparing a 500-base sequence against a data base of 40 million bases (using a Smith-Waterman dynamic programming algorithm)

Computer	Time
Sun Sparcstation 1	5 hours
Cray 2	12 minutes
Connection Machine 3	1 minute
BISP	3.5 seconds

biology and then develop and apply a tool in computer science that could be applied to it, thus bringing computer science into biology through the student. This program would develop interdisciplinary scientists, those with expertise in biology and other disciplines and the ability to forge interdisciplinary collaborations. Moreover, these students will be the channels through which biologists and scientists from other disciplines may actively collaborate on developing biologically oriented techniques. I believe interdisciplinary scientists will play a major leadership role in biology and medicine of the twenty-first century.

Interdisciplinary collaboration will be essential to the progress of biology in the next century. The future of biology will depend upon the analysis of complex systems and networks that may involve molecules, cells, or even arrays of cells. If we are ever to understand such systems, the individual elements in the network must be defined, as must the nature of their connectivity. Computer models will be required to explore network behavior when individual elements are perturbed. Finally, the modeled behavior will then have to be tested on real biological systems. The living systems may be whole organisms or appropriately reconstructed subsystems of organisms. The human genome project will take a big step forward in identifying key elements of the complex system responsible for human growth and development by delineating the elements of the 100,000 human genes.

Once the sequence of the entire human genome is known, various computational and biological approaches can be taken to determine the location of the 100,000 genes. Several computer programs have combined the various general features of genes so that they may be identified among raw DNA sequence data—by looking for special base compositions of coding regions, for example, or for special sequences at exon-intron boundaries. Another approach will compare newly analyzed sequence data against all preexisting gene sequences from humans or model organisms, the idea being that sequence similarities may help reveal gene boundaries. Finally, the sequence of the human and mouse genomes will be compared. The mouse contains most human genes. The coding regions (and regulatory elements) are far more highly conserved during evolution than the intervening DNA that surrounds the genes. Accordingly, an important element in the genome project will be the comparative sequence analysis of human and mouse DNA to aid in the identification and analysis of genes. The identification of coding regions is made more difficult by the fact that many genes show alternative patterns of RNA splicing; from the same gene sequence on DNA, several different messenger RNAs may be transcribed that splice together different combinations of exons or place particular exons at different sites. In the end, to define all the alternative forms of particular genes, one may have to study carefully the messenger RNAs in appropriate tissues. Nevertheless, the identification of most of the 100,000 human genes will provide biologists with an enormously powerful tool for exploring many aspects of contemporary biology.

Some biologists have argued that the copy DNAs (cDNA) of the mRNA should be sequenced rather than the DNA of the genome. The cDNA provides a direct read-out of the coding regions of genes, and these sequences, termed *expressed sequence tags* (ESTs), could also be used as markers spread throughout the genome to facilitate obtaining DNA fragments for a physical map. Since each automated DNA sequencer could readily define 5,000 ESTs per year, ESTs for many of the 100,000 or so human genes could readily be obtained early in the genome program.[6] This truly represents a biological gold rush—and it has many fascinating implications. ESTs will permit a rapid assessment of the human genes through similarity analyses and raise fascinating patent questions (see Chapter 14). The regulatory elements will not be expressed in ESTs, nor will many other sequences important for general chromosomal functions. Moreover, for a variety of technical reasons, not all human genes can be identified by the EST approach (see below). Hence genomic and cDNA sequencing are both important for the genome program.

Each gene has regulatory elements or special DNA sequences that generally fall within 500 to 5,000 base pairs of the boundary of the gene itself (Figure 19). The regulatory elements function by virtue of the fact that specific DNA binding proteins interact with them. Called transactivating factors, they have three distinct functions. They control the temporal (developmental time) and spatial (tissue site) modes of expression and thereby coordinate a gene's expression in particular cells with that of thousands of other genes. They also control the amplitude of expression. For exam-

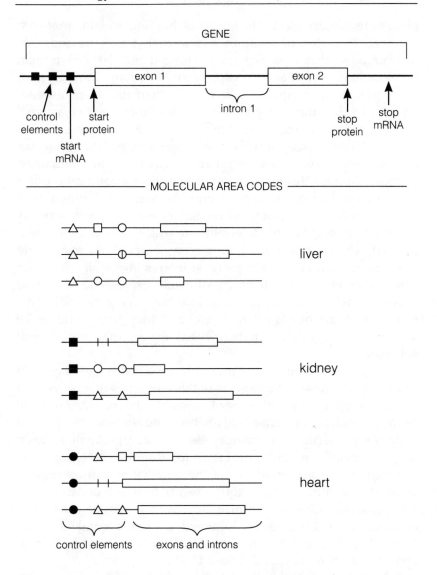

Figure 19 Only some parts (the exons) of the sequence that makes up a gene are transcribed into mRNA, the actual instructions for making the gene products. In addition, the gene also has noncoding regions (the introns) and control elements that regulate the expression of each gene. Also indicated here are the special control elements that serve as start and stop sites for mRNA and protein. The control elements for each gene constitute a molecular area code address. These regulatory addresses will someday permit us to identify, by computer analyses of their regulatory sequences, the temporal and spatial sites of expression as well as the amplitude of expression for each gene. Hence, they will permit us to determine in which organ or cell type particular genes are expressed.

ple, the regulatory elements and DNA binding proteins that control the expression of albumin dictate that it is expressed only in liver cells, that it is expressed late but not early in human development, and that it is expressed at perhaps thousands of times greater concentrations of mRNA than the average gene. These three functions may be written as a "molecular area code" (see Figure 19). The idea is that the three elements of gene expression will be dictated by specific DNA sequences and that these can be deciphered, just as we decipher a regular telephone number: suppose the first three digits determine the spatial location of a particular sequence; the second four the temporal location, and so on. In other words, defined regulatory elements may serve as molecular area codes to determine in which cells a gene is expressed, when it is expressed during development, the magnitude of its expression and, perhaps most interesting of all, the other genes with which it will be coordinately expressed. Molecular area codes will be an important tool for identifying the individual members of the biological network, and they will therefore be a part of the regulatory network that the genome project will delineate.

The regulatory elements or molecular area codes will in general be found in precisely the same way that the genes themselves are found. Comparisons with other known regulatory elements will be carried out by computational analysis and, in time, the general sequence properties of regulatory elements can probably be used to create specific computer programs to recognize these elements. In addition, cross-comparisons of the putative regulatory regions in mouse and human DNA sequences will be useful in delineating the regulatory elements, because they, as with their gene counterparts, will be highly conserved. Indeed, the first mammalian regulatory element ever identified was found because it is highly conserved in both human and mouse DNA.

The study of individual proteins in biology has traditionally started with the identification of a particular function, the development of an assay for this function, and the use of the assay to purify the protein that carries out the function. After sequencing the protein (that is, determining the order of its amino-acid subunits), the genetic code dictionary is used to translate the protein into DNA sequence; DNA probes are then synthesized and the gene is cloned by conventional recombinant techniques. The genome project will reverse this approach. In the future, we will know the 100,000 human genes and will have to develop new approaches and tools for ascertaining their functions. Indeed, the genome project will empower us to analyze genes that are inaccessible to the contemporary techniques of molecular biotechnology. For example, more than half of our genes are expressed in the brain, and many of them are expressed for such a short time during development and in so few cells that virtually no contemporary techniques would permit us to identify them. Perhaps we will be able to identify some of them only through direct sequence analyses of genomic DNA.

How might one go about ascertaining the function of newly discovered genes? First, one can do a search through existing data bases to see whether other genes of known function exhibit similar sequence characteristics. Second, the molecular area codes provided by the regulatory elements will generate insights as to the temporal, spatial, and coordinate expression of genes that may be useful in speculating on gene functions. Third, the genes may also provide information as to where in the cell the functions of corresponding genes are localized, once again providing insight into function. Finally, many genes may be present in model organisms whose genomes will be sequenced by the genome project. Thus, if a gene that corresponds to an unknown human gene, is found in the fly or the nematode, the model organism may be used for experimentation to discover the function of the gene in humans.

The sequences of all human genes will permit us to identify the corresponding proteins. This information will in turn allow us to find the motifs and domains that are the building blocks of proteins (Figure 20). Domains are the individual functional units within the protein; motifs are the building-block components for each domain. Indeed, a protein might be likened to a train. The domains would be the individual cars in the train, each different type of car—the flat car, the engine, the caboose—carrying out a different function. The motifs of a domain would be the individual components of cars, such as the wheels, the car sides and windows. A protein may have from one to fifteen or even more domains. For example, the antibody molecule that protects humans against foreign invaders (such as viruses and bacteria) folds into six domains, two of which are involved in recognizing the invad-

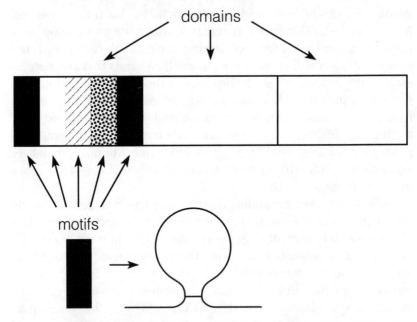

Figure 20 Proteins are composed of smaller building blocks, motifs and domains. Different domains have different functions, such as identification or elimination of a particular molecule. Motifs are the structural components of the protein molecule.

ers and four of which are involved in destroying or eliminating them. Each domain is composed of smaller motifs, called "β-pleated sheets." Having the sequence of all human proteins will allow us to use computational techniques to define domains and motifs. Indeed, if we identify the possible 100 to 500 motifs that are the fundamental building-block components of proteins, we will have a valuable tool for understanding the functions of the protein and how the order of the amino-acid subunits determines its three-dimensional structure. This is the so-called protein-folding problem.

The protein-folding problem is one of the major unsolved mysteries of contemporary biology. Within the next fifteen to twenty years it may be possible to decipher the folding rules so that, given the primary amino-acid sequence of a particular protein, we can predict what its three-dimensional structure would be. It is clear that the protein motifs may play a fundamental role in this process: that is, once the structure of a particular motif has been determined, then all variant forms of this motif expressed in different proteins will have very similar structures. If we could determine the 100 to 500 basic structures of the protein motifs, then we would have a structural alphabet for understanding how proteins are assembled in three dimensions. Other approaches will also facilitate solutions to the protein folding problem. I have in mind theoretical calculations such as energy minimization, *in vitro* mutagenesis to alter a gene's DNA sequence rationally so as to determine how the corresponding protein structure changes, as well as an examination of many additional proteins with high-resolution three-dimensional structures.

Once we can predict how a protein will fold in three dimensions, still another task will remain: to predict, from first principles, the function of the protein and understand how its structure and function are related. It is interesting to note that there is still no protein in contemporary biology for which we understand completely how its structure enables it to carry out its function. The step from structure to an understanding of function is a challenging one. Once again, new tools and approaches will have to be developed to take it.

The genome project in the twenty-first century will have a profound impact on medicine, both for diagnosis and therapy. The development of automated instrumentation for examining DNA polymorphisms raises the possibility of identifying the polymorphic forms of genes that cause disease or predispose individuals to disease. The ability to recognize particular DNA sequences by molecular complementarity between probe and target DNA is known as DNA diagnostics (Figure 21). This technology will figure in the diagnosis of genetic diseases whose single-gene defects have been identified; in determining the presence of dominant or recessive oncogenes that may predispose an individual to cancer; in the identification of infectious agents, such as the AIDS virus; and in forensics—that is, the use of DNA fingerprints to identify the donor origins of any tissue or blood sample. Perhaps the most important area of DNA diagnostics will be the identification of genes that predispose individuals to disease. However, many such diseases—cardiovascular, neurological, autoimmune—are polygenic; they are the result of the action of two or more genes.

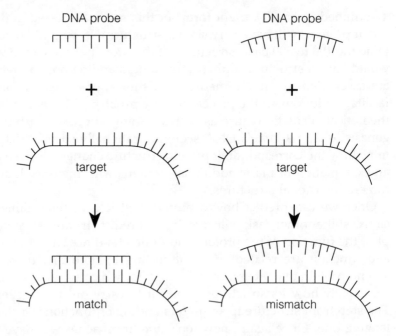

Figure 21   DNA diagnostics employs a stretch of known DNA sequence (a probe) to determine whether a gene from a patient's (target) DNA is complementary or not. If the probe is complementary to a normal gene, then a mismatch will indicate a mutation in the patient's DNA.

Human genetic mapping will permit the identification of specific predisposing genes and DNA diagnostics will facilitate their analysis in many different individuals.

To illustrate, Table 11, abstracted from a recent review in the *New England Journal of Medicine*, gives the factors that predispose one to cardiovascular disease, the leading killer in the United States today. These factors fall into two categories—modifiable and nonmodifiable. The majority of the nonmodifiable factors are genetic in origin. In the future, we will be able to identify the various genes that code for cardiovascular function (Table 12) and, through DNA diagnostics, to identify whether an individual possesses genes that predispose him or her to cardiovascular disease. A second example is the observation made by ourselves and others that two and possibly three different immune receptor genes—one on chromosome 6, one on chromosome 7, and one on chromosome 14—predispose certain humans to the autoimmune

Table 11   Risk factors for cardiovascular disease

Genetic predisposition Male Aging	⎫ ⎬ Unmodifiable ⎭
Elevated levels of low-density lipoprotein cholesterol Low levels of high-density lipoprotein cholesterol Smoking Arterial hypertension Physical inactivity Obesity Type II diabetes	⎫ ⎬ Modifiable ⎭

Table 12   Polygenic factors potentially contributing to cardiovascular disease

Cells	Factors
Endothelial Platelets Monocytes/macrophages Vascular smooth muscle Fibroblasts	Variety of growth factors and chemoattractants

Genetic differences

- Production of growth factors or chemoattractants
- Responses to these factors
- Host of paracrine and autocrine factors
- Production of thromboxane in platelets or prostacyclin in endothelial cells

disease multiple sclerosis. Moreover, therapeutic approaches will be designed to circumvent the limitations of these defective genes. Approaches to circumvention might include new techniques in molecular pharmacology, special manipulations of the immune system (immunotherapy), appropriate avoidance or manipulation of environmental factors such as smoking, and, in the future, genetic engineering to replace defective genes in certain tissues with good genes.

The diagnosis of disease-predisposing genes will alter the basic practice of medicine in the twenty-first century. Perhaps in twenty years it will be possible to take DNA from newborns and analyze fifty or more genes for the allelic forms that can predispose the infant to many common diseases—cardiovascular, cancer, auto-immune, or metabolic. For each defective gene there will be therapeutic regimens that will circumvent the limitations of the defective gene. Thus medicine will move from a reactive mode (curing patients already sick) to a preventive mode (keeping people well). Preventive medicine should enable most individuals to live a normal, healthy, and intellectually alert life without disease.

It has been estimated that the identification of the gene for cystic fibrosis cost approximately $150 million dollars. If the genetic and sequence maps of the human genome were known, it would be possible to identify a particular disease or predisposing genes for perhaps $200,000. In the future, we will use our detailed genetic maps to localize a particular disease or predisposing gene to a specific chromosome; indeed, we will find it within a two-centimorgan region within that chromosome. Then we will use the sequence information for this smaller region to identify the specific sequence responsible for the particular disease gene. Thus, the identification of disease genes will become a simple, straightforward, and inexpensive process.

Once the human and mouse genomes have been determined, we will be able to model human gene defects in the mouse. Techniques are now being developed whereby genes can be precisely placed in their appropriate location in the chromosomes of embryonic stem cells, and these cells in turn can develop into mice. Accordingly, once the mutation for Huntington's disease has been identified, the corresponding gene defect could be created in the homologous mouse gene. The mouse would be a model for studying means of circumventing the disease, at least until genetic engineering might be able to correct the consequences of this tragic mutation. In this manner, we will be able to model a variety of different human diseases in mice to determine appropriate therapeutic approaches.

Once the 100,000 human genes have been identified, they will be used as therapeutic reagents for dealing with all aspects of human disease. If we can use molecular area codes to identify all of the genes expressed in a particular cell, such as the lymphocyte, then we can begin to model, to experiment, and hence to understand in some detail the interactions of genes that generate this unique cell phenotype. These studies are beyond the genome program, but as with protein folding, the identification of all human genes will provide key insights for subsequent analyses. Likewise, if one can query the computer for "heart" and obtain a listing of the genes that are expressed in the heart, so one can begin to model, experiment, and understand in detail the physiology of this organ and its pathology as well. Likewise, the genome project should have a major impact on our understanding of the brain. Our ability to understand how networks of neurons interact with one other and transmit information may be facilitated by an understanding of the most basic building-block components of these networks, the genes that specify the proteins that are active in the brain. Once we comprehend the normal physiology of various organs and organ systems, then we can begin to comprehend in detail the consequences of subtle disease pathologies and design appropriate therapeutic responses.

The benefits of the human genome project to industry are likely to be enormous, both through the information available from sequence and genetic maps and from the development of new techniques and instrumentation. Knowledge of the 100,000 human genes will provide a vast therapeutic repertoire with which the pharmaceutical industry can attack fundamental aspects of human disease. The spectacular success of erythropoietin (EPO), a hormone that promotes the development of red blood cells, and granulocyte-colony stimulating factor (G-CSF), a hormone that promotes the development of white blood cells to fight infections, is evident in therapies for chronic anemia and cancer, respectively. In the future, we can expect to have literally hundreds, if not thousands, of additional proteins that will facilitate the development of therapeutic approaches to a variety of different diseases.

DNA diagnostics and the identification of genes that cause and predispose to disease will place enormous pressure on the pharmaceutical industry to come up with appropriate therapeutic strategies. The gap between the ability to diagnose and the ability to treat genetic diseases could well be five to twenty or more years.

One striking new approach to the control of gene expression is the use of anti-sense nucleic acids. This entails the use of nucleic acid probes that can bind to nuclear RNA and block its processing or exit from the nucleus or that can bind directly to the gene to prevent its transcription into RNA. These approaches are in the earliest stages of exploration, but if they are successful anti-sense therapy will be strikingly specific, in that it will enable the regulation of specific genes to be precisely controlled. These approaches may have important implications for many major human diseases including cancer, cardiovascular disease, immune diseases such as allergies, and autoimmune diseases. Clearly the delineation of the 100,000 human genes will provide vital DNA sequence information for the anti-sense strategies.

As the protein-folding problem is solved, exciting new possibilities for therapy will arise. It will be possible to design new therapeutic proteins of virtually any desired shape. For example, the genes in tumor cells often may express unique tumor-specific molecules, or antigens (Figure 22). Once a particular tumor antigen (or gene) has been sequenced, its three-dimensional structure can be deduced. A recognition unit can then be designed that is complementary to the tumor antigen and that has a killing domain attached to the recognition unit, which will function to destroy any tumor cell when the recognition unit attaches. In this manner, individually specific therapeutic reagents can be designed for many different tumors. If it is to succeed, this strategy requires the identification of unique or highly specific tumor antigens, an objective that should be reached within the time required to solve the protein-folding problem, which will likely come within the next fifteen to twenty years. The ultimate molecular engineering objective for the pharmaceutical industry is to design small organic molecules that have long half-lives and that may be taken orally as a replacement for protein therapeutic reagents. What the genome project will provide is 100,000 three-dimensional shapes (proteins) that execute the functions of life; these shapes can be used to engineer appropriate small molecules with diverse therapeutic potentials.

New industrial opportunities will arise from DNA diagnostics. They will encompass those aspects of medicine previously discussed and many additional applications. DNA fingerprinting may be used to identify personnel in the armed services. Applied to animals, DNA diagnostics will unequivocally delineate the parentage of prize dairy cattle or race horses. Genetic maps will be created for the major crop plants and be employed to identify and

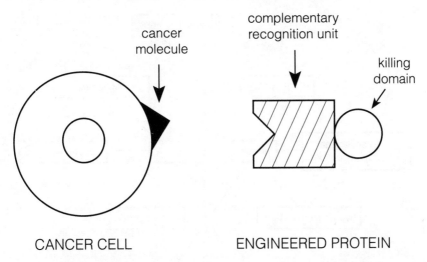

Figure 22 A solution to the protein-folding problem will offer great therapeutic bene-fits. Once the DNA sequence that codes for a cancer molecule is known, for example, the three-dimensional structure of the molecule (shown here as a triangle) may be discovered. Then it will be possible to engineer a protein that will attach to the mole-cule (obviously a round shape would not work in this example) and destroy it.

ultimately to engineer for desirable polygenic traits, such as a higher protein content or better taste.

By spawning the development of new technologies and instru-mentation, the genome project will obviously create opportunities for companies that now produce biological instrumentation. For example, chemical and biological robots will be needed for routine tasks such as cloning, mapping, or sequencing. Opportunities will arise for companies to offer commercially many services that are currently provided primarily by molecular biologists. These in-clude genetic mapping, DNA sequencing, cloning, and gene transfer to cells or organisms, to name only a few.

There will be striking future industrial opportunities in biocom-puting. New software will be needed for the signal processing and image analysis associated with a wide variety of analytic and preparative instruments: DNA sequencers, chemical and biologi-cal robots, DNA mappers, mass spectrometers, NMR machines, X-ray crystallography, and so on. The combinatorial problems of biology—string matching, for example—will require the develop-ment of new algorithms, the development of new hardware such

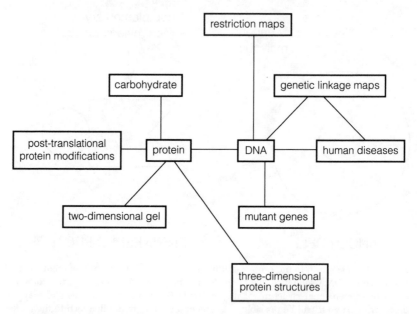

Figure 23   A model of some of the biological data bases and network connections that will be needed in the future.

as specialized coprocessors, and, increasingly, the use of parallel computers. In the future, there may be more than one hundred distinct biological data bases (Figure 23). It will be an enormous challenge to maintain these data bases as well as to make all of them readily accessible to the biologist or physician user. The development of new object-oriented data bases, which can organize information in keeping with its functional attributes, provide interesting new possibilities for instantaneous accessibility. It is also clear in the future that biologists are going to be quite dependent on the computer modeling of complex systems and networks to create new hypotheses that can then be tested in biological systems or living organisms. The opportunities in biological computer modeling will be enormous.

The United States is now the undisputed world leader in biotechnology. The genome project will help to ensure that we retain this lead. An important question is, To what extent can U.S. industry take advantage of this leadership? Without a national commitment to the support of long-term research endeavors and of their possible commercial opportunities, the outlook is uncertain.

The human genome project is unique in several regards. Since it is one of the first major biological initiatives that has technology development as a centerpiece, the need is tremendous for an inter-disciplinary attack on challenging mapping, sequencing, and informatics problems. These problems will require the application of leading-edge techniques and instrumentation from applied mathematics, applied physics, chemistry, computer science, and biology. Moreover, the genome project, if successfully executed, will enormously enrich the infrastructure of biology by providing biologists and physicists with computer access to genetic, physical, and sequencing maps. For example, identifying the molecular addresses encoded in the regulatory elements of human genes will make available powerful data for attacking fundamental problems in developmental biology. Likewise, identifying the lexicon of perhaps 100 to 500 protein motifs may lead to valuable insights for attacking the protein-folding problem. Neither developmental biology nor protein folding is a problem inherent to the genome project; rather, the genome project will provide new tools for attacking these fundamental problems in other areas of biology. This infrastructure will fundamentally alter the practice of biology and medicine as we move into the twenty-first century. It will also secure the leadership of the United States in biotechnology and present U.S. industry with a wealth of opportunities.

To some, much of this discussion may appear fanciful science fiction. Yet the pace of biological discovery and technological advances continues to accelerate. This is truly the golden age of biology. Twenty years ago few of us could have imagined where we would be today. I suspect that, if anything, I have greatly underestimated the magnitude of the changes that will come, in part, as a consequence of the human genome project. I believe that we will learn more about human development and pathology in the next twenty-five years than we have in the past two thousand.

A Personal View of the Project

7

When I was going into science, people were concerned with questions of where we came from. Some people gave mystical answers—for example, "the truth came from revelation." But as a college kid I was influenced by Linus Pauling, who said, "We came from chemistry." I have spent my career trying to get a chemical explanation for life, the explanation of why we are human beings and not monkeys. The reason, of course, is our DNA. If you can study life from the level of DNA, you have a real explanation for its processes. So of course I think that the human genome project is a glorious goal.

People ask why *I* want to get the human genome. Some suggest that the reason is that it would be a wonderful end to my career—start out with the double helix and end up with the human genome. That *is* a good story. It seems almost a miracle to me that fifty years ago we could have been so ignorant of the nature of the genetic material and now can imagine that we will have the complete genetic blueprint of man. Just getting the complete description of a bacterium—say, the five million bases of E. coli— would make an extraordinary moment in history. There is a greater degree of urgency among older scientists than among younger ones to do the human genome now. The younger scientists can work on their grants until they are bored and still get the genome before they die. But to me it is crucial that we get the human genome now rather than twenty years from now, because I might be dead then and I don't want to miss out on learning how life works.

Still, I sometimes find myself moved to wonder, Is it ethical for me to do my job? A kind of backlash against the human genome project has cropped up from some scientists—good ones as well as not so good ones. What seems to have outraged many people was that, in 1990, against the proposed increase of 3.6 percent in the president's budget for all NIH funds, the human genome project was proposed for an increase of 86 percent—from roughly $60 million to $108 million. Feeling dispossessed, some scientific groups have begun to behave like postal workers' unions. The biological chemists, the molecular biologists, and the cell biologists have hired a lobbyist, a former congressman from Maine, to get the overall NIH appropriation increased. If such moves succeed, then maybe we won't have this terrible situation of really good scientists claiming that they are not getting funded because all the money is going to the human genome project.

In the meantime, hate letters have made the rounds, including the rounds of Congress, contending that the project is "bad science"—not only bad, but sort of wicked. The letters say that the project is wasting money at a time when resources for research are greatly threatened: If good people are failing to get grants, why go ahead with a program that is just going to spend billions of dollars sequencing junk? In 1990, someone in my office tried to get a distinguished biologist to help peer-review a big grant application. The biologist said, "No, not the human genome!" as though he were talking about syphilis.

The biologist sent me a FAX asking me to explain why he should not oppose the human genome program. I called him up and said that, though I couldn't prove it, Congress actually seemed to *like* the human genome program because it promised to find out something about disease. Congress was excited that maybe we scientists were worried about disease instead of just about getting grants. The primary mission of the National Institutes of Health is to improve American health, to give us healthier lives, not to give jobs to scientists. I think that the scientific community, if it wants to be ethically responsible to society, has to ask whether we are spending research money in a way that offers the best go at diseases.

The fact is that understanding how DNA operates provides an enormous advantage over working only with proteins or fats or carbohydrates. The best illustration of this advantage has been tumor viruses. If we had not been able to study cancer at the level of the change in DNA that starts it, the disease would still be a hopeless field. Every time a new enzyme was discovered, hope would rise that it was the cause of cancer. Cancer used to be considered a graveyard for biochemists, even good ones, many of whom wanted to cap their careers by solving cancer but failed. Not until the genetic foundation for cancer was identified could you really begin to say what goes wrong to make this terrible human affliction.

A similar example is Alzheimer's disease. Are we going to find out what Alzheimer's is and why it causes brain failure without getting the genes that we know predispose certain people to the disease? Maybe we will, but I would not bet on it. But if we can get the gene or genes implicated in the disease, I am confident that we will save hundreds of millions of dollars, if not billions, that would have been spent on worthless research.

Every year, Congress passes a bill for even more money to study Alzheimer's. Congress is voting for good goals, but we do not really know how to use the money. It is not as if all the federal budget for health and all the basic research grants add up to good research. All the study sections in the National Institutes of Health do not receive applications of equal value; they often endorse research projects or programs because they address important problems. The programs themselves are not terrible, but they often have a low probability of paying off. I am sure that half the NIH budget is spent on good intentions rather than on a realistically high probability that a research program will have a direct impact on one of the major human diseases.

The pressure is enormous to do something about mental disease because it can be terrible, as anyone knows who has a friend or family member suffering from it. We do spend a vast amount of money studying mental diseases, yet the effort yields very little. Manic-depressive disease leads to great moments of mania—perhaps the successful careers of a number of scientists can be attributed to it—but it also leads to depression, tragedy, and suicides. Lithium relieves some of the symptoms, but a drug is not the complete answer, as any psychiatrist will tell you. It is pretty clear that manic depression has a genetic cause. Several scientists thought they had located the gene on a chromosome. But then it got lost, and so long as it is lost, we are lost.

It is also pretty clear that alcoholism bears some relationship to genes. This view comes from studies on identical twins adopted and raised by different families. There *are* alcoholic families. It is not likely that their members are morally weak; they just cannot tolerate alcohol chemically. But no one has found the gene or genes for susceptibility to alcoholism, and the chance of finding the genetic sources are probably low until a much more sophisticated human genetic community exists—plus the money to get the pedigrees and all the genetic markers.

Some diseases are not going to be easy to crack. For a long time, people have been trying to discover the cause of schizophrenia by looking for chemical differences in the urine or the blood, a research strategy that has not been successful. It is not going to be easy to find the genes behind schizophrenia either, because reliable pedigree data are difficult to compile and the condition is hard to diagnose. Thus both directions offer low probabilities, but it is still better to waste your money doing genetics because genetics lies at the heart of so much. Of course scientists should find out what the brain is. I believe in neurobiology and have tried to help raise money to support the field. But I do not believe that its current approaches will necessarily lead to the real, deep cause of manic-depressive disease.

In 1989 Congressman Joe Early said to me, "I'm tired of putting fingers in dikes!" In combating disease, genetics helps enormously if it is a bad gene that contributes to the cause. Ignoring genes is like trying to solve a murder without finding the murderer. All we have are victims. With time, if we find the genes for Alzheimer's disease and for manic depression, then less money will be wasted on research that goes nowhere. Congressmen can only feel good if they are spending money on good things, so we have to convince them that the best use for their money is DNA research.

The human genome project is really trying to push a little more money toward DNA-based research. Since we can now produce good genetic maps that allow us to locate culprit chromosomes and then actually find the genes for disease (as Francis Collins found the gene for cystic fibrosis), genetics should be a very high priority on the agenda of NIH research. We are extremely lucky that when James Wyngaarden was director of NIH, he saw to the establishment of what is now a permanent division within NIH called the Center for Human Genome Research. I doubt that I convinced the biologist who sent me the FAX, but I may eventually, since he is very bright. I want to convince as many people as I can of the merits of the human genome project, but not to cap my career and have something that sounds good in my obituary. I can make best use of my time by trying to mobilize the country to do something about diseases that have hit my family and many others. I am sort of a concerned parent for whom things have not gone completely right. So, I am trying to enlist a group of people who will help us get these genes, and do what I think Congress wants us to do.

The ultimate objective of the human genome program is to learn the nucleotide sequence of human DNA. We want the program completed in roughly fifteen years. By completed we do not mean every last nucleotide sequence. If we get 98 percent of the regions that are functional, that will probably be the end of it. We will not worry about spending infinite amounts of money trying to sequence things we know probably contain little information. We could define the end of it to be the identification of all the human genes—that is, we will be done when we have located the coding sequences and can declare that human beings on the average contain, say, 248,000 genes, with variations such that some individuals, for example, have a gene present in four copies and some in three, and that for some the gene is nonessential. It has recently been learned that only a third of yeast genes are essential. Knock out two-thirds of them and the yeast still multiply. Studying things that are not essential will keep the people in the yeast world going for a long time. I think we can safely say the project will be over when we can identify the genes.

We probably will be unable to identify the genes until we get most of the DNA sequenced, because we will not know where they are. It would be nice if the whole program could be done by copy DNA (cDNA)—that is, by purely functional DNA—so that we would not have to sequence all the junk, but we will never know whether we have all the cDNAs. This is not to say we should not do cDNA; we will actually fund grants for people trying to find better techniques for getting rare cDNA in tissue-specific places. But I think that we have got to sequence the whole thing.

In the first five years, we will push to achieve three major objectives. First, we will try to get good genetic maps, so that each chromosome has enough genetic markers on it actually to locate a gene if a pedigree is available. Currently, we have only about 150 markers that are sufficiently informative for assigning the location of genes. We have started a crash program to persuade people to make a lot of markers and to put them into a public repository made available to the whole world. We want to change the current practice among researchers of not sharing their markers because they want to be the first to find a gene and encourage everyone to make markers available to everyone.

The second objective is to make overlapping fragments of DNA available so that anyone looking for a gene in a particular piece of a certain chromosome will be able to get it by paying some nominal sum. The fragment will not be totally free, but it will certainly be there for anyone who seriously wants it. Techniques for doing this seem to be available now; it should not require more than $10 million to stockpile overlapping fragments of a given chromosome. To put this figure into perspective, Francis Collins has said that finding the cystic fibrosis gene was expensive— between $10 million and $50 million. If all the markers had been available, it would have cost only $5 million. I think we can establish an overlapping fragment library for the entire human genome for a couple of hundred million dollars, which will certainly reduce the costs of subsequent disease hunts. We will end up with a map of overlapping fragments, each one identified by three or four DNA sequences along it called sequence tag sites. With PCR, researchers will be able to pull out all the human DNA that may be wanted.

The third major objective is to support scientists trying to do megabase sequencing in one place in a reasonable period of time. An example of this type of project is a proposal from Walter Gilbert to sequence a mycoplasma, which is really a small (800 kilobases) bacterium. Gilbert's proposal, whether he lives up to it or not, is to do a million bases a year within two years. We want to encourage people to do sequencing of megabases with the aim of reducing the cost—so that within a couple of years it will fall to about a dollar a base pair, and then perhaps even to fifty cents. We will not accept a grant application from someone who proposes to sequence some DNA the old fashioned way, with graduate students or postdoctoral fellows, at the current cost—five to ten dollars a base pair—just out of curiosity about it.

People continue to work in the old-fashioned way, but I have my doubts that it advances careers. It used to be that you could get a job if you could sequence DNA. Now, if you sequence too much, you probably cannot get a job because you have done nothing interesting. We human genome projecteers are actually *good* people; we want to save graduate students and postdocs from ever having to sequence by giving them a tool. We want sequencing to be done by much cleverer ways—by machine or by multiplexing and with automatic gel readers—so that researchers will not have to go crazy just doing the same sequencing procedures over and over.

A Japanese scientist told me a very unlikely story—one so unlikely that it must be true. He was describing the Japanese effort to sequence a chloroplast DNA, which was about 120,000 base pairs. Two groups were in a competitive race in Japan to get the sequences of a few different chloroplasts. Both came out successful, but mutiny broke out in one of the teams. It is imaginable that an American graduate student might tell his supervisor to go to hell; it is unimaginable that a Japanese graduate student might do the same. In the face of the extraordinary mutiny, the Japanese supervisors decided that forced-labor sequencing was too inhumane and resolved to change the system.

We hope to spend 10 to 20 percent of our total money trying to develop sequencing methods that could make the life of future students more humane. We face the problem of convincing NIH study sections—those peer-review bodies that assess and approve research proposals—to take a sufficiently adventurous attitude toward the development of fast-sequencing technologies. They tend to be willing to fund something only if they know it can be done. What we have to fund are projects whose outcome is uncertain, and we know no way to proceed other than to trust the investigator who has a good idea and to give out the money. Since we have not yet done a megabase in a single project, that makes for a problem in obtaining study-section approval. In contrast, mapping will breeze through peer review because many scientists have shown it can be done. I am confident that with all the brains in our field, we can reduce the cost of sequencing by a factor of ten.

The NIH genome project will also try to get some real data on model organisms. I will be happy if we get ten quite different bacteria sequenced up through yeast. We are now supporting a joint program between the Medical Research Council, in England, and the Laboratory of Molecular Biology in Cambridge, and the group in St. Louis that has developed yeast artificial chromosomes to sequence the genome of a roundworm. The roundworm community is eager to do it because they've already got the overlapping DNA fragments. We hope to get the sequence out in ten years. It's about the equivalent of an average human chromosome—about a hundred megabases—but with less repetitive DNA, and so probably with fewer problems. There is also an effort to sequence a plant genome, arabadopsis, which we hope will be led by the National Science Foundation with help from other agencies, including ourselves. This is roughly seventy megabases, and the project should be a real boon to botany. Except for perhaps one bacterium, none of this probably would ever have been funded in the absence of the human genome program.

Among the reasons for wanting to find bacterial genes is to help find the human ones. People ask, How are you going to identify a gene if it is interspersed with so much junk and you lack a cDNA? How are you going to know you have it? That is obviously going to be hard in some cases, but if you have obtained the corresponding bacterial gene without many repetitive sequences and if you are clever, you ought to be able to spot the differences. I can imagine that typical work for undergraduates will be to find the gene once all the sequence has been obtained. Professors could tell their students: If you can identify a gene, we will let you go on to graduate school and do real science.

The human genome project is sufficiently justifiable so that if no other country wants to help fund it, the United States should do the whole thing. We are rich enough to do it. But I doubt that we will be allowed to do it alone, because others are going to worry that it might actually be commercially interesting, and they will worry that we will be disinclined to distribute the data very fast if we have paid for it ourselves. It is my hope that we can spread out the cost of sequencing and data distribution over many countries. As soon as a gene has been identified, it should be thrown into an international data base.

But there are problems that I don't see how to get around. If a stretch of DNA is sequenced in an academic laboratory, a university lawyer will say, "That looks like a serotonin receptor. Patent it!" Mutant forms of the cystic fibrosis gene have been patented by the universities of Toronto and Michigan. They will get some royalties and maybe build better student unions with the revenues. I am at a loss to know how to put valuable DNA sequences in the public domain fast when a lot of people want to keep them private. I just hope that other major nations come in. The Japanese will not let anyone who doesn't pay for it see their work. I figure that strategy might work. People might actually pay for sequence information if that is the only way to get to see it. So I have to seem a bad guy and say: I *will* withhold information that we generate if other countries refuse to join in an open sharing arrangement. But, in truth, it would be very distasteful to me to get into a situation where we were withholding the data for reasons of national advantage.

The acquisition of human DNA information has already begun to pose serious ethical problems. I think that somehow we have to get it into the laws that anyone's DNA—the message it gives—is confidential and that the only one who has a right to look at it is the person herself or himself. Still, the ethics get complicated if you can spot a gene in a newborn child that produces a disease for which no treatment exists. Sometimes these defects will be hard to spot, but sometimes, as in muscular dystrophy, they can be very easy to detect. As we begin to get data of this kind, people are going to get nervous and some are going to be violent opponents of the project unless they can feel that they or their friends will not be discriminated against on the basis of their DNA. If someone can go look at your DNA and see that you have a deletion on one of your anti-oncogenes and that you will be more liable to die of cancer at an early age, then you might be discriminated against in, say, employment or insurance coverage.

Laws are needed to prevent genetic discrimination and to protect rights that should not be signed away too easily. If you are poor, it will be highly tempting to say, "Yes, look at my DNA because I want the job in the asbestos factory." If you have no money, a job in an asbestos factory is better than no job. Issues like these demand a lot of discussion, at least so that DNA-related laws are not enacted prematurely. For that reason, we are putting more than 3 percent of the genome project money into an ethics program; and we will put more into it if we find that it needs more.

We have faced up to this challenge already with DNA finger-prints. The National Center for Genome Research has given $50,000 to the National Research Council–National Academy of Sciences study on DNA fingerprinting, which has lawyers and judges advising it. The police want a DNA register of sex offenders; other people may want one of dishonest accountants. People will want DNA fingerprints to prove that a politician's children are really his. At a meeting in Leicester, England, Alec Jeffries showed a slide of a letter from a woman who runs a small hotel in Wales and who wrote that it would be a good idea to have a DNA fingerprint register of bedwetters. Different people will want different information—the possibilities are unlimited. I don't think *anyone* should have access to anyone else's DNA finger-prints.

We need to explore the social implications of human genome research and figure out some protection for people's privacy so that these fears do not sabotage the entire project. Deep down, I think that the only thing that could stop our program is fear; if people are afraid of the information we will find, they will keep us from finding it. We have to convince our fellow citizens somehow that there will be more advantages to knowing the human genome than to not knowing it.

ETHICS, LAW, AND SOCIETY

III

DOROTHY NELKIN

The Social Power of
Genetic Information

8

The testing of human characteristics is a ubiquitous trend in American society. Our enchantment with diagnostics was nicely captured in a recent *New Yorker* cartoon: a drive-through testing center on a busy highway advertised tests for "emissions, drugs, intelligence, cholesterol, polygraph, blood pressure, soil and water, steering and brakes, stress and loyalty."[1] Besides identifying the current preoccupation with measurement, the cartoon suggests many of its characteristics: like hamburgers or gasoline, testing is offered as a commodity; you can simply drive through, get tested, and then drive away. The center does not differentiate people from machines—both are objects that can be reduced to parts, examined, and assessed. Nor does it distinguish people's physical health from their behavior; blood pressure, deviance, intelligence, integrity, political loyalty—all can be routinely tested, just like the brakes of a car. Finally, and perhaps most important for the purpose of my analysis, most of the tests available in this drive-in station are not intended simply to diagnose manifest symptoms of illness or malfunction; their purpose is to discover the truth behind appearances, to detect conditions that are latent, asymptomatic, or predictive of possible future problems.

These are precisely the characteristics of tests that are emerging from research in genetics and in the neurosciences—tests that detect ever more subtle biological differences among individuals and predict disease before the manifestation of symptoms. Genetic tests are emerging from new methods of scanning the sequences of DNA that form the basis of biological inheritance. Through the location of markers that cosegregate with genes for illness in members of a family in which a hereditary disease occurs, geneticists can identify predispositions to a growing number of hereditary diseases. Tests now exist for about thirty disorders, and as more genes and markers are identified (there are jokes these days about the "gene of the week"), it is anticipated that tests will be available to indicate predisposition not only to purely genetic diseases but also to very complex disorders suspected of having a genetic component. These include mental illness, hyperactivity, early-onset Alzheimer's, certain forms of cancer, and alcoholism and addiction. In other words, tests will predict behaviors as well as disease.

Another type of predictive test is emerging from imaging in the neurosciences. These tests are not usually discussed with reference to genetics, but they form a complementary area of research based largely on genetic assumptions. Positron emission tomography (PET) and related technologies—some with remarkable acronyms such as SPECT, BEAM, and SQUID—are essentially brain-mapping devices. Experiments using these techniques seek to visualize not the structures in the brain but the way the brain functions under different conditions in order to study the relationship between brain functioning and particular behaviors. The PET-scanning laboratories are experimenting on the diagnosis of behavioral disorders or potential behavioral disorders—predispositions to violence, learning disabilities, and psychiatric diseases—*before* symptoms are expressed. For example, PET studies of violent patients suggest that particular abnormalities in the brain can be used to "anticipate" an individual's outbursts of rage and inability to control violent impulses—all of obvious interest to the criminal justice system.

Many of the more sophisticated genetic and neurological tests are still limited to experimental investigation, but according to the National Institute of Mental Health, presymptomatic detection of psychiatric disease will become routine.[2] Moreover, as has often been the case in the history of medical invention, diagnostic techniques are far ahead of therapeutic possibilities. For the short term, perhaps the most important social consequences of these new diagnostic tests will arise less from their actual use than from their bearing on the definition of deviance and disease. They are providing the theoretical models to explain very complex behaviors in simple biological terms.

Charles Scriver, a past president of the American Society for Human Genetics, has observed that genetics has made inroads into the medical mind and that genetic information is increasingly appearing in medical records.[3] For example, psychiatrists, inclined in any case toward deterministic explanations, appear to be increasingly committed to behavioral genetics. Medical receptivity has been encouraged by the utility of genetic testing in clinical contexts, where genetic information may help define therapeutic modalities. Genetic therapies for most diseases are far in the future, but knowledge of genetic defects can still be medically useful. Postnatal tests for phenylketonuria (PKU), compulsory in many states, have allowed control of this disease through rather simple dietary measures. A severe genetic disease that can result in mental retardation, PKU can be controlled by removing phenylalanine from the diet of afflicted children.

The most familiar use of genetic information today comes from prenatal testing. Amniocentesis is used to detect the presence of chromosomal abnormalities, including trisomies such as Down's syndrome, unbalanced translocations, mosaics, or sex-chromosome abnormalities. It can also be used to find biochemical abnormalities at the genetic level, detecting up to 180 genetic disorders, including Huntington's disease (HD), sickle cell anemia, Tay-Sachs disease, and (through alpha-fetoprotein measurements) neural tube defects. A newer technique, chorionic villus sampling, can detect genetic abnormalities at only ten weeks. Genetic testing is encouraged by legal pressures—such as wrongful-birth and wrongful-life suits against physicians who neglected to offer their pregnant patients tests that could predict fetal disorders. If tests are available, they will be used.

Prenatal tests, of course, raise troubling issues of abortion. But tests that identify adults who are carriers of a hereditary disease, or who themselves are at risk, have even greater psychological implications. Imagine the reaction of a person who discovers that he or she is sure to contract a devastating disease like Huntington's or early-onset Alzheimer's. Imagine the impact on family members who learn they too are at risk. While HD tests are available to those who know they are at risk because a parent has the disease, relatively few people have chosen to be tested.

The acquisition of genetic information also generates clinical dilemmas. Genetic counselors are beginning to ask questions such as, "Who is the patient—the individual? The family—the spouse, the sister, the brother, or the child?" The debatable issue of partner notification, of so much concern in the case of AIDS, is problematic here as well. Should a spouse or a child be informed if a person is a carrier of a genetic disease? It has been argued that compelling social interests call for compulsory genetic testing of people at risk of genetic disease and for informing family members about the biological status of their relatives.

Beyond its psychological effects, knowledge that one is presymptomatic has social and economic implications. In *Dangerous Diagnostics* Laurence Tancredi and I explored the increasing pervasiveness of biological assumptions in nonclinical settings.[4] All institutions—employers, insurers, schools, and the courts—seek strategies that will increase economic efficiency, lower costs, and reduce or minimize future risks. Organizational needs can be served by tests that predict how the body functions and can be expected to function in the course of a person's life.

Testing is not only a medical procedure but a way of creating social categories. It may be used to preserve existing social arrangements and to enhance the control of certain groups over others. This is not a new idea. Michel Foucault, for example, termed pedagogical examinations a strategy of political domination, a means of "normalization." He described the examination as a normalizing gaze that "introduces constraints of conformity, that compares, differentiates, hierarchizes, homogenizes and excludes."[5] A psychiatrist, Walter Reich, developed a similar analysis of psychiatric tests that were used for many years in the Soviet Union to reinforce political and social values.[6] An anthropological literature scrutinizes the tendency to use biological arguments to shape individuals to institutional values. As the anthropologist Mary Douglas has put it, "Institutions bestow sameness; they turn the body's shape to their conventions."[7]

The increased preoccupation with testing in American culture also reflects our tendency to approach problems with an actuarial mindset. Actuarial thinking requires one to calculate the cost of future contingencies, taking into account expected losses, and to select good risks while excluding bad ones. All this entails an understanding of the individual with reference to a statistical aggregate. In this context, the information derived from tests becomes a valued resource. Reflecting the actuarial mindset, the gathering of personal information by government agencies, employers, and schools has greatly increased over the past two decades. Testing is part of this trend. The screening of prospective employees for drug use, for example, continues despite concerns about the accuracy and legitimacy of the tests. The pressure to test for the AIDS virus is relentless, despite the discriminatory implications. And the use of standardized tests in schools is growing despite questions about their validity as a measure of intelligence and predictor of performance. Indeed, the faith in facts and the numbers derived from testing has obscured the uncertainties intrinsic to measurements of individual "aptitude." Standardized tests are widely accepted as neutral, necessary, and benign.

Just as the value of facts is part of the actuarial mentality, so too is the tendency to reduce social problems to measurable biological dimensions—that is, to dimensions that can be revealed through a test. Our popular culture widely assumes an ideal of biological normality or perfection against which individuals can be measured. One finds in the press, for example, broad acceptance of stereotyped sociobiological assumptions, and a pervasive belief that complex human behavior can be reduced to biological or genetic explanations.[8] Among the traits attributed to genetics have been mental illness, homosexuality, dangerousness, job success, exhibitionism, arson, stress, risk-taking, shyness, social potency, traditionalism, and even zest for life. Such complex conditions are attributed to biological determinants with minimal reference to social or environmental influence. When E. O. Wilson's *Sociobiology* was published in 1975,[9] *Business Week* ran a series of articles on "The Genetic Defense of the Free Market." ("Competitive self-interest, the bioeconomists say, has its origins in the human gene pool.") News coverage of the Baby M case included a story in *U.S. News and World Report* entitled, "How Genes Shape Personality," which took the "solid evidence . . . that our very character is molded by heredity" to question whether Baby M's future really hinged on which family would bring her up. Family magazines recommend genograms and family health pedigrees as ways to predict the future characteristics of children. You can be sure that genetic ideas have been popularized when you see a button saying, "Gene Police! You—Out of the Pool!"; or a Mother's Day card, to a daughter who is herself a mother, that says on the front, "What a good Mother you are," and on the inside, "It's all in the genes." Even the advertising industry seems to have assimilated genetic concepts: an ad for a BMW boasts its "genetic advantage."

Media stereotypes are not just an invention of journalists; they reflect the images projected by scientists in their own communications. Much has been written on the jaded history of genetics and on the eugenic assumptions that shaped both scientific thought and social policy up until World War II.[10] Less has been said about the implications of the latest scientific discourse, as some scientists themselves suggest the social meanings inherent in their work. Until recently, most scientists, with the notable exception of Arthur Jensen, have been reluctant to extend their ideas into the realm of social values. The advances of the 1980s, however, seem to be inspiring more frequent references to the applications of genetic understanding to social policy. Geneticist Marjorie Shaw, for example, has asserted that "the law must control the spread of genes causing severe deleterious effects, just as disabling pathogenic bacteria and viruses are controlled."[11] In effect, she is adapting a public health model to genetic disease, referring to biologically vertical rather than horizontal contagion. She calls for the police powers of the state to prevent genetic risk by controlling reproduction in families informed about potential genetic disorders.

The editor of *Science* magazine, Daniel Koshland, asserted in 1987 that in the warfare between nature and nurture, nature has clearly won, with all that this implies for the idea of genetic determinism and the immutability of inherited traits.[12] References are appearing in scientific discourse to the pollution of the gene pool, to genetically healthy societies, or to "optimal" genetic strategies.[13]

The broad cultural appeal of genetic concepts needs to be kept in mind if we are to understand the social power of genetic information. Social policies are mediated through institutions such as schools, courts, insurers, and employers. For such institutions, biological tests are but an extension of early pedagogical and psychiatric tests; conceptually, they are not completely new. Like earlier tests, they serve as gatekeepers, controlling access, for example, to employment or insurance. In effect, the apparent predictive power of biological tests allows an organization to select its clients on the basis of its own economic and administrative needs. Genetic tests are particularly powerful tools. Because of their apparent certainty, they are credible, and because of their specificity, they imply that institutional decisions are implemented for the good of the individual.

Tests can be used to redefine socially derived syndromes as problems of the individual, placing blame in ways that reduce public accountability and protect routine institutional practices. The availability of biological tests, in effect, gives an organization a scientific means to deal with failures or unusual problems without threatening its basic values or disrupting its existing programs.

For example, when public schools face pressures for accountability, it is convenient for educators to explain learning difficulties or behavioral problems in terms of individual disabilities. Educational failures were once explained in terms of cultural deprivation or nutritional deficiencies; over the past decade they have been redefined as learning difficulties—that is, problems located in the student's brain. Hyperactivity was once interpreted as a problem in classroom dynamics. Some thirty years ago, children's problems were even blamed on an alleged deficit in the home environment called "working-mother syndrome."[14] Now these problems have been defined as attention-deficit disorders, a difficulty intrinsic to the child.[15] I hardly want to denigrate the notion that behavior or learning disabilities exist; they certainly do. But by removing blame from the schools or other social institutions and dissolving demands for accountability, diagnostic labels can too easily become an institutional convenience. The consequences are not necessarily all bad; removing guilt from families, for example, can be useful. But labels can stigmatize slow learners as inherently, and therefore permanently, disabled, and they can draw attention away from the social interactions that clearly influence learning.[16]

Testing can also be exploited to sanction routine institutional practices in the workplace. Biological tests may identify the susceptibility of particular workers to harm from toxic exposures.[17] Although justified in the first instance as a way to protect workers' health, they can be used to exclude those who are most vulnerable and thus avoid costly changes in the workplace environment. It is the employee who is burdened with responsibility; it is the employee who can be expected to fit the environment or to move to another job.

The predictive capacity of biological tests is also useful to organizations as a means of facilitating efficient long-term planning. Companies are not only employers; they are also insurers, and as insurers they are increasingly reluctant to employ those whose life-styles or genetics may predispose them to a future illness. About half of American employers require preemployment medical exams. These include various predictive tests ranging from psychological tests for future executives to lower-back X rays for construction workers, from tests for drug use to screens for AIDS. In the context of growing economic competition, screening techniques that identify those predisposed to genetic disease can become a cost-effective way to control absenteeism, reduce compensation claims, and avoid future medical costs for workers and their families.[18]

Efficient planning is imperative in the administration of prepaid medical plans. The financial dilemmas of insurance companies and government policies that link reimbursement decisions to specific diagnostic categories encourage medical administrators to predict and control future risks.[19] These pressures have converged with the threat of malpractice suits to create powerful incentives to back up health care decisions with objective and predictive information. They encourage what is called "skimming"—competition among health maintenance organizations (HMOs), for example, for the so-called profitable patient, the person who has predictable and reimbursable illnesses. Diagnostic technologies help to categorize patients; they provide the technical evidence to support controversial decisions and the patient profiles to control access to health care facilities.

The need for efficiency and cost containment can influence the use of prenatal tests. Genetic disorders are believed to occur in about 5 percent of all live births and to account for nearly 30 percent of all pediatric hospital admissions and 12 percent of all adult admissions in the United States. While the presumed beneficiary of fetal testing is the prospective parent, health care providers and insurers also benefit from genetic information that may have consequences for future medical obligations. Some insurers have threatened not to cover the medical expenses of a child with a genetic disease once the mother has been warned through a prenatal test that the fetus may be afflicted. She may choose either to abort or to give birth to a child who may have extraordinarily costly, uninsured medical expenses. The "choice" in such cases is surely limited.

The biological status of a person's body, indicated by tests, may also be grounds for exclusion from insurance. Today about 37 million people in the United States have no public or private health insurance. About 15 percent of those who are insured are covered individually (that is, not by a company group plan) and must meet underwriting standards by providing their health history, information on family illnesses, and evidence of their current state of health. Sometimes tests are required. In 1987, 20 percent of applicants were issued policies that excluded preexisting conditions unless they paid higher premiums; 8 percent were denied coverage for diseases such as obesity, cancer, schizophrenia, and AIDS.[20] Similarly, HMOs in 1987 denied membership to 24 percent of individual applicants.

The medical directors of insurance companies expect access to information from genetic tests in order to make judgments about coverage and to calculate rates. Given that insurance rates are based on risk predictions, it is not strange to find the industry expecting to be informed. It is standard behavior for insurers to insist on all available health information about insurance applicants. In recent years, insurability classifications have been multiplied. The Prudential Insurance Company, for example, increased its classification categories from ten in 1980 to nineteen in 1986. Its vice-president announced that the company will be making more, and more subtle, distinctions on the basis of biological—that is, predictive—tests.[21]

Finally, tests are used to support controversial decisions about the disposition of those who will not or cannot conform to institutional norms. Consulting psychiatrists seek technical support for decisions that are often highly controversial. For example, they are frequently asked by hospitals to evaluate the competence of patients—those who, say, fail to comply with decisions about recommended treatment or about placement in a nursing home. Biological data help to buttress the evaluations, for they appear more solid than phenomenological psychiatric opinion.

In the legal system, genetic information is becoming more than just a source of evidence. It is also influencing traditional legal concepts. Assumptions about the importance of genetics are increasingly the basis of legal decisions in a wide variety of fields, including torts (wrongful-life cases), family law (custody disputes), trusts and estates (distributing an intestate's assets), and criminal law (determining responsibility).[22] The courts are especially receptive to admitting hard evidence to sort out conflicting psychiatric opinions in decisions about responsibility, culpability, and disposition of criminal defendants. One recalls the popular disaffection with the court's lenience in the case of John Hinckley, the man who attempted to assassinate President Ronald Reagan. The outcome of the Hinckley trial rested on the requisite burden of proof, and the prosecution, relying on the assessments of psychiatrists, failed in the jury's view to prove Hinckley's responsibility for the act. Thus, Hinckley was found to be not guilty by reason of insanity and committed to a mental institution.[23]

The Hinckley case came along at a time of mounting debate over the weakness of the rules for the insanity defense, and it strengthened a growing outcry for reforms in determinations of criminal responsibility. Many courts prefer to limit psychiatric opinion and to substitute more objective data that would reduce bias and disagreement. In keeping with this trend, a neurologist in California has introduced the use of PET scans into the courts as a scientific basis for sentencing decisions that require consideration of the responsibility and the rehabilitation potential of convicted criminals. Pilot studies of violent patients have linked deviant behavior to specific abnormalities in tbe brain.[24] Purportedly, PET evidence can establish mental illness, and help the courts predict recidivism and make decisions about sentencing. In the view of many people, the predictive reliability of this technology is limited, yet legal scholars, writing on biological psychiatry, anticipate that the courts will increasingly use information from brain scans to evaluate responsibility and to predict the likelihood of future dangerousness.[25] It used to be that textbooks on criminology devoted no more than a chapter to predictions of dangerousness. Now some criminology textbooks are organized around the notion of biological prediction of criminal behavior.[26]

Because they are grounded in science, genetic diagnostics are compelling. Images on a screen convey precision. Statistical findings processed by computers appear, at least to nonscientists, to be objective, neutral, beyond refutation, somehow equivalent to truth. But the results of these tests are subject to many interpretive fallacies. The evidence produced by most diagnostic tests is only inferential, and interpretation rests on statistical definitions of the normal. Moreover, interpretation often assumes causation where there is only correlation. And in the present state of the biological testing art, the error rates—false positives, false negatives—are high. Even reliable tests cannot predict when or how the disease will strike, for in many cases the manifestation of symptoms—their severity and moment of onset—rests on random events or intervening factors, such as diet, life-style, or environment.

The interpretive assumptions underlying the use of biological tests become especially critical when they are used to screen large populations—for example, in testing people for AIDS or screening workers for genetic susceptibility to toxic substances. The purpose of screening, unlike clinical testing, is not to discover the cause of manifest symptoms in an individual but to deduce statistical levels of disease in a population. In such circumstances, inconsistencies may remain undetected and the actual behavior of individuals ignored. Inevitably, some people tested will be mislabeled and will suffer problematic consequences, such as loss of a job.

Despite the technical limits of biological tests, the testing industry is growing rapidly in anticipation of an enormous market.[27] Biotechnology companies are competing to develop probes for determining genetic diseases, for they assume that genetic testing will be mandatory in many organizations. Similarly, neuroscience firms are the latest targets of venture capital.

As advances in genetics and in the neurosciences provide powerful instruments for predicting disabilities and behavioral abnormalities, tests can be developed that are efficient, inexpensive, accurate, and, above all, nonintrusive. It would be simple to test every newborn child. The American Society for Human Genetics has questioned proposals for sampling the umbilical cord of newborn children and storing the DNA. What are the implications for privacy? For later discrimination? It is feasible to establish national DNA data banks to store information about a person's parentage and predisposition to disease. Every person could have a genetic map on file. Some private biotechnology companies are advertising genetic repositories and urging families to deposit specimens of their DNA for future analysis. Some firms are predicting that most people will eventually have their genetic profiles on record.

Such predictions reflect a naive optimism—that the medical benefits of anticipating genetic disease, the social benefits of records that would facilitate control of criminal elements, and the economic benefits of having data for rational planning will justify the development of DNA data banks. Data banks today already contain considerable personal information about a large number of people. In some states information on hospitalized mental patients is stored in state files, and genetic registries record birth defects. A dispute took place in New York State over tagging state records on birth defects with identifying information, such as Social Security numbers. The state wants these identifiers in order to do research on genetic abnormalities caused by occupational hazards. But genetic counselors do not want to release this information on grounds of privacy; they distrust the state's willingness to limit its use of this information to this specific research.

Nevertheless, DNA data banks are growing. Crime control agencies have a special interest in genetic records.[28] The FBI has a data bank containing the DNA fingerprints of paroled ex-convicts. The British allow the police in Northern Ireland to take a lip scrape without consent from anyone suspected of future terrorism, in order to create a data bank that can be used for later identification. The practice includes a kind of respectful nicety: mouth scrapes are prohibited, since they would invade the region inside the teeth.

The possibilities for genetic discrimination are obvious, and many cases are beginning to appear. The asymptomatic ill— people who have no symptoms but who are known to have a genetic disease—have been barred from insurance and employment, and even refused driver's licenses. In effect, genetic risk for a disease has been reified as the disease itself, even in the absence of obvious manifestations of illness.[29]

The significance of biological profiles rests, of course, on how they will actually be used. Groups besides schools, employers, insurers, and law enforcement agencies have an interest in the genetic or neurobiological condition of the people in their domains. The Department of Motor Vehicles, immigration authorities, creditors, adoption agencies, organ transplant registries, professional sports teams, sexual partners, the military, even university tenure committees—all may have reasons for wanting access to diagnostic information about the present and future health of individuals. One can imagine a kind of Jonathan Swift scenario—families demanding information about their genetic roots, adoption brokers probing the genetic history of children in order to find appropriate matches, or commercial firms storing genetic profiles and selling them to interested agencies.

The rising tide of biological testing poses a broad range of challenges to standards of civil liberty—particularly the right of privacy to medical information. Numerous organizations, such as insurers and crime control agencies, may insist, with legal and policy support, that their access to medical information is a necessity and a right in view of their responsibilities. But their insistence is at the least debatable, given the exclusion, stigmatization, and discrimination that may ensue from the use and abuse of testing.

The growing availability of biological tests are also challenging standards of professional responsibility—particularly the obligations of confidentiality. As the biological and genetic underpinnings of disease are exposed, the role of the medical expert in nonclinical contexts becomes increasingly important. The company doctor, the school psychologist, the forensic psychiatrist—all professionals with conflicting interests—have assumed greater responsibility in their respective institutions. Sometimes called "double agents," physicians in these ambiguous roles have dual loyalties—to the company they work for and to their patients.[30] As the responsibilities of physicians in nonclinical settings increase, so too do dilemmas of professional ethics.

The most serious implication of biological testing is the risk of expanding the number of people who simply do not "fit in." The refinement of food-product testing in the 1960s and 1970s allowed greater sensitivity in the detection of carcinogens, and the number of products defined as problematic greatly increased. In very similar ways, improved diagnostics are refining our capacity to identify deviations from the norm, and, as in the case of product testing, more precise tests are expanding the number of persons defined as diseased. Morever, by allowing anticipation of future problems that may not be symptomatically expressed for years, tests are, in effect, creating a new category of persons, the "presymptomatic ill."

Even as tests improve in certainty and extend the range of what they can predict, questions of interpretation will remain. What degree of correlation will be necessary between existing markers and subsequent physical or behavioral manifestations before social action—such as exclusion from work, tracking in special education programs, or establishing competency to stand trial—may be taken? How do we balance the institutional need for economic stability against the rights of the individual? What is to be defined as normal or abnormal? And whose yardsticks should prevail? In all, we risk *increasing* the number of people defined as unemployable, uneducable, or uninsurable. We risk, in other words, creating a genetic underclass.

ERIC LANDER

DNA Fingerprinting: Science, Law, and the Ultimate Identifier

9

Throughout the twentieth century, basic advances in the study of human genetics have consistently provided new tools for the analysis of evidence samples in criminal cases and paternity disputes. Forensic genetic typing began with the discovery of the ABO blood group and soon expanded to other blood groups, serum proteins, and red blood cell enzymes. Because these proteins each occur in several alternative forms, forensic scientists could compare a suspect's proteins with those found in an evidentiary sample to determine whether the suspect is "included" among or "excluded" from the set of individuals who might have contributed the sample. Typically, a randomly chosen person would have a 95 percent probability of being excluded, but this still leaves a substantial possibility of a match occurring simply by chance. Thus, a test that yielded a result of "inclusion" based on such procedures could never be definitive proof of guilt.

Forensic scientists recognized the need for genetic markers with greater discriminatory power. The human leukocyte antigen (HLA) proteins offered one promising candidate: these cell surface proteins are extremely diverse—it is this variability, in fact, that is the basis of tissue transplant rejection. Unfortunately, HLA proteins proved to be too fragile to be reliably typed from dried evidentiary stains; its use has thus been limited primarily to paternity cases, in which fresh samples can be obtained.

The situation changed dramatically with the recognition of a much richer source of variation than proteins—namely, DNA sequences, particularly the restriction fragment length polymorphisms (RFLPs). Some 2,000 RFLPs have been identified to date, spanning all the human chromosomes. Forensic scientists were quick to realize that DNA is the ultimate identifier. It has all the requisite properties: DNA shows abundant variation (there is about 1 site of variation per 1,000 DNA nucleotides, in a genome of about 3 billion nucleotides); DNA is present in all cells (excepting red blood cells); a person's DNA is identical throughout the body and is invariant throughout life; and DNA, a fairly stable molecule, is likely to be preserved in dried stains.

The basic methodology of DNA fingerprinting is relatively simple. First, DNA is extracted from an evidence sample and from a suspect's blood. The DNAs are then chopped into millions of fragments by a restriction enzyme that cuts at defined sequences, and the fragments are fractionated by gel electrophoresis: each sample is loaded at the top of its own lane on the gel and is subjected to an electrical field that runs along the length of the gel, causing the DNA fragments to migrate at speeds depending on their size (small fragments move faster than large fragments). At the end of the process, the DNA fragments in each lane are separated according to size along the lane. The DNA is then transferred onto a special piece of paper called *membrane* and fixed in place. It is now ready for analysis.

To visualize the DNA fragments corresponding to any particular *locus* on the chromosome, one must use a *radioactive probe* containing a short DNA sequence from the region. The radioactive probe is washed over the membrane and binds to the complementary sequences. One then exposes the membrane to X-ray film overnight to see where the radioactive probe has bound; these locations are marked by dark horizontal *bands* on the film, which is called an *autoradiogram*. The bands constitute the sample's DNA pattern for the locus in question.

For each locus, one compares whether the DNA patterns of the sample (the precise number and positions of the bands) match those of the white cells in the suspect blood. If the patterns do not match at every locus, then they must have come from different sources (barring technical errors). If the patterns *do* match at every locus, then they *may* have come from the same source—that is, they are consistent with having come from the same source, although it is also possible that they came from different people who simply happen to have the same patterns at these particular loci. If one finds matches at *enough* loci, one eventually concludes that the samples must have come from the same person.

How many matching loci are enough? The answer depends on the degree of variability at each locus—that is, the chance that two randomly chosen people would show the same pattern at the locus. To maximize the discriminatory power, forensic scientists have opted to worked with highly polymorphic RFLPs called *variable number of tandem repeat* (VNTR) loci. (See Figure 24.) As their name suggests, the loci contain variable numbers of adjacent DNA sequence repeats. Some chromosomes may carry 30 tandem copies, others 31 tandem copies, and so on. At many such loci, there may be scores of alternative lengths. Provided that one can accurately discriminate among the length differences, one has a powerful system for DNA fingerprinting. Currently, most DNA fingerprinting laboratories examine four VNTR loci. These loci constitute only a tiny fraction of the variability present in the human genome, but forensic scientists feel it is enough to provide a great deal of information about identity.

In the mid-1980s, several private companies were founded to commercialize DNA fingerprinting for the identification of criminals; the most prominent were Cellmark Diagnostics of Germantown, Maryland, and Lifecodes Corporation of Valhalla, New York. In 1988, DNA fingerprinting was first admitted as evidence in court in the case of *Florida v. Tommy Lee Andrews*.[1] In January 1989, the Federal Bureau of Investigation (FBI), having cautiously studied the technology in its own laboratories, began to accept casework from state forensic establishments. Since then, DNA fingerprinting has been used in more than a hundred cases in the United States and has been formally allowed in at least one jurisdiction in about two-thirds of the states.

As a new technology, DNA fingerprinting had to be found in each of the courts to satisfy well-established standards for the admissibility of novel scientific evidence. The most common standard is known as the Frye rule—in recognition of its establishment by a federal court in the 1923 trial of James Frye, a young black man who was accused of having murdered a white man in Washington, D.C. Frye's lawyer urged the court to admit as evidence

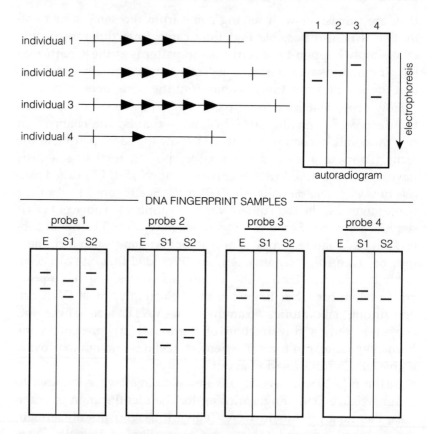

Figure 24 DNA "fingerprints" are a record of polymorphisms in a person's DNA. Forensic scientists analyze a kind of RFLP called variable number of tandem repeat (VNTR) loci. In a particular chromosomal region, for example, a DNA sequence (denoted by an arrow in the sketch at the top) may be repeated 3, 4, 5 times, or only once, on chromosomes from different individuals. When the DNA from these chromosomes is cut with a particular restriction enzyme (whose cutting site is indicated by the vertical bars), the fragments produced differ in size because of the variable number of tandem repeats. After radiolabeling and electrophoresis, the resulting autoradiogram shows "bands" indicating the size of the fragments (ranging from largest at the top to smallest at the bottom).

For illustration, autoradiograms are shown for three samples probed with four different probes. Sample E is an evidence sample, samples S1 and S2 were taken from possible suspects. For each probe, the DNA yields one band from the paternally derived chromosome and one from the maternally derived chromosome. Two bands are typically seen, although there may only be one band if the two copies of the chromosome have fragments of the same size. Sample S2 matches the evidence sample E at the four loci considered, whereas sample S1 matches it only once. Suspect S2 would be said to be "included" in the set of individuals that might have given rise to the evidence sample, and suspect S1 would be definitively "excluded" as the source of the sample.

the results of a "systolic blood pressure test"—an early form of lie detector—on the grounds of the general rule allowing experts to testify on matters of specialized experience or knowledge. Because the polygraph was a new technology, however, the court imposed a more stringent evidentiary rule, holding:

> Just when a scientific principle or discovery crosses the line between the experimental and demonstrable stages is difficult to define. Somewhere in this twilight zone the evidential force of the principle must be recognized, and while courts will go a long way in admitting expert testimony deduced from a well-recognized scientific principle or discovery, the thing from which the deduction is made must be sufficiently established to have gained general acceptance in the particular field to which it belongs.[2]

The court, believing that the polygraph did not yet command general acceptance, declined to admit the lie-detector results. (Some jurisdictions apply a somewhat different standard, the relevancy test, based on the Federal Rules of Evidence, but the question revolves around the same basic issues.)

"General acceptance in the field in which it belongs" is a vague standard—sufficiently vague, perhaps, to have permitted the courts quickly to decide that DNA fingerprinting satisfied the Frye rule simply because DNA analysis was widely accepted in medical applications. Running through most of the court decisions was a common set of declarations: that DNA was absolutely invariant in every cell in the body; that, therefore, evidence samples would be identical to samples from the guilty party; and that falsely positive identifications would be essentially impossible.[3]

The rapidity with which the courts accepted DNA fingerprinting is understandable. In theory, the procedure is flawless: if enough sites of genetic variation are examined, it is certainly possible to determine whether two samples come from the same source.

In practice, however, DNA fingerprinting can be quite problematic. The difficulties become readily apparent from a comparison of DNA forensics with DNA medical diagnostics. DNA diagnostics can be conducted under optimal laboratory conditions: the samples are fresh, clean, and from a single individual. If uncertainty arises in the results, new samples can be taken and the test can be redone, which makes for a high standard of accuracy in the procedure. By contrast, DNA forensics compels the biologist to work with whatever samples happen to be found at the scene of a crime. Samples may have been exposed to numerous environmental insults: they may be degraded; they may be mixtures of samples from different individuals, as happens in a multiple rape. The forensic biologist often has only a microgram or less of sample DNA, enough to do perhaps only one test. If the test has ambiguous results, it often cannot be repeated because the sample will have been used up.

Moreover, DNA diagnostics usually asks only a simple question: which of two alternative RFLP alleles has a parent passed on to his child? Because there are only two possible alternatives, there are natural consistency checks to guard against errors. By contrast, DNA forensics resembles analytical biochemistry. Given two samples about which we know nothing in advance, we wish to determine whether or not they are identical. We first need to determine if the band patterns match, a decision which requires us to make fine judgments about whether small differences between patterns are meaningful. If we decide that the patterns match at a few sites of variation, we must then assess the probability that the match might have occurred by chance. For this purpose, we must know the distribution of band patterns in the general population.

For these reasons, DNA forensics is considerably more challenging than DNA diagnostics. At the time that DNA fingerprinting was first introduced as evidence, however, the courts were not troubled by these potential problems—and neither were many other people, myself included. These issues have become obvious only in retrospect, as the result of experience.

I became involved in DNA fingerprinting as the result of attending a conference on DNA forensics that was held at the Banbury Center of the Cold Spring Harbor Laboratory, in Long Island, in November 1988.[4] As a human geneticist who had worked with RFLP markers in medical diagnostics, I had been invited to provide an outside perspective on the forensic applications. After the conference, I was asked to examine evidence in a Bronx, New York murder case, *New York v. Castro*.[5] After much reluctance, I eventually agreed to testify as an expert witness—without compensation—in the pretrial hearing on the admissibility of the DNA evidence.

Vilma Ponce and her two-year-old daughter Natasha were bru-
tally murdered in their Bronx apartment. The victim's common-
law husband cast suspicion on the building's janitor, José Castro.
When the police interviewed Castro, they noticed a small blood
stain on his wristwatch. They confiscated the watch and sent it
for DNA testing to Lifecodes Corporation. Lifecodes compared
DNA from the blood stain to DNA from the victims, using three
autosomal probes and one probe on the Y chromosome to determine sex. The company reported that the blood on the watch
matched the blood from the deceased mother and stated that the
frequency of the DNA band pattern was one in 100 million among
the U.S. Hispanic population.

In the pretrial hearing on the admissibility of the DNA evidence,
numerous problems became manifest.[6] According to the testing
laboratory's report, the bloodstain on the watch and the sample
from the deceased mother both showed three DNA bands when
analyzed with the probe for the locus DXYS14. (Locus names reflect the chromosome and order of discovery; this one is on both
the X and Y chromosomes and was the fourteenth to be found.)
However, all the experts at the pretrial hearing—including those
from Lifecodes—agreed that there were two extra bands in the
DNA from the watch. (See Figure 25.) Why were these additional
bands not mentioned in the laboratory's report? Why didn't they
prove that the samples did not match? Lifecodes argued that the
extra bands were nonhuman contaminants, although the conclusion was based on casual speculation and no experiments were
carried out to determine the bands' origin (which would have
required only the simple step of repeating the hybridization with
an uncontaminated preparation of the DNA probe). Lifecodes argued that the two extra bands could not have come from the locus
in question, because their pattern was not consistent with the
known properties of the locus. This argument was specious, as
became evident in the testimony of Howard Cooke, the scientist
at the Medical Research Council laboratory in Edinburgh who had
discovered the locus and had provided the probes to Lifecodes.
He stated that there was simply no way to determine whether the
extra bands were or were not human on the basis of the pattern—
the experiment should have been repeated when the ambiguity
had been found. (By the time of trial, the DNA on the filter had
been exhausted and it was too late to repeat the work.)

Why had Lifecodes dismissed the two extra bands? In all likeli-

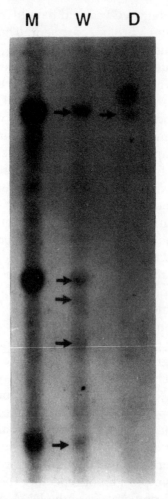

Figure 25 This autoradiogram from *New York v. Castro* shows samples from the deceased mother (M), the bloodstain on the defendant's watch (W), and the deceased daughter (D), all hybridized with a probe for the locus DXYS14. This locus is somewhat more complex than the usual VNTR loci; it can produce from 1 to 8 bands in a single individual's DNA. The mother's DNA shows three bands. Witnesses from both sides agreed that the DNA from the bloodstain on the watch showed five bands (indicated by arrows). Although the lowest three bands had equivalent intensity, the testing lab decided that the nonmatching bands were artifacts (variations due to unknown effects, such as testing at an incorrect temperature) and could be ignored. Accordingly, the lab declared a match between M and W. In addition, the testing lab's formal report stated that the daughter's sample (D) showed three bands in exactly the same positions as those in the mother's sample. At the hearing, no scientists could find these bands. These data suggest that the testing lab's interpretation of the DNA patterns may have been overly influenced by lane-to-lane comparisons; accuracy demands independent assessment of each pattern.

hood, the examiners had not identified the pattern in each lane independently, but rather had allowed themselves to be influenced by sample-to-sample comparison. Seeing three matching bands, the eye tends to ignore the extra bands. It is a perfectly natural tendency, especially since sample-to-sample comparison is an acceptable practice in scientific experiments for which there are built-in scientific controls to avoid errors. Unfortunately, it is a dangerous practice when comparing unknown DNA samples—since one will tend to discount precisely those differences that would exonerate an innocent defendant.

The hypothesis that Lifecodes had been influenced by sample-to-sample comparisons was reinforced by the conclusions it drew about samples from the daughter. According to Lifecodes' laboratory records, the laboratory found three bands in the daughter's blood in exactly the same positions as in the blood of the mother and the blood on the watch. In fact, the daughter's DNA did *not* show three such bands! At least, no witnesses could find them in court—including the Lifecodes scientist who had originally recorded the three bands. Rather, the daughter's lane showed only a single band. Again, Lifecodes' original findings had likely resulted from unconscious bias due to lane-to-lane comparison.

As noted above, one must make fine measurements about band positions when using hypervariable VNTR loci since there are many possible alternative sizes that must be distinguished from one another. When setting up its system, Lifecodes measured the accuracy of its system and defined a precise matching rule that it printed at the bottom of its forensic reports: two bands should be declared to match if their measured sizes were within 1.8 percent of one another.

Remarkably, the laboratory records in the *Castro* case showed that bands at two autosomal loci, D2S44 and D17S79, fell outside Lifecodes' own matching rule. By the company's own quantitative standard, these bands should have been declared non-matches. Why were they called a match? Notwithstanding the declared matching rule, it emerged at trial that Lifecodes never actually used the quantitative matching rule to determine whether samples matched. Instead, the judgments were made by eyeball.

One of the key propositions for which the *Castro* case now stands is that the eyeball is not enough; if hypervariable VNTR systems are to be used, quantitative measurements must be made.

To Lifecodes' credit, it changed protocol following the hearing and now uses quantitative measurements in declaring matches.

Another example of confusion arose in the sex test done on the samples from the mother, the daughter, and the watch. The sex test is very simple: it involves a probe for a highly repeated sequence present some 2,000 times on the Y chromosome, a substantial fraction of the entire chromosome. DNA from males should give a very intense band of a characteristic size, while DNA from females should show no signal. In the *Castro* case, the mother, the daughter, and the watch all yielded no signal. Accordingly, Lifecodes concluded that the blood on the watch came from a female.

There was one problem with the test, however. Standard laboratory procedure would require the inclusion of a positive control—that is, a sample from a male. Otherwise, one could not know whether the male banding pattern did not appear because the blood on the watch came from a female or because the test had been performed incorrectly. Lifecodes' DNA test did have a sample lane labeled "Control." Curiously, however, the control lane also showed no signal! What was the control sample?

This simple question proved to generate tremendous confusion. Originally, Lifecodes' laboratory director testified that the control DNA was from a female-derived cell line—an odd control to use when performing sex tests, since it would not show a positive signal. Two weeks later, however, Lifecodes' laboratory technician testified that the control DNA was not from a female cell line but rather from a male scientist at the company. When the laboratory director returned to the witness stand, he was asked to explain why the now-male control lane failed to show a positive signal on the Y-chromosome test. He offered a remarkable explanation—that the male scientist had a rare genetic abnormality, a short Y chromosome lacking the sequence in question. Several scientific witnesses, including myself, subsequently pointed out to the court that such deletions were known to be extremely rare (they have a frequency less than 1 in 10,000), and they are almost invariably associated with medical abnormalities. A normal individual with such a deletion would be so unusual as to be worthy of a publication in a scientific journal. A week later, the laboratory director returned to report that the control DNA was not from the male scientist after all but rather from a female laboratory technician. According to his testimony, Lifecodes had determined the identity of the control sample by its DNA fingerprint pattern—surely an unexpected use of the technology. Remarkably, Lifecodes had apparently never recorded the identity of the DNA samples that it used as a scientific control.

Shortly after I testified but while the hearing was still in progress, I happened to attend a scientific meeting co-organized by Richard Roberts, a scientist at Cold Spring Harbor Laboratory who had testified for the prosecution at the beginning of the *Castro* hearing about DNA fingerprinting in general. Reviewing the evidence that had come to light since his testimony, he recognized the problems with the evidence and proposed a promising course of action: he would get together all the outside scientists who had testified for the prosecution; I would gather all the outside scientists who had testified for the defense; and we would all meet without the lawyers to discuss the evidence. The meeting took place on the morning of May 11, 1989, in a borrowed office in Manhattan. After examining the autoradiograms, the population data base, and other records, we all agreed that the evidence was seriously flawed. We issued a joint statement outlining the most serious problem, concluding that "the DNA data in this case are not scientifically reliable enough to support the assertion that the samples . . . do or do not match," and added, "If these data were submitted to a peer-reviewed journal in support of a conclusion, they would not be accepted. Further experimentation would be required."

Faced with unanimity among the independent scientific witnesses, the judge eventually ruled that DNA fingerprinting evidence was admissible in principle but that the analysis in this case had not been conducted in accordance with generally accepted principles. He ruled that the DNA evidence of a match between the blood on the watch and the blood of the victims was legally inadmissible.

DNA fingerprinting requires a critical mind about what a DNA band pattern might mean. Otherwise, one runs the risk of being blinded by the technology's stunning power and ignoring simple alternatives. In the case of *Pennsylvania v. Shorter*, a man was accused of the rape and murder of his daughter.[7] A washcloth with sperm on it was found in his house and was sent to Cellmark for DNA typing. The laboratory identified a DNA pattern in a sample of the washcloth that matched the DNA of the father. It also identified, by differential extraction of vaginal epithelial cells, a second pattern that matched neither the father's DNA nor the daughter's. Cellmark explained that the second pattern must have come from someone else but did not make much of this finding, and the prosecution intended to use the washcloth as proof positive of the rape. However, a scientific expert hired by the defense noted something unusual about the second pattern: it shared half of its bands with the daughter, one band at each locus. This is precisely the result one would expect from the mother's DNA. In short, the bodily fluids on the washcloth were probably the residues of marital intercourse, not of the daughter's rape. Confronted with this observation, the prosecution withdrew the DNA evidence entirely. (This is not to say that the man was innocent; in fact, he pleaded guilty in return for a reduced sentence.)

DNA fingerprinting also requires rather precise information about the nature of technical artifacts, as was shown by a child rape case, *Maine v. McLeod*.[8] In this case, the suspect's DNA and the semen sample appeared to be similar, but the banding patterns were shifted vertically relative to one another, according to the analysis performed by Lifecodes. Such a difference could indicate that the samples came from different individuals, or it could have been due to a phenomenon known as *bandshifting* (see Figure 26). Occasionally one sample migrates more rapidly than another in the electric field (due to differences in sample concentration, salt concentrations, contaminants, or other reasons), and the bands thus appear to have shifted to a different position. To decide between the two possibilities, one can analyze the samples with a DNA probe for a constant or *monomorphic* locus—one that does not vary in the population but rather is the same in every individual. If the monomorphic patterns were in the same place, then it would be clear that bandshifting had *not* occurred and that the difference between the polymorphic patterns would be properly interpreted as an exclusion. If the monomorphic patterns were shifted by the same amount as the polymorphic patterns, one might conclude that bandshifting had occurred and attempt to correct for its effect.

The McLeod case dramatized the problems in correcting for

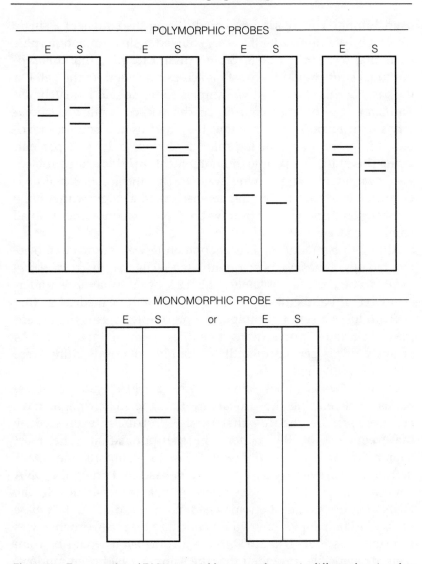

Figure 26 Two samples of DNA may yield patterns that are in different locations but are simply shifted relative to one another. There are two possible explanations: the DNAs are actually from different sources, or the DNAs are from the same source but one sample has migrated faster for technical reasons (such as a difference in salt concentration). The second alternative is called "bandshifting." To distinguish these possibilities, one can use a monomorphic probe that detects a DNA fragment of the same length in all individuals. If this constant-length fragment is in the same place in both samples, one can conclude that there has been no bandshifting and the samples must be different. If the "constant" fragments are not in the same place, then band-shifting has occurred. One must then attempt to correct quantitatively for the extent of bandshifting.

bandshifting. The hearing occurred during the course of a single week. On Wednesday, Lifecodes showed a single monomorphic locus that had shifted by 3.15 percent and testified that this proportional shift should be constant across the length of the gel. On the basis of this shift, the two samples could be said to match. On Thursday, the defense confronted the witness with the laboratory's own records showing that they had used a second monomorphic probe that had indicated a shift of only 1.72 percent. Using this shift, the sample did not match! By Friday, the problem was clear to the judge, who was, to say the least, concerned. Over the weekend, before the defense called a single witness, the prosecution decided to withdraw the DNA evidence and drop all criminal charges.[9]

Although bandshifting is a phenomenon well-known to molecular biology, it has not been quantitated well enough for measurements to be corrected reliably. This lack of information is understandable: if bandshifting occurs in a research or medical setting and produces a serious ambiguity, one simply repeats the experiment. In forensic applications, one does not have this luxury. As of this writing, serious quantitative studies of bandshifting were still to be published.

Even if two samples show matching DNA patterns, it remains to determine the probability that the match might have occurred just by chance—that is, the probability expressed by the frequency of the pattern in the population. The most straightforward approach would be to compare the DNA pattern to a previously assembled data base containing the DNA patterns of a randomly selected population sample. If the DNA pattern in question matched no pattern in a data base of, say, 1,000 people, one would conclude that its frequency was probably less than 1 in 1,000. Such a conclusion would be completely defensible, provided that the sampling had been random. (In fact, the sampling schemes used by testing laboratories leave much to be desired in this regard: the FBI's original Caucasian data bases consist of samples from FBI agents, which hardly is a random sample with regard to ethnicity.)

DNA typing laboratories, however, claim much more extreme probabilities. They cite odds that range from 1 in 100,000, to 1 in 100 million, to 1 in 739 trillion in one case. How are such probabilities calculated? The explanation is simple: the laboratory assumes that each allele (that is, each band) in the DNA pattern is statistically independent and then multiplies the population frequencies of the alleles to produce the sometimes astronomical odds reported. Of course, the key is whether the assumption of statistical independence is correct.

The law has confronted the issue of statistical independence in the past. In a famous case called *California v. Collins,* an eyewitness reported seeing an interracial couple, the woman being a blonde, the man being black, leaving the scene of a crime in a yellow automobile.[10] The police subsequently picked up such a couple. At the trial, a mathematician testified about the probability of finding a couple that matched this description. As part of his calculation, he multiplied (a) the frequency of blonde women in the population times (b) the frequency of black men times (c) the frequency of interracial couples. Using arithmetic of this sort, he reached the conclusion that the frequency of such couples was about 1 in 12,000,000; the jury convicted the defendants. The California Supreme Court overturned the conviction, because of this improper statistical testimony. Among several problems, the court noted that the three categories described in (a), (b), and (c) are certainly not statistically independent and so their frequencies cannot be multiplied.

In population genetics, the question of statistical independence comes down to whether the general population is randomly mixed or whether it contains genetically differentiated subgroups.[11] If the latter is the case, then bands will not be statistically independent: if a band common among southern Italians is found at one locus, for example, it is more likely that the suspect has southern Italian heritage and thus it is more likely that a band common among southern Italians will be found at another locus.

The correct way to calculate population frequencies for a heterogeneous population is a complex question. Remarkably, DNA testing laboratories initially paid little attention to this point. Lifecodes stated in a scientific paper that it saw no evidence of heterogeneity in its data bases, but it did not present the evidence itself. When its evidence was made public in the *Castro* trial, the data actually supported the opposite conclusion. Given Lifecodes' assertion about its measurement accuracy—the laboratory director testified that the system could distinguish two-band patterns from one-band patterns provided that the fragments differed by 0.6

percent in molecular weight—there were far too many single band (or homozygous) patterns in the study. The high proportion of homozygous patterns indicates a correlation between the allele inherited from father and the allele inherited from mother in the population studies that Lifecodes had performed to test the assumption of statistical independence required for calculating population frequencies. In fact, one of the *Castro* witnesses was a scientist who had served as peer reviewer for the scientific paper describing the population study. After reviewing the data, he testified that the data contradicted the assertions made by Lifecodes and that he would never have accepted the paper for publication had he seen the data at the time of submission.

Population genetics has proved to be a very thorny and confusing issue for the courts. Some courts have accepted the calculations wholesale; other courts have ventured into correcting the arithmetic; still others have thrown out DNA fingerprinting altogether pending a resolution of this issue. The last category includes the supreme courts of Minnesota and Massachusetts. Clearly, this important issue requires prompt and rigorous resolution.

———————

Several years of experience with DNA fingerprinting have taught a number of lessons.

First, developers of the technology have perhaps been too clever by half, permitting the better to be the enemy of the good. By using the most polymorphic systems, those consisting of variable number of tandem repeats, they have certainly increased the discriminatory power of the technology but at the expense of requiring measurements of fragment position much more accurate than those routinely used in medical applications. If they had instead used a larger number of polymorphic systems each with only a handful of discrete alleles, the discriminatory power would have been slightly lower but the analysis much less demanding. Inasmuch as prosecutors report that odds of 1 in 1,000 are usually more than adequate for getting convictions, the latter choice may have been wiser—at least in the early stages of this new technology.

Second, new technology also has a tendency to give rise to new needs. Even though HLA testing provides virtually a 99 percent chance of exclusion in paternity suits and is quite inexpensive, some authorities have called for much more costly DNA fingerprinting in all paternity cases. In a March 1990 editorial, *The Lancet* suggested that DNA fingerprinting was entirely superfluous for this purpose; that it was simply a frill, a method that had turned into a need simply because it was possible. Similarly, the FBI is eager to establish a national computerized DNA fingerprint data base (analogous to the automated fingerprint data bases), although no one has done a careful cost-benefit analysis to ascertain the value of such a system for criminal investigations. DNA fingerprinting has many benefits, but they must be evaluated carefully.

Finally, the more powerful a technology such as DNA fingerprinting is in principle, the less likely is it to be closely scrutinized and challenged in practice. This is a Faustian bargain. We should be most cautious precisely about those technologies that are most powerful and most valuable—otherwise our tolerance of substandard work will return to haunt us. (Indeed, this is beginning to happen with DNA fingerprinting: some defense attorneys are trying to turn prosecution-commissioned DNA analyses to their client's advantage by arguing that small differences and artifacts exclude their client.) It is always wise to remember John Gardner's apothegm: "The society which scorns excellence in plumbing because it is a humble activity and tolerates shoddiness in philosophy because it is an exalted activity, shall have neither good plumbing nor good philosophy. Neither its pipes nor its theories will hold water."[12]

Fortunately, DNA fingerprinting is steadily improving—partly because the technology is getting better, partly because efforts are being made to resolve the problems that experience has exposed. In response to a joint suggestion of the experts in the *Castro* case, the U.S. National Academy of Sciences established a committee to recommend standards for DNA fingerprinting. Perhaps such a committee should have been constituted at the outset, but the truth is that if we had attempted to write down a set of rules prior to any trial experience, we would have been far off the mark. As with case law, scientific "case" wisdom develops only through the accretion of examples. In the interim until standards are in place, individual defendants require vigorous legal advocacy by counsel competent in the evaluation of scientific evidence.

In the long run, the reliability of DNA fingerprinting would undoubtedly benefit from independent, blind proficiency testing of DNA laboratories. Some proficiency testing has been conducted by the California Association of Crime Laboratory Directors (CACLD), but this organization is not independent of the laboratories and the testing has not been blind. Still, it is far better than nothing and has turned up several problems. On the first round of testing, Cellmark Diagnostics declared one false match out of fifty samples. The error turned out to result from sample mix-up: someone had unwittingly loaded the same sample in two different lanes! Cellmark thereafter required that every DNA transfer had to be witnessed and signed off by two people. In a second proficiency test, completed in March 1990, Cellmark declared another false match. The reason for this error is still not clear; it may have been due to a contaminated reagent.

The occurrence of false matches should come as no surprise: clinical laboratory errors occur in all areas at rates estimated at between 1 percent and 5 percent. Mistakes inevitably get made. Proficiency testing compels us to find our errors and confront their causes—with the result that procedures improve. It should also remind us that it makes no sense to report DNA fingerprinting results as accurate to one part in a hundred million, if the rate of laboratory errors is on the order of even 1 percent.

To ensure high standards, DNA fingerprinting laboratories are very likely—and quite properly—headed for governmental regulation. There are a number of possible routes being explored, including laboratory accreditation by a professional organization, state regulation, or federal regulation.

It may prove easier to solve the technical issues in DNA fingerprinting than some of the legal problems that arise from it. For example, defendants in a case involving DNA evidence should be entitled to expert counsel. Even though DNA fingerprinting has been excluded in only a few cases (such as *Castro*), there have been dozens of cases in which prosecutors have chosen to withdraw the evidence when the defense has hired experts to scrutinize it. Still, an indigent defendant's right to expert assistance is not a settled matter in American law and, if it is recognized, it will surely prove expensive to ensure.

Another issue is that a number of legislatures have passed laws making DNA fingerprinting automatically admissible, without bothering to define what DNA fingerprinting is. This is a worrisome trend. DNA fingerprinting is not a single technology, but rather a diverse collection of methods for assessing DNA differences—each at different stages of development. Providing blanket admissibility for any type of DNA analysis whatsoever (as these loosely drawn statutes might appear to do) is an invitation for mischief.

With growing acceptance of DNA identification and growing interest in DNA data banks, it is reasonable to ask whether our DNA might become our "social security number." Already, the military is interested in using DNA identification for all its personnel (as an aid, for example, in the identification of remains); only cost prevents the Department of Defense from pursuing the option at the moment, and this barrier is expected to decrease with newer and more efficient DNA-typing technologies. Some authorities have proposed DNA typing of all newborns, to facilitate the identification of a child who is kidnapped and later found. Once begun, newborn typing could lead to the creation of a national DNA data base that specifies not only each individual's identifying pattern but also his or her medical and possibly behavioral traits. Scenarios of this kind are not in immediate prospect, but neither are they far-fetched.

The proliferation of DNA data bases will surely challenge the individual's right of privacy in connection with criminal justice systems. Given a sufficiently important case (say, a serial killer), police someday will seek to search DNA patterns stored in medical data bases to find the criminal. Innocent people will have nothing to fear from such a search, they will argue. And, assuming that the technical problems have been cleared up by then, they will be substantially correct. Nevertheless, such searches seem intrusive to many observers.

Even without data bases, police investigations will pose similar problems. If the police have identified twenty possible suspects, they will someday ask a judge to require DNA samples to be taken from each to determine whether any of them match. Today, a court would probably deny the request because it would involve taking blood samples, which is rather intrusive. However, improved technology (such as the polymerase chain reaction, a powerful method for amplifying DNA *in vitro*) is making it possible for DNA typing to be carried out on a wad of spit or a single hair. Courts may not view the request as particularly intrusive, and may be persuaded by the argument that the innocent have nothing to fear from the test. Perhaps courts will allow testing of everyone in the vicinity of crime. If so, the notion of requiring "probable cause" may begin to erode.[13]

DNA fingerprinting is a major contribution of molecular biology to the criminal justice system. To be sure, current DNA fingerprinting procedures will be replaced by far more sensitive and less expensive techniques—including amplification methods such as the polymerase chain reaction, detection methods that allow direct inference of DNA sequence, and automated devices that reduce the possibility of human error. The prospect of improvement is no reason to forestall the use of DNA fingerprinting today, however. To wait would, perhaps, allow some innocent people to be convicted and some guilty ones to go free.

That said, it would be wise to guard against an overweening confidence in the technology, not to mention biased used of it— the tendency to see in the bands what prosecutors and defense attorneys want to see. It would also be wise to think hard and deeply about how to meet the challenges to civil liberties that DNA typing will pose.

NANCY WEXLER

Clairvoyance and Caution: Repercussions from the Human Genome Project

10

The natural trajectory of human genome research is toward the identification of genes, genes that control normal biological functions and genes that create genetic disease or interact with other genes to precipitate hereditary disorders. Genes are being localized far more rapidly than treatments are being developed for the afflictions they cause, and the human genome project will accelerate this trend. The acquisition of genetic knowledge is, in short, outpacing the accumulation of therapeutic power—a condition that poses special difficulties for genetic knowing.

Our expectation is that the characterization of a disease-instigating gene will greatly assist our understanding of how and why it causes a malfunction in the body. It makes good sense to go to the root of the problem. But to learn a gene's secret, first you must find it. And finding it is not so simple. It is much easier to locate the neighborhood in the genome where a gene resides than it is to determine its exact address.

Lilliput and Brobdingnag: Beyond Gulliver's Travels. The magnitude of the challenge arises from the vast amount of DNA contained in the diploid human genome, which includes all of a person's genetic material. If strung out, the DNA in a single human genome would stretch to about two meters, but the diameter of the strand would amount to only about two billionths of a meter, 20 angstroms, a span a hundred times smaller than a wavelength of light. If the DNA from a single cell from every human being on the planet—6 billion people—were stitched end to end, the resulting string would girdle the earth about 300 times. If the genomes from *every* cell of the 6 billion people were laid out end to end, they would extend 700 billion, billion miles—enough to wrap around our galaxy more than 700 times.

To understand the enormous problem of finding a gene somewhere on an individual's strand of DNA, imagine that a single human genome is long enough to circle the globe. On this scale, the amount of DNA in a chromosome would extend for a thousand miles. A gene would span just one twentieth of a mile, and a disease-causing defect—a point mutation, a change in only one DNA base pair—could run as short as one twentieth of an inch. What we are thus searching for is comparable to a fraction of an inch on the circumference of the globe! In this immense morass of DNA, finding the exact address of a gene and pinpointing its fault makes for extremely tough going, and it requires all of the creativity and ingenuity of everyone engaged in the quest.

Hunting for Huntington's. The spectacular difficulty of the problem has been made painfully clear by the search for the gene causing Huntington's disease (HD). Huntington's disease is a movement disorder—causing uncontrollable jerking and writhing movements of all parts of the body, called chorea. Even more distressing to patients and families than the obvious movements, it is preceded or accompanied by cognitive changes leading to profound intellectual deterioration and frequently severe emotional disturbances, usually suicidal depression and occasionally hallucinations and delusions. The disease runs a course of about fifteen to twenty-five years and is inevitably fatal. Its usual onset is between the ages of thirty-five and forty-five, but it can start as early as two and as late as the early eighties, an age when it can be hard to detect. The later the onset, the milder the symptoms. If the diagnosis is missed in an elderly person, manifestation of the disease in the next generation may appear erroneously to be due to a new mutation. No treatments are known beyond some marginal and temporary palliation for the movements and antidepressants for the psychiatric symptoms.

Huntington's disease is the product of a gene transmitted in an autosomally dominant inheritance pattern—in other words, a gene that occurs on one of the twenty-two non-sex human chromosomes and whose effect dominates its normal partner. It is entirely penetrant, which means that if a gene carrier lives long enough, the disease is inexorably expressed.

One peculiarity of Huntington's disease is that the sex of the parent transmitting the abnormal gene seems to play a role in determining the age of disease onset in offspring. Children, both male and female, who fall ill when twenty years old or younger almost invariably have inherited the disease from their fathers. Whether a gene is passed on through an egg or sperm sometimes affects its level of expressivity, a phenomenon called "imprinting." (One possible explanation for imprinting is that the number of methyl groups added to a gene vary, depending on the sex of the parent passing on the gene.) This differential expression may, in turn, influence the timing of disease onset. Or other modifying genetic factors may alter the timing and expression of the HD gene. The identification and manipulation of these factors may lead to early therapeutic measures: if disease onset could be pushed until later in life, the illness might not be so onerous.

The fact that the development of new therapeutics for Huntington's disease and other hereditary disorders may require the pursuit and characterization of many normal as well as abnormal genes underscores the need for a unified and concerted effort such as the human genome project. For Huntington's disease therapeutics, genes that determine critical chemical pathways affected by the gene defect may prove more amenable to intervention than the HD gene itself.

"Riflips" to the rescue. In looking for the fraction of an inch responsible for HD on the globe of DNA, we get some very clever help from restriction enzymes that identify small, normal variations in DNA called restriction fragment length polymorphisms—the RFLPs that geneticists pronounce as "riflips." Whenever a restriction enzyme sees its unique recognition site, it cuts the DNA right at that spot, like a miniature pair of scissors. The locations of these sites vary among individuals, and, as a result, the DNA fragments between two sites differ in length. When DNA is cut with restriction enzymes, these differences in fragment sizes can differentiate

one person from another, one chromosome from another, and they are inherited, just as genes are. "Riflips" act as markers in a person's DNA, a telltale indicator of genetic identity. (There are now many new kinds of very informative markers that do not require restriction enzymes but still serve the same function of identifying specific regions in the DNA.)

When we began the search for the HD gene, we were looking for a RFLP marker that was close to it. We can get an idea of the relative closeness of a marker and a gene on a chromosome because of a process called recombination—the tendency of segments of paired chromosomes (one from the father, the other from the mother) to change places during the creation of gametes, a kind of genetic "do-si-do." The further apart the marker is from the gene, the more likely it is that one of these "recombination events" will separate them; the closer together, the less likely. For every one million base pairs of DNA, there is a 1 percent chance that a recombination event will take place. Counting the number of recombination events found gives you a fairly good estimate of genetic distance between two markers or a marker and a gene. (I explain recombination probabilities to myself by imagining an earthquake at the North Pole where thousands of penguins occupy a huge ice floe; when the ice breaks up, two penguins sitting next to each other are more likely to stay on the same little piece of ice while two penguins far away from each other will drift away on separate pieces down each half of the globe.)

If one of these penguins is a DNA marker and the other the HD gene on the same floe, the two will travel together. If the "penguins" are close on the same chromosome, they will be transmitted to offspring in a Mendelian fashion with a high degree of regularity. So if a mother with Huntington's disease has a pattern-A RFLP next to her HD gene, and the father, who is not affected with Huntington's disease, has a pattern-B RFLP next to his normal gene, then their children with the B pattern will most likely not inherit HD and those who have inherited the HD gene will show the A pattern. (Because of the possibility that a recombination event will separate and rearrange markers and genes, we can only say "most likely." See Figure 27.)

In 1979, when the search began for the HD gene, the idea of "mapping" genes using RFLP markers was totally new and thought to be whimsical, if not heretical. No one had actually

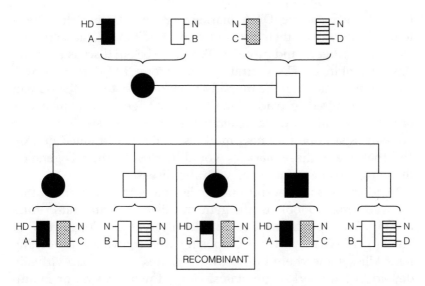

Figure 27 In this pedigree, women are denoted by circles, men by squares. Shaded symbols indicate that the person has Huntington's disease. Shown next to each person are the top portions of the short arms of both chromosome 4's. The Huntington's gene locus is sitting very close to and above a marker locus that has four different variants, or alleles (A, B, C, D). The mother, who is affected by Huntington's disease, carries the A allele of the marker near the disease gene (marked HD). Her normal gene (denoted as N), on her other chromosome 4, is sitting next to the B allele at the marker locus. The father's two chromosome 4's can be differentiated because even though he has two normal alleles at the HD gene locus, one is near the C allele of the marker while the other is near the D allele. All four chromosomes in the parents can be distinguished by the fact that each carries a distinct allele at the marker locus.

When the HD gene was passed on to the eldest daughter, it was accompanied by the A allele at the marker locus. She and her three brothers are not recombinants because the copies of chromosome 4 that they inherited from their mother are exactly the same as her chromosome 4. If one looked only at the marker type of the younger sister, one would conclude that she had escaped HD. However, she shows clear symptoms of the disease, an unequivocal indication that the gene is also present. This indicates that a recombination event or "crossover" has occurred during the formation of the ova from which she developed. The HD gene and the normal gene switched places to create two new "recombinant" chromosomes.

Recombination events are a measure of genetic distance: the further apart the gene and marker are, the more likely it is that they will recombine. For research purposes, discovering a recombination event can be beneficial in pinpointing the precise location of the gene. On the other hand, recombination events can lead to disastrous clinical misdiagnoses, as the diagnosis is based only on inspection of marker patterns and not on the presence of the gene itself. In a late-onset disorder, it is inadvisable to conclude that people who have the alleles that *usually* indicate the presence of the gene but are asymptomatic have a genetic recombination because they may not have yet developed the illness. Only those with unequivocal Huntington's disease and the marker alleles that typically travel with the normal gene should be utilized for research into recombination events.

located a gene using DNA markers, although genes had been found by virtue of their proximity to "traditional" markers, red blood cell antigens and proteins. When we began discussing using this recombinant DNA strategy, only one RFLP marker was known; the claim that the myriad of markers existed in the human genome required to place some near every gene of interest was only theoretical, an extrapolation from other species. We had to hope against hope that new markers could be fashioned quickly and that one of these markers would lie close to the HD gene on the chromosomes of persons with the disease.

Needless to say, several knowledgeable scientists told us that we were crazy to look for the gene in this haphazard, hit-or-miss fashion. They predicted it could take fifty years or longer to find our target. What we were proposing was equivalent to looking for a killer somewhere in the United States with a map virtually devoid of landmarks—no states, cities, towns, rivers, or mountains, and certainly no street addresses or zip codes—with absolutely no points of demarcation by which to locate the murderer. Our critics said "wait until a more detailed genetic map is available, one with many more regularly spaced markers." This is, of course, a much better strategy if you have the time to wait. But we are in a race against the Huntington's disease killer and have no time to spare.

Venezuela bound. In 1979, despite such sensible advice, we began hunting for the Huntington's disease gene. We knew that just finding the location of the gene would not tell us anything about the nature of the gene defect itself. But we reasoned that if we could close in on the gene from either direction, using markers more and more closely spaced until we finally honed in on the gene itself, we could then direct all our energies and resources toward identifying the gene defect and developing therapeutic interventions. If we were incredibly lucky, the markers could tell us, "your murderer is living in Red Lodge, Montana," and then we could continue the hunt door-to-door.

The only way you can tell if a marker is close to an unmapped gene is to observe if the two consistently "travel" together in a family. We know certain people have the HD gene because they are sick. We were looking for people who had the disease to have one form of a marker and their unaffected relatives to have another form of that same marker. We needed to study large families, as the HD gene itself might vary in its locale from family to family, and the markers, having nothing to do with disease itself, would certainly vary across families as to which form of the marker traveled with the gene.

So we were looking for a large extended multigenerational family in which we could observe many instances of the Huntington's disease gene or its normal counterpart being passed on—and we knew of just such a family, although when we began we had no idea how huge and important this family would eventually turn out to be. Members of the kindred live in Venezuela in three rural villages—San Luis, Barranquitas, and Laguneta—on the shores of Lake Maracaibo. Because it is situated in the northern region of Latin America and Lake Maracaibo is actually a huge ocean gulf, Venezuela has long communicated directly with Europe, and many European genes have appeared in the local population. Story has it that some sailor with Huntington's disease came over to trade and left his legacy, but we do not know if this is apocryphal.

We have been able to trace the disease as far back as the early 1800s, to a woman appropriately named María Concepción. We know that María lived in the "pueblos de agua," villages built on stilts in the water next to shores too marshy, jungly, and inhospitable to accommodate human life. Laguneta, where many of María's descendants live, remains such a stilt village.

María was the founder of a kindred now numbering close to 11,000 people, living and deceased. In the pedigree, there are 371 persons with Huntington's disease, 1,266 at 50 percent risk and 2,395 at 25 percent risk for the disease. Of the 11,000, 9,000 are living and the majority are under the age of forty. In these small and impoverished towns, we estimate that there are over 660 asymptomatic gene carriers who are too young to show symptoms, but as years pass—if no treatment is found—they will surely die. It is crushing to look at these exuberant children full of hope and expectation, despite poverty, despite illiteracy, despite dangerous and exhausting work for the boys fishing in small boats in the turbulent lake, or for even the tiny girls tending house and caring for ill parents, despite a brutalizing disease robbing them of parents, grandparents, aunts, uncles, and cousins—they are joyous and wild with life, until the disease attacks. Every year we add more people to the pedigree who will suffer, diagnose more new cases, and watch helplessly as more and more begin that sad journey toward deterioration and death. It is impossible to be immune to their plight. It is only possible to be passionate and driven and work desperately to save as many as we can before it is too late.

My original interest in this family was sparked by its pockets of consanguinity. The kindred is now so numerous and extensive, spanning eight generations, that it is not particularly inbred. It also has a tumult of Hispanic and other European genes intermingled with indigenous Indian genes to make for a very rich genetic mixture. But in some branches of the families, gene carriers intermarried and produced offspring with a 25 percent chance of inheriting the HD gene from both parents, that is, being homozygous for the disease. We were hoping to find more direct evidence of the biochemical cause of the disease by studying a homozygote, who would have no normal gene product to mask the workings of the defective gene. When I first went to Venezuela, in 1979, it was only to look for such families. We did find a large family in which both parents were affected, drew blood samples from these family members, and went home thinking we would have a very small study.

Anatomy of a gene search. By 1981, however, we had changed our rationale for research with the Venezuelan family from searching for homozygotes to a full scale genetic linkage project using DNA markers. I went to Lake Maracaibo with a small research team in March 1981 for what proved to be the first of an annual month-long expedition ever since. We were assisted in all aspects of our work by an extraordinary physician, Dr. Américo Negrette, who first correctly diagnosed Huntington's disease in this population and began constructing the pedigree, and two of his students, Dr. Ramón Ávila Girón and especially Dr. Ernesto Bonilla, who continues to be an active investigator. Our first task was to extend the pedigree that our Venezuelan colleagues had begun. Most couples represented in the pedigree are not legally married. Our pedigree is entirely composed of oral record, but just asking people to name their relatives has worked out pretty well. We check for possible nonpaternity but have found the rate lower than in the United States because people readily identify the father, even if he is not the mother's current partner.

Mapping the Huntington's disease gene required obtaining blood samples from as many relatives as possible in families in which the disease was traveling, according to certain guidelines for collection. Our Venezuelan colleagues warned us that the family members might be reluctant to give blood or refuse altogether; for many it was the first time ever. We have been privileged to receive extraordinary cooperation, despite people's fears. Most gave on behalf of their children. Although they may not know the exact odds, kindred members are experts on this disease and have a keen sense of the threat to their offspring. As birth control has become more available to them, it is being used increasingly.

I felt it was important for the Venezuelan family members to know that Huntington's disease was also in my family and in many others in the United States but that our families were not large enough to offer the gift to research that their family was capable of providing. We needed their help to find a cure. At that time we were doing skin biopsies, which I also had done for research. The family members were dubious of my story until I pointed out my skin biopsy scar and my wonderful colleague and friend, Fidela Gomez, a Florida nurse, grabbed my arm and dragged me around the room shouting, "She has the mark, she has the mark!" My mark and I became something of a passport for our research team and its activities.

Since the blood had to be sent immediately to James Gusella's laboratory in Boston, we could draw blood only when a team member was leaving and could hand-carry the samples. Those days became known as "draw days"—chaotic days in boiling hot, deafeningly noisy rooms jam-packed with people of all ages, days spent going to sweltering homes where throngs of children would shout out in gleeful horror the number of tubes of blood we were drawing. The men tended to be more recalcitrant than the women, fearing they would lose some vital bodily fluid and be weakened, or unable to work or drink, if they gave blood.

Genetic jackpot. At Massachusetts General Hospital, DNA was extracted from blood samples from the Venezuelan family members.

Jim Gusella was also studying a large American family with Huntington's disease from Iowa. He searched the DNA from these
two families for a telltale marker, helping to develop what were
to become standard laboratory procedures in such ventures. Jim
sliced up each person's DNA with restriction enzymes. He then
developed markers, RFLPs, which he made radioactive. These
markers were called anonymous because he did not know on
which human chromosome they were located, only that they were
in one unique spot in the genome, just like a gene, and they came
in several forms so that individuals could be differentiated from
one another. The fragments of chopped-up DNA from the family
members were put on a gel that separates fragments on the basis
of size. The radioactive probe (denatured, or single-stranded) was
then added. When the probe met its counterpart single-stranded
DNA on the gel, the two strands would reanneal, a process called
hybridization. As the probe is radioactive, it would "light up"
where it was stuck on the gel, revealing distinctive bands. One
would then need to check if a certain pattern of bands appeared
only in individuals who had the disease and another pattern in
their relatives who were healthy. If this difference was true more
often than would be expected by chance, it would be very likely
that the marker and the gene were close together on the same
chromosome.

We all expected that the detection of a marker linked to the
Huntington's disease gene would require thousands of tests and
probes, but the third probe that Gusella characterized and the
twelfth one he tried hit the jackpot. He began with the Iowan
family, whose samples were the first to be collected, and the
probe, called G8, was weakly positive, but not significantly so.

This finding gave him the crucial push, however, to try G8 in
the Venezuelan family—and it was the only probe he needed! It
immediately showed odds far better than 1,000 to 1 that it was
very close to the HD gene. P. Michael Conneally at Indiana University performed the computer linkage analyses that definitely
proved that this probe and the HD gene were close neighbors.
Almost all the Venezuelan family members with HD had one form
of the marker while their healthy relatives had another. At the
time linkage was discovered, the chromosomal location of the
probe was unknown, but it was quickly mapped, using *in situ*
hybridization and other techniques, to chromosome 4. By infer-

ence, the position of the gene was mapped as well. Out of 3 billion possible base pairs on 23 chromosomes, we now knew we were a mere 4 million base pairs below the culprit gene way on the very top of the short arm of chromosome 4. We triumphantly announced the feat in an article in *Nature*, in November 1983.[1]

It had taken us just three years—an astonishingly short time—to localize the HD gene. Our critics and even our supporters said, rightly, that we had been incredibly lucky. It was as though, without the map of the United States, we had looked for the killer by chance in Red Lodge, Montana, and found the neighborhood where he was living.

An elusive prey. Next we needed to find the exact location of the HD gene, isolate it, and learn its secret. Since January 1984, the Hereditary Disease Foundation has supported a formal collaboration of seven scientists around the country and the world searching for this gene: Francis Collins, at the University of Michigan; Anna Maria Frischauf and Hans Lehrach of Imperial Cancer Research Fund, London; Peter S. Harper, University of Wales College of Medicine; David Housman, Massachusetts Institute of Technology; James Gusella, Massachusetts General Hospital/Harvard Medical School; and John Wasmuth, University of California, Irvine. The task has been arduous in the extreme in this inhospitable terrain at the top of chromosome 4. It has been like crawling up Mt. Everest over the past eight years. First we thought the gene was at the telomere, the very end of the chromosome. Now recent work has indicated that we probably jumped over the gene in our rush to get to the top and it actually is not quite to the end. I used to say confidently that we would be right on it within six months for sure, but I don't say that anymore.

Homozygotes for Huntington's. I mentioned earlier that our first research interest in the Venezuelan kindred was to find a homozygote for the illness. Once a marker for the gene was found, we immediately used it to learn which offspring might have inherited the HD gene from both affected parents. This included the family which originally drew us to Venezuela, a family of fourteen children and more than seventy grandchildren and great-grandchildren. Over the last decade of work, we have also found two other families in which both parents have Huntington's disease and many more in which both are at risk or one affected, the other at risk. Eight probable homozygotes have been identified from these families and more undoubtedly exist.

Even though a dominant gene is defined as "dominating" its normal partner, the few homozygotes who have been found for human dominant genetic diseases have been described as more severely affected than heterozygotes for that same illness. This would suggest a dose effect, even for dominant disorders. The normal gene plays an ameliorative role, even in one dose, and two doses of the defective gene makes the illness worse.

Huntington's disease provides the only exception to this clinical experience: it is the first completely dominant human hereditary disorder that has been genetically documented. Those who are most likely homozygous are not different clinically from their heterozygote relatives. Some putative homozygotes are symptomatically normal, presumably too young yet to develop signs, others have minor neurologic abnormalities, while some have definitive Huntington's disease but no earlier or more severely than anyone else. One tragedy that homozygotes face uniquely: all of their children will be afflicted with HD, as the homozygous parent has no normal genes at that locus to contribute. This is especially agonizing in Venezuela, where the family sizes are so large. The cells of the homozygotes may hold clues that will unlock this devastating disorder but until an intervention is found their families barely sustain themselves in widening pools of suffering. Every year we examine more family members and every year our hearts sink.

The Venezuelan Reference Collections. Over the past decade during which we have been doing field work in Venezuela, it has become increasingly apparent that this excellent kindred, so helpful for studies of Huntington's disease because of its size, geographical proximity, and cooperativeness, among other qualities, is also uniquely superb for gene mapping in general, called reference mapping. Mapping the human genome requires that we track many markers and genes, just like the probe G8 and the HD gene, traveling through generations. The Venezuelan family is unparalleled for determining who is giving what to whom. We can follow eight generations (the grandmothers in this community are only in their thirties). Family studies from this population have been used to make a marker map of chromosome 21, which was helpful in locating genes for Alzheimer's disease and amyotrophic lateral sclerosis (Lou Gehrig's disease); maps of chromosome 17 and 22, where genes causing two forms of neurofibromatosis are located; and a map of chromosome 11, which was used to search for a possible site for manic depression. We are also trying to investigate other genetic or polygenic afflictions in the community—for example, obesity, diabetes, and hypertension—for which large study populations are needed.

In this work, it is, of course, imperative to get clinical diagnoses right, because if they are wrong, the genetic analyses are going to be totally incorrect. We are fortunate to be able to maintain contact with the family and return every year. A member of our team, a superb Venezuelan physician named Dra. Margot de Young, attends the family members year round. We try, as much as possible, to collect blood samples from an individual only once. But sometimes you find someone in whom a critical recombination event has taken place that can help localize the HD gene more precisely. It is essential to be able to return to that person to reconfirm the diagnosis and reanalyze a new DNA sample to eliminate the possibility of a laboratory error. Our continuing access to family members makes this reference collection additionally valuable; investigators specified that they would not recontact individuals who have contributed to other major family collections for genetic reference mapping and are thereby precluded from gathering clinical information or checking mistakes.

A new era: Prediction outstrips prevention. While the search for the Huntington's disease gene goes on, painstakingly, the discovery of *markers* linked to the gene has opened a new, exciting yet troubling era: presymptomatic and prenatal diagnosis of Huntington's disease with no cure in sight.

Immediately after localizing the Huntington's disease gene, we confronted the question of genetic heterogeneity: Is the HD gene in the same chromosomal locale in all families with Huntington's disease throughout the world? Many other genetic disorders manifest genetic heterogeneity—the causative gene may be on several different chromosomes, even though phenotypically the symptoms of the illness appear to be the same in all the affected families. Was our chromosome 4 home for Huntington's disease unique to the Lake Maracaibo kindred and an Iowan family, or was it universal? Over one hundred families have been tested from throughout the world—in Europe, North and South America, even Papua, New Guinea—and in all of them the HD gene is in the same chromosomal locale on the top of chromosome 4. The actual mutations at that spot may turn out to differ, but the region is the same. Given its universality, we can now use G8 and other markers subsequently found closer to the gene to test whether an individual has the HD gene before any symptoms appear, even before birth. So here we confront our worst fears: our scientific success puts us on the threshold of an era of unknown but imaginable dangers. We can predict the flood but cannot leave or stop the tide. We can tell people that they possess the gene and will eventually come down with the disease, but we have no cure or even therapy to offer to soften the blow.

Cystic fibrosis as a model. Whether or not a disease shows chromosomal heterogeneity (more than one chromosomal location), or allelic heterogeneity (more than one mutation in the same gene at the same chromosomal locus), makes a big difference in genetic counseling. A current case in point is cystic fibrosis (CF), the most common hereditary disorder of Caucasians. Those who suffer from it have pancreatic enzyme insufficiencies and severe lung abnormalities; they cannot clear fluids from their lungs, which become inviting parks for bacteria. Children with cystic fibrosis now often survive until early adulthood, but the disease eventually proves fatal. About one in twenty-five Caucasians carries a single abnormal gene for this condition but is not clinically affected. CF is a recessive illness; each child of two parents who both have one gene for the disorder has a one in four chance of inheriting two genes for cystic fibrosis and expressing the disease. There are about 30,000 individuals with CF in the United States. For a Caucasian person in the population with no family history of the illness, the likelihood of having a child affected is about one in two thousand. If a reliable test were available to detect carriers, people might choose to use it.

In September 1989, Francis Collins, Lap-Chee Tsui, and Jack Riordan isolated a mutation that is found in 70 percent of all individuals with cystic fibrosis.[2] Screening people for this particular three-base-pair deletion known as delta 508 would identify 70 percent of all CF carriers. If, however, you test positive but your husband does not, you still will not know whether he really is negative or just has a different mutation, one of the remaining 30 percent not yet known. Calculations indicate that screening couples who are both CF carriers would detect one partner but not the other in more than half the cases. To put the matter another way, if you assume that because your husband or wife has tested negative he or she is free of the gene for cystic fibrosis, half the time you would be wrong. Eminent geneticists on behalf of the American Society of Human Genetics issued a statement, with which other experts convened by the National Center for Human Genome Research, the National Institutes of Health (NIH), the Department of Energy (DOE) Human Genome Program, and other Institutes of the NIH concurred, recommending against population screening for cystic fibrosis until additional mutations were found and the test was more accurate.[3] Although testing could be beneficial for those with a known family history of CF, geneticists stated that any other testing was premature and certainly not the standard of care. They advised that widespread use of the test should await two essential elements: identification of a larger proportion of all mutations, and putting in place the service infrastructure to give the test with adequate counseling provided. (Over 100 new mutations have been found since 1989, and the requirement for a higher degree of accuracy of detection is being met. Adequate counseling services are still insufficient, however, and providing these may be even more difficult to implement than achieving the scientific goals.)

In the case of cystic fibrosis, it appears to be useful for parents to know when a child is born that it is affected. Earlier and more aggressive interventions with antibiotics, pancreatic enzyme therapy, and physical therapy can definitely assist the child. Prospective parents and prospective partners might also want this information for family planning purposes. When should you give it?

After conception, when options are limited to keeping or terminating the pregnancy? Before conception? When a couple applies for a marriage license? Should testing be obligatory before marriage? Should screening on a large scale be done at the school level? At what age? Should all children be screened at birth for CF carrier status? Each of these scenarios have very different economic, medical, psychological, and social repercussions.

Genetic illiteracy. In all of these screening programs people must understand the difference between being a carrier of one abnormal gene for a recessive condition, in which the carrier usually has no symptoms, as opposed to an affected individual who has two copies of the abnormal gene. People must equally understand that carriers for a dominant disorder, a late-onset illness such as Huntington's disease or polycystic kidney disease, will, in fact, get sick. In recessive diseases the carrier is only a carrier, but in dominant disorders the carrier will become a patient. How do we explain technically complex and emotionally charged information to ordinary people, many of whom never heard of DNA and barely of genes, who have hardly a clue about probability, and whose science education never equipped them to make choices regarding these matters? How do we ensure justice in access to counseling services and make them available to more than the white middle and upper classes who typically utilize them now? Genetic diseases cross ethnic and class boundaries, but access to services, unfortunately, does not.

How do we guarantee that the doctors who test individuals or populations provide adequate genetic counseling when doctors themselves have minimal training in genetics and often fundamentally misunderstand its principles? What should we do about doctors who say to a couple with one child affected by cystic fibrosis and who are contemplating having another, "Don't worry, lightning never strikes twice in the same place." Or—the ultimate in confusion about a genetically dominant disease— "Don't worry about Huntington's, just tell 'em to *marry out* into families that don't have it!" Such medical mistakes have increasingly been addressed through malpractice suits, including wrongful birth and wrongful life cases. In *wrongful birth* cases, parents of a seriously impaired child bring an action claiming that the child should never have been born. The parents argue that they were deprived through the negligence of a health care provider of the information they needed to decide whether to initiate or continue a pregnancy. Had they known, they claim, they never would have had the child. *Wrongful life* is an action brought by the child claiming that it should never have been born.[4] Must we resort to the threat of lawsuits to ensure that good medical practices will be followed? Or should we have sufficient ingenuity and imagination to be able to introduce new genetic findings into medical practice without increasing the litigiousness of our already embattled society? I believe we can figure out how to offer people genetic information in a way they can understand and assimilate. We can resolve these difficulties if we start working on them *now*, before the deluge of new tests the human genome program will bring.

All in the family. A major problem in presymptomatic and prenatal testing using linked DNA markers is that the whole family must be involved. When we have the gene in hand and can detect directly the specific mutation in the gene, we will only need to look at an individual's DNA. But for tests using linked RFLPs, the marker patterns of all the relatives must be traced to determine which pattern of the marker in that family is consistently traveling with the appearance of the HD gene. For example, in the Venezuelan family it is the C variant of the marker G8 that tracks with the HD gene, but in the Iowan family it is the A variant. Over time, the process of recombination will gradually change which particular pattern of a marker is near a gene, unless the marker is exceedingly close to the gene. If a gene and a marker are such close neighbors that they are virtually never separated, they are described as being in "linkage disequilibrium." Within a family, however, the same pattern of the marker will tend to travel with the gene and the few recombination events—the random exchange of segments between two homologous chromosomes—that may have taken place tend to be obvious. This is why diagnostic testing using linked markers *must* be done in families: it is imperative to determine what pattern of markers prognosticates HD in that particular family. It is a tedious way of doing diagnostic testing, but until the gene itself can be found it is the only way of doing it and it is how tests must be carried out right now, not only for Huntington's disease but also for polycystic kidney disease and others. (Anybody in a family with a genetic disease—this probably includes everybody—should think about storing samples of DNA from relatives whose genotype would be essential to know for diagnostic testing. This can easily be done by freezing a DNA sample extracted from blood. DNA can also be taken from the brain, skin fibroblasts, or almost any other tissue, even after it has been frozen for a long time. The most important relatives to you are those in the family with the illness and those clearly unaffected, parents of these individuals and your own parents. If you have a genetic disorder, banking your own DNA could be critical to your descendants. Each family might have its own genetic variation, its own "genetic fingerprint" of the gene in question, and it is best to preserve a sample of the particular gene that plagues your family rather than extrapolate from the genes from other families.)

There are many families in which an insufficient number of genetically informative people are living (or have banked DNA samples) to permit diagnostic testing for Huntington's disease. And many people would prefer not to know their own genetic status—whether or not they will develop HD. Can anything be offered for these people? One kind of test—called a nondisclosing prenatal test—allows couples at risk to gather some information about a fetus. This test can tell, almost definitively, if a fetus *is not* going to have HD but cannot tell definitely if the fetus *is* carrying the gene.

A person at risk has one chromosome 4 from a parent with HD, the other from a parent without the disease. The chromosome 4 from the affected parent may or may not be carrying the HD gene. The at-risk individual will pass on only one chromosome 4 to the fetus; the other comes from the partner. If that chromosome is the one from the unaffected parent, the fetus has a negligible risk (there is always some risk due to the possibility of recombination). If that chromosome comes from the affected parent, the fetus has the same risk as the at-risk parent: fifty-fifty. In a nondisclosing prenatal test, the risk status of the at-risk parent is not altered at all. The only new information that is acquired is whether the fetus has the chromosome 4 that came from its affected grandparent, in which case it has a fifty-fifty risk, or from its nonaffected grandparent, in which case its risk is minimal. And all that is required for this test is a DNA sample from the fetus, obtained through amniocentesis or chorionic villus sampling, both its parents, and one or both parents of the person at risk—a minimum of four people. (If the affected grandparent has died, his or her genotype can usually be inferred from the other individuals available.)

When we first began offering testing for HD, many of us involved in providing the test thought that nondisclosing prenatal tests would be a preferred option. It offers a chance to ensure that children would be free of the disease while at the same time protecting individuals at risk from learning potentially traumatic information. But comparatively few have utilized the test. Its worst aspect is the possibility of aborting a fetus with a 50 percent probability of not having Huntington's disease, the same risk as the at-risk parent. Just imagine—you're pregnant or you've fathered a baby, you're attached emotionally, your fantasies are engaged, and now you're confronted with the choice of aborting a baby who might be perfectly normal. How easy will it be for you to become pregnant again? How fast is your biological clock ticking? What if it happens again? A one in two chance is high. Some people feel that aborting a fetus with a 50 percent risk of HD is equivalent to aborting themselves, a rejection of who they are and of their legitimate place in the world. This sentiment is sometimes voiced by those who are disabled, those who object to genetic testing on the grounds that it is designed to eliminate people like them. Because of these particular difficulties, nondisclosing prenatal testing must be offered in a context of intensive counseling and support. If the couple is willing and interested, it is extremely valuable for research purposes to study any tissues resulting from terminations, particularly with respect to learning more about the timing of the HD gene expression. It is possible that the gene is expressed only *in utero*.

Considerations for genetic counseling. There are many factors that influence the nature and quality of genetic services. An important problem is timing—when should you give genetic information? For a late-onset disorder like Huntington's disease, timing issues are complex. We are often faced with providing presymptomatic test information to someone whose affected parent is in the last stages of the disease or has just died of it. Etched in their mind's eye is the disease at its devastating worst. And now you are telling someone who is perfectly healthy and normal, "You have a 96 percent probability of having the HD gene"—which they hear as, "You are going to be just like your mother or father." Wrenching news.

An opposite problem can occur if people coming for testing have never seen the disease. The affected parent may have drifted away from the family, died in a remote hospital somewhere, and the children know only that they are at risk but have never encountered anyone with Huntington's disease. Or a parent may have just been diagnosed and has only minimal symptoms. Newly confronting the ambiguity of being at risk, these people find it intolerable and run to the nearest testing clinic. But what happens if you test these "disease-naive" individuals and shortly after receiving the information that they are probably gene-positive, they happen to turn on *NOVA* and there on television is a graphic display of Huntington's disease from beginning to death. They think, "Oh my God! I had no idea this is what people at the testing center were talking about!"

If you decide to educate potential testees about the disease to enable them to make more informed choices with respect to being tested, how much education is appropriate? What if someone whose father has just been diagnosed comes to you for testing? You don't want to drag him or her down to the local state hospital, where many patients with HD live, or to a nursing home and say, "This is what is in store for your father, and maybe you." You cannot crack open a person's very beneficial shell of denial too radically, and yet you also cannot allow someone to be tested without having some fundamental appreciation of the meaning of the results. It is difficult to shatter denial and shoal it up at the same time; denial is a critical component of coping and must be treated with respect. Information should be carefully titrated, and intensive counseling over time is essential.

My fear is that when presymptomatic and prenatal laboratory tests become more rapid and accurate—for example, when polymerase chain reaction (PCR) techniques can detect the mutation itself—there will be a temptation to short-circuit the testing process, to make it faster and offer less counseling. But no matter whether the test is easy or not, the impact of the information is equally crucial for one's life. There is still no interdiction we can offer, no treatment or preventive. And even if testing can be done on a single individual without DNA samples from relatives, Huntington's disease is a *family disease* and the test results for one member reverberate through all.

At the moment, relatives must donate a DNA sample for linkage testing to be done. They sign an informed consent form when they give blood indicating that they give their permission for the sample to be used for the presymptomatic testing of someone in the family. Usually the at-risk person requesting testing must arrange with relatives for samples to be sent to the laboratory or for neurological examinations to be conducted on critical relatives whose clinical status must be accurate; with all these preparations going on, the at-risk individual's desire to be tested is generally known and discussed in the family. Relatives have an opportunity to convey their feelings and, in some instances, they try to dissuade the individual from continuing with the test. Some parents have gone even further to exert their influence and have refused to give a DNA sample, thereby halting the test. One parent refused because the testing program in that locale did not provide adequate counseling and follow-up. Many of the testing centers have encountered other situations in which parents might be willing to give a DNA sample for one offspring's test but not for another: "Jane can take the news, John can't." Of course, once you know what pattern of the marker the disease travels with in that family, you do not need to retest the parent's sample for each child. Whose rights take precedence—John, who says "I can take it, and besides, I want to get married"; or Jane, who says "Your arguing is depriving me of my test, and besides, I want to have a baby"; or the father who says, "I own my genetic profile—you can't rob me of my genetic information and use it without my permission for purposes of which I don't approve."

A similar problem arises when identical twins arrive at the testing center and one wants to be tested and the other does not. Now who should hold sway? One center said, "We'll test you but don't tell your twin." It doesn't work. If you are free of the disease it would be almost inconceivable not to run to your twin with the good news. And if the outcome is HD, it's hard to explain uncontrolled crying as a chronic cold to people who know you well. Other testing centers confronted with this predicament consulted ethicists who gave them pronouncements that autonomy is higher on the scale of ethical virtues than privacy, so the centers decided to proceed with the test. But to my mind, autonomy or privacy may be irrelevant if the twin who is not part of the testing process and has not even had the benefit of counseling learns the truth and commits suicide. The immediacy of each individual's psychological reality *must* take precedence over abstract, theoretical values and issues. One cannot consult a guidebook for who should be tested and under what circumstances. The professionals giving presymptomatic test counseling must be trained psychotherapeutically to help determine the best solutions for individuals and families as a whole.

Another factor that is insufficiently appreciated is that when one person in a family is tested, the entire family is tested and all must live with the outcomes. Many parents of persons at risk feel guilty about and responsible for their children's risk status, even though they may have known nothing about HD when the children were born. In some families, three or four children simultaneously may be diagnosed presymptomatically. A parent who has spent fifteen or twenty years caring for an ill spouse now has a grim preview of the future: the prospect of caring for children as well, or knowing that the children may have to depend on the mercy of strangers. One woman said, "When my husband died after twenty-five years of illness, I felt like a light had finally come on at the end of the tunnel. Now I watch my daughter and see her movements and the light has extinguished."

Testing minors. Because presymptomatic testing for Huntington's disease is difficult to undergo for all and potentially devastating for some, given the absence of any therapy, those professionals who are offering the test, including myself, have decided to restrict it to individuals who can give informed consent and who are age eighteen or older. This is not a legal requirement but it has been accepted as part of the HD testing protocol for centers throughout the world. Until we know more about the impact of having this information on adults who are choosing to learn it of their own free will, the testing center professionals are reluctant to test minors, at their own request, or to give information to parents about their minor children, with or without their children's knowledge. This conviction, on my part, was reinforced when a woman told me she wanted to test her two minor children because she only had enough money to send one to Harvard.

But parents do advance cogent arguments for testing their minor children—they need the information for making financial and other life plans. It would certainly make a difference to know that any or all of your children are going to develop HD. Withholding this information from parents goes against the typical situation found in family case law, in which parents are entitled to medical information and the only instances in which the courts are likely to intervene is when parents are *not* providing medical attention for religious or other reasons.[5]

One of the complexities of providing nondisclosing prenatal testing is that it sometimes forces you to give information about a minor child despite your protocol. Prenatal testing for HD is not offered to couples who have no intention of terminating the pregnancy because there is no medical advantage to the parents to have this information, and because prenatal testing does pose a slight medical risk to the fetus and entails testing a minor without its consent. But if a couple finds that their fetus has a 50 percent risk of having HD, they still may change their minds about a termination and carry the pregnancy to term. If the at-risk parent later develops the disease, the genetic identity of the fetus is also revealed. This violation of the privacy of the minor must be endured because parents are entitled to change their minds regarding the termination of a pregnancy. But very careful counseling must be provided so that the couple knows exactly what the test involves and the nature of their options.

Another potential controversy among test providers is the decision made by those currently offering presymptomatic testing for Huntington's disease not to test children who are in agencies awaiting adoption. The adoption agency personnel requesting testing argue that children shown to be free of the disease will be more eligible for adoption. To date, test providers have responded that such testing still violates the privacy of a minor without informed consent and, furthermore, may consign those found to be gene-positive to permanent placement in an orphanage, eliminating any hope that a child still only at-risk might hold for adoption.

Some of these decisions regarding who and when to test may be altered as we learn more about the meaning of genetic diagnostic information to those who are receiving it. During the 1970s, in Canada, several studies by Charles Scriver and colleagues showed that secondary-school children who learned they were carriers for the recessive gene causing Tay-Sachs felt stigmatized and somehow inferior to their classmates even though being a carrier was in no way injurious to their health.[6] It was an emotional stigmatization. Will such responses be common? Some people insist, "Make genetic testing mandatory when couples get married." Others advise, "Integrate it into the genetic services, so that couples can be tested when they are contemplating pregnancy or are already pregnant." However, people disinclined to choose abortion might want to have genetic information before selecting their mate. A sickle cell screening program in Orchemenos, Greece, undertaken in the early 1970s before prenatal diagnosis was possible, found that 23 percent of the population carried the trait.[7] Those discovered to be carriers were stigmatized and, as a result, they sometimes concealed their carrier status in order not to jeopardize marriage prospects. The net consequence was that the same number of affected babies was born as before the screening program. In two of the four matings resulting in the birth of an affected child, the women hid their carrier status, and in the other two the couples had children although cognizant of the risks. Once prenatal testing for sickle cell disease and thalassemia became widely available, carrier status became less of an impediment to social acceptance, even in predominantly rural, Catholic countries like Sardinia.[8]

Genetic misunderstandings and their implications. I am always surprised by the imaginative ways in which people can misinterpret genetic information. One common and very understandable mistake is the belief that at least one person in every family will be sick. In the Huntington's disease testing programs, people often arrive with the conviction that whether they will or will not get the disease depends on the fate of siblings. If my siblings are sick, my risk goes down; if they are old and healthy, my risk goes up. This is a perfectly reasonable misinterpretation given the way in which genetic inheritance is usually explained. Most genetic textbooks and consumer pamphlets teach principles of inheritance by showing a family of four children and two parents in which two children are affected and two are not. And doctors often explain risks by saying "half your children" or "one-quarter of your children will become sick." You must always say, "*Each* child has a fifty-fifty or 25 percent risk, regardless of the rest of the family." The day people in Venezuela became really confused about inheritance was the day the newspapers published such a diagram.

It is difficult to teach someone that "chance has no memory" and that whatever may have happened in a previous conception has no bearing on whether or not a child carries the HD gene. Each person has his or her *individual* risk and whatever happens to the rest of the siblings does not matter. I often ask people to flip coins during the counseling sessions to see, concretely, how it is possible to flip ten heads in a row. If they flip a coin which says that they will get HD, it also gives that dire possibility some hard reality.

The idea that one's life or death is controlled in such a random way as a flip of a coin is appalling for most people. We try to make sense, make meaning of our lives. We try magically to control that coin toss by inventing rules governing who gets sick and who does not, but the fact remains: it is totally a random accident of fate which gametes meet, and in that moment the future is sealed.

Gazing into the crystal ball. There are approximately 22 HD testing centers in the United States, 14 in Canada, 5 in Great Britain, and several more in Europe and Scandinavia. Probably fewer than 1,000 people have been tested in all these centers. This skewing of test results toward those who test negatively is probably a result of the arduous and intense counseling required by the test protocol. As people come to appreciate on an emotional level, not merely intellectually, the magnitude of a positive diagnosis on their lives, those who have some inkling that this is to be their fate may decide not to be tested.

It has also been the experience of most centers that some individuals coming for presymptomatic testing are clearly already symptomatic but show no awareness of these symptoms. They mostly want to learn if they will have HD in the future, not in the present. If they are ready psychologically for the information, they should be diagnosed clinically, not by DNA analysis, or else encouraged to come back at a later date.

One group of individuals who come for testing I call the "altru-istic testees." These are people who would prefer not to be tested but are doing so to clarify the risk for their children who are getting to be dating or marrying age. The genetic risk for these people is lower because they are older, but many of them really do not want to be tested and would prefer not to "rock the boat." We know very little about the response of this group to a diagnosis of probably gene-positive.

Some clients who learn that they do not carry the disease gene find that this knowledge does not make all of their problems dis-appear. They may still have trouble finding the "right person" or managing their careers. Before the test, being at risk became a convenient excuse for putting decisions on hold, for postponing and avoiding issues that may have nothing to do with being at risk but become entangled in the risk situation. People at risk say, "Well, if I just weren't at risk, I could figure this all out." Then, suddenly, they are no longer at risk, but they still can't figure things out. The problems have become too entrenched, too much a part of their characters.

This is not to say that good news does not also lead to ecstasy and joy. Some people alter their lives—have children, move, or change jobs—and feel wonderful about the change. Others expe-rience a kind of survival guilt and sense of unease with respect to other family members who may not know their genetic status or who tested positive for the gene.

Learning by experience in Venezuela. It is this latter group—those who test most likely gene-positive—who preoccupy us the most, who capture our imagination and concern. As there are only a very few such people, less than 100 people in the United States, we know little about how this new "presymptomatic group" will react to the bad news. Some clues are provided by our experience in Venezuela, where we are just determining the clinical status of the study population. Only in rare instances does anyone ask us, after a neurological examination, what we have learned. On one occasion, however, a woman in her twenties, mother of several children and pregnant again, came to be examined. We concluded that she did have Huntington's disease and were taken by sur-prise when she asked us what we thought. We asked her how she felt about herself and she said, "just fine—no HD so far," which gave us the immediate clue to find out more about her and what this diagnosis would mean in her life. We told her we would like to get to know her better and see her over time—we would be there all month and again the following year and encouraged her to spend time with us. A moment after she left, her friend came running into the clinic at top speed, looking panicked, and asked, "What did you say to her?" When we told her, she sank back in relief. "Oh, thank God," she said, explaining that her friend had announced, "I'm going to ask the American doctors if I have Huntington's and if I do I'm committing suicide." Just as in the United States, suicide attempts and completions do occur in the Lake Maracaibo community.

The advent of presymptomatic testing presented us with a dilemma: Should we try to make it available to the Venezuelan families? Scientifically it is essential for all of us working in the field to be blind to the genotypes of the people whom we are following. It is impossible to do an unbiased clinical examination once you know the genotype, and those not doing examinations must remain blind to prevent nonverbal or accidental revelation of the information. It is particularly crucial not to know genotypes when assessing individuals with putative recombination events.

We were uncomfortable, however, with our research requirements depriving anyone of necessary and desired diagnostic information, so we visited with professionals at the University of Zulia in Maracaibo to see whether, if we were to perform the laboratory genotyping, they could arrange for faculty to provide the necessary genetic counseling that is indispensable to accompany genetic information. The "barrios" or neighborhoods where we work are quite poor and have a reputation for violence. To our concern, we felt that those providing this momentous information would give insufficient time and attention, and we were unwilling to release genotype data under these circumstances. We even worked with scientists in Caracas to try to set up the necessary laboratory and counseling structures we felt were imperative to have on site, but it turned out to be impossible.

We also arranged a meeting with family members who had heard that there had been a critical discovery and were disappointed that it was not the cure. We tried to explain what had been found and what it might mean to them. One man pointed to the bridge over the lake near where we were meeting and said succinctly, "If you tell me I am going to have this disease and I do not have someone to talk to about this, I am going to run to the nearest bridge and jump off!" We felt that if we gave diagnostic information and then left for a year, we would be acting like hit-and-run drivers. We are also limited by the options available to people who request genetic information: abortion is illegal in Venezuela, which eliminates the value of prenatal diagnosis for the HD gene since minors are not tested. We finally decided to provide clinical diagnoses, if people ask for them, and genetic counseling based on age of onset of HD in the family, but not to give genotype information.

Presymptomatic testing: preliminary outcomes. Our experience with genetic diagnostic testing for Huntington's disease in our own country suggests that, with very intensive counseling, the few people who have tested positively tend, by and large, to do pretty well. In Canada, one person who came for presymptomatic testing and discovered she was already symptomatic made a suicide attempt, and there has been one hospitalization in the United States that I know of for severe depression following a presymptomatic positive diagnosis. Most people who have tested most likely positive for the presence of the gene have had this news for only a year or two, and we have no idea how this group will respond once they find themselves becoming symptomatic. One woman told me she was often asked if she regretted her decision to be tested. She added, "You know, I don't think so, but I really can't afford to think about that question too long because I'm afraid it's going to take up housekeeping in my mind."

We are unable to tell people when the disease will start; we can just say that they most likely have the gene. In follow-up interviews some time after testing, people who have tested positive were asked if they think they will develop the disease; some reply, "I don't think so, because God will cure me, or science will cure me, or the test was wrong."[9] It is traumatizing to be totally healthy and know with almost 100 percent certainty that Huntington's disease is in your future.

The gain/loss quandary. We know very little about how people decide whether or not to be diagnosed presymptomatically. Two

psychologists, Daniel Kahneman and Amos Tversky, have studied how people assess risks and make decisions based on those
assessments. One of their scenarios involves imagining that you
command six hundred soldiers in battle and you must choose
between two possible routes.[10] If you take the first route, you will
certainly save two hundred soldiers; if you take the second, the
odds are one in three that all will be saved but there is a two-thirds
chance that all six hundred will die. Another scenario is as follows:
You are a commander of six hundred troops. If you take the first
route four hundred of your soldiers will certainly die; if you
choose the second, the odds are one in three that none of your
soldiers will die and two in three that all six hundred will die. Of
course, the two scenarios are exactly the same, but one is phrased
in terms of most certainly saving two hundred soldiers and the
other in terms of losing the lives of four hundred men. Kahneman
and Tversky have found, by and large, most people are not *risk*-
averse but *loss*-averse. In the first battle scenario, which emphasizes saving two hundred men, most people will take the first
route, preferring the sure thing of saving some lives rather than
gambling with all. But when the same choice is rephrased to emphasize that four hundred men will certainly die, more will choose
to gamble for the possibility of saving all six hundred soldiers and
will choose the second route. Faced with a certain gain, people
tend to be conservative and maintain what they have, but faced
with a certain loss, people are more willing to gamble. If you are
given a certain amount of money, people tend to keep what they
are given rather than gamble for the prospect of winning more.
If you must give up money, you would be more inclined to gamble
and even risk losing more for the chance of not losing any. Kahneman and Tversky argue that people dislike losses and will guard
against them. But if the loss is *certain*, then people are willing to
risk an even bigger loss if there is a chance of avoiding any loss
at all.

Kahneman and Tversky's findings emphasize why it is so important to explain genetic information both in terms of gain and
loss. Telling clients they have one chance in four of having an
affected child conveys one psychological message, saying that
they have three chances in four to have a normal baby conveys a
different one—even though the statistics are the same. Clients
need to be given both constructions of the genetic information.

A person taking a genetic test makes a terrific gain-loss calculation. The gain, obviously, is to learn that you do not have the genes for Alzheimer's, cystic fibrosis, Huntington's, or any number of other diseases. The loss is to learn that you do. Is learning the good news worth risking hearing the bad? Many who come for testing already feel they are in a loss position; they consider being at risk just as bad as knowing they will be affected. They assume they cannot do one thing or another because they are at risk, even though they could do what they want. Nothing is really stopping them, but they are paralyzed by their risk situation: because certain things are unknown, everything becomes impossible. I asked a woman why she wanted to be tested. She replied, "If I find out I'm going to have HD, I'll take my son to Hawaii; but if I'm OK, then I'll wait." I said, "If you want to take your son to Hawaii, why are you waiting until you get diagnosed with HD to do it? Because by the time that you're ready to take him to Hawaii, he's going with his girlfriend, not his mother."

Since many people at risk already feel themselves to be in a *loss* situation, they are more willing to take a test which may throw them for a greater loss—learning that they do, in fact, carry the gene. If they see their lives more or less in a *gain* situation—they have chosen their careers, had their children—then they are more conservative about maintaining that gain and not as willing to take the gamble. Some people will take the gamble for the sake of their children, because the only way to clarify their risk is to take the test.

What makes these problems so difficult is the absence of recourse to treatment. If you can do something about the disease, then there will be an incentive to undergo presymptomatic testing and the catastrophic nature of a positive diagnosis will be tempered. If the treatment in only marginal, the choice to be tested will still be difficult. Attitudes toward HIV testing changed when people learned that the drug AZT could retard disease onset in individuals positive for HIV.

The human genome project: Road map to health. The human genome project should eventually point the way to preventions and cures. The project should, within the next few years, establish an "index" map of the human genome, with markers spaced about every 10 million base pairs, placing at least one marker close enough to every gene of interest to locate it. This map should get us in the neighborhood of most disease-causing genes. Then will follow the construction of a "high-resolution" map, one with thousands of markers spaced every million or so base pairs. With this detailed map, it should be possible to localize genes more rapidly and precisely, and finding the genes should lead to their sequencing and characterization. This map will guide the "gene hunters" as they navigate along the genome.

Some people protest that simply knowing the molecular lesion in a gene does not guarantee the development of new treatments for the disease it causes. The mistake in the gene causing sickle cell disease has been known for twenty-five years and there is still no effective therapy or cure. But this discovery may have been ahead of its time; with newer technologies today, the defect may be more amenable to intervention. Of course, it may be true that knowing the cause of a disease at a molecular level may do nothing toward advancing palliation or cure. But it makes intellectual sense to go after the gene, the cause of the disease, as one possible avenue of interdiction. It is not the only route but it is a reasonable one in studying the etiology of a disorder. If you want to stop the damage of the Nile constantly overflowing its banks, you could either build protections along the length of the river shore or go to the source of the Nile and try to control the flow before damage occurs.

There is preliminary evidence that the identification of the abnormal genes causing cystic fibrosis, α_1-antitrypsin deficiency, neurofibromatosis, and other hereditary disorders may lead to promising new avenues of research on therapeutics. Scientists have been able to place a normal human α_1-antitrypsin gene in the epithelial lining of a rat lung, using a virus that causes the common cold, an adenovirus, to transfer in the gene.[11] The rat lung tissue, both in the test tube and in the animal, produced normal human gene product for some time. Two other scientific groups have inserted the "normal" cystic fibrosis gene into lung and pancreatic tissues in culture taken from people with cystic fibrosis, also using a virus to transport in the gene. In tissue culture, the normal gene was able to reverse the effects of the disease and blocked channels functioned normally.[12]

Some investigators have contemplated using an aerosol spray

to deliver the normal gene into diseased lung tissue. Dr. James Wilson of the Univeristy of Michigan, leader of one group that corrected CF in culture, commented to the *New York Times*, "My tendency is to be very conservative . . . at this point, it's impossible for me not to be optimistic about CF."[13]

These are all very preliminary findings and much work is ahead to demonstrate that genetic therapy, therapy based on using the normal gene itself as a treatment, is effective. But they are intriguing leads. And at a minimum, identifying the gene focuses all the energy and resources that had been scattered in analyzing irrelevant regions of the chromosome while the gene search was under way onto the aberrant gene itself.

Ethical, legal, and social issues. There are many social, psychological, ethical, legal, and economic problems awaiting us that I have not even mentioned. Once we have an improved capacity to diagnose disorders presymptomatically, many more individuals and families may face the loss of health and life insurance. They may be exposed to discrimination from employers and stigmatization and ostracism from friends and relatives. Predictive information can be fraught with dangers to individuals and to society. To address these concerns, the National Center for Human Genome Research, the National Institutes for Health, and the Human Genome Program of the Department of Energy have established the Joint Working Group on Ethical, Legal, and Social Issues associated with mapping and sequencing the human genome. It is the mandate of this working group to support research in these critical arenas and develop policy recommendations for the necessary protections that must be put in place as new genetic tests are being developed.

If there are so many personal, social, and economic hazards and successful cures are not assured, some people ask, why proceed with the project? How can we *not* proceed? Many who suffer from hereditary diseases already make huge economic sacrifices, already pay exorbitant psychological and social costs. I could not go to Venezuela and say to those expectant people, "Sorry, we've called off the search for the Huntington's disease gene because having the gene in hand is too dangerous and there is no guarantee of a cure."

I am an optimist. Even though I feel that this hiatus in which we will be able only to predict and not to prevent will be exceedingly difficult—it will stress medical, social, and economic systems that were already under a severe strain before the advent of the human genome project—I believe that the knowledge will be worth the risks. We are learning from our experience with Huntington's disease and other disorders about the power of clairvoyance and the need for caution. We are preparing for the future when tests for breast cancer, colon cancer, heart disease, Alzheimer's disease, manic depression, and schizophrenia might well be available. For a while we may have the worst of all possible worlds—limited or no treatments, high hopes and probably unrealistic expectations, insurance repercussions—everything to challenge our inventiveness and stamina. But these ingredients will be, I hope, catalysts for change. The stakes are high; the payoff is high. I am reminded of a line by the poet Delmore Schwartz: "In dreams begin responsibilities."

RUTH SCHWARTZ COWAN

Genetic Technology and Reproductive Choice: An Ethics for Autonomy

11

If we think of the human genome project as a technological system (an altogether appropriate way of thinking of it), then we can see that, like all technological systems, the project produces something: the product is not widgets or chips or ingots, but information. Actually, two types of information: for heuristic purposes we can call one type "internal" and the other "external." Internal information feeds back into the system itself, helping to improve either equipment or maps and thereby making the whole enterprise both more efficient and more productive. Internal information is important and significant to those who work on the project, but it is not the information most relevant for moral and social decisionmaking, and therefore it is not the information on which this essay will focus.

In its social characteristics, external information differs markedly from internal information. All technological systems exist in a particular social world. The technological system, for example, that produces ingots exists in the social world that we call the "iron and steel industry" and the technological system that produces chips exists in something we could call an "offshore multinational." Most of the human genome project lies within the social world that we call "science," or sometimes "medical science," but—and this is a crucial point—some of the information that it produces, external information, leaves that social world and enters a different one, what we call "medicine," or sometimes "clinical practice." This is the genomic information we have to worry about when we worry about the social and ethical implications of the genome project: not the information that stays within the system but the information that gets out.

The major actors in the medical social world differ from the major actors in the scientific social world. Even the same individual changes social role when crossing the boundary from one world to another. The scientific world is populated by "principal investigators" and "technicians" and "suppliers" and "laboratory directors"; the medical world has "practitioners" and "patients" and "technicians" and "families" and "nurses" and—let us not forget them—"third-party payers." When a person who is a lab director in one social world becomes a patient in the other social world, that person may remain physically intact (let us hope so!) but her or his social role, and social power, changes profoundly. There are at least two things that people in the medical system have more power over than people in the scientific system: (1) individual decisions—moral decisions, relating to life and death—and (2) group decisions—policy decisions, also relating to life and death.

For the foreseeable future, when genomic information enters the medical world it is going to do so through the practice of prenatal diagnosis. Gene therapy is, in many ways, a wonderful goal, but it lies in the future. Prenatal diagnosis is already with us, here and now. Indeed, it has been with us, in most countries of the developed world (and, to a very restricted extent, also in the underdeveloped world) for more than twenty years. Prenatal diagnosis is a technological system that is delivered to patients either through an obstetric practice or through a practice that specializes in medical genetics. Either way it requires that a sample of fetal tissue be removed for analysis; chorionic villus sampling is a way of taking a biopsy of fetal membranous tissue during the ninth or tenth week after conception; amniocentesis is a way of obtaining fetal cells that have been sloughed off into amniotic fluid between the fourteenth and the sixteenth week. Once they are removed fetal cells (and fetal biochemical products, which can also be found in amniotic fluid) can be analyzed: something can be learned both about the number and morphology of the chromosomes; and something can also be learned about the presence or absence of certain genes. This diagnostic process—discovering whether a gene is present or absent—is the process by which knowledge gained from the human genome project will flow into the medical system, as the discovery of the gene for cystic fibrosis has already indicated. We need to be very clear about what therapy is currently available for most diseases or disabilities that can be diagnosed prenatally: *none*. The only recourse for patients whose fetuses are diagnosed as having Down's syndrome, or spina bifida, or Turner's syndrome, or Tay-Sachs disease, or sickle cell anemia, or one of the thalassemias is abortion, a process that can hardly be described as therapeutic. This means that, for the foreseeable future, the ethical and social implications of the human genome project are going to be inextricable from the ethical and social implications of abortion.

Since prenatal diagnosis has been with us for more than two decades we already know a good deal about how it has developed and what its social and ethical implications have been. I propose, therefore, to examine the history of prenatal diagnosis from two perspectives—from the perspective of the history of technology and from the perspective of feminist ethics—by way of exploring some of the social and ethical implications that the genome initiative has created for us in the present and will likely continue to create in the future.

The history of technology is about the same disciplinary age as molecular genetics; the Society for the History of Technology was founded and began publishing a quarterly journal, *Technology and Culture*, in 1958. Although many historians are loathe to speak of "historical laws" (there being far more variables in a historical system than in any other), I believe many would acknowledge that historians of technology have isolated at least three characteristics of technological systems that are generally applicable to most modern examples that have been studied in great detail.

First: *motives matter*. In technology, as opposed to science, the intentions of innovators can be incorporated, literally, into the structure of an artifact—built into the plumbing, as it were. A simple, illustrative example is the Northern State Parkway, a beautifully landscaped, four-lane highway built to accommodate east-west traffic on Long Island in the late 1930s.[1] The highway overpasses, intended as part of the design, are faced with field-stone. They are also very low; buses and trucks cannot pass under them (as errant drivers still periodically discover). The highway was planned and designed by New York's master builder, Robert Moses. In the course of doing the research for his painstaking biography of Moses, Robert Caro discovered that the overpasses were deliberately, consciously built too low for buses.[2] The Northern State Parkway is the feeder road that connects New York City to another of Moses's brilliant creations, Jones Beach, and Moses did not want his spectacular beach overwhelmed by poor people from New York City, people too poor to own their own cars, people who rely on public transport. The subway to Coney Island was good enough for them; Jones Beach was to be reserved for their "betters." Motives matter.

Motives, on the other hand, are *not wholly determinate*; this is a corollary of the first principle. The Long Island Expressway (which runs parallel to the Northern State Parkway) foiled Moses's plans for Jones Beach, albeit twenty years later. People, infinitely ingenious, can find ways to get around an innovator's intentions: turbines can be redesigned to handle gases rather than liquids; internal combustion engines can be modified to burn wood or coal rather than liquid fuel; swords can even, let us hope, be beaten into plowshares. The trouble with all this modification and redesign is that it is expensive: time, trouble, and money have to be expended to subvert or subdue the original intention—and sometimes that time, trouble, and money are not easy to come by. Better, all things being equal, to have designed the thing according to the "right" motives first; in the real world of technological systems, motives may not be wholly deterministic but money sometimes is.

Second, technological systems usually have *embedded objectives which are different from their overt objectives*, overt objectives being the ones that are used to sell technologies to potential consumers. Consider, for a moment, the development of the household refrigerator.[3] Two types of refrigerators were being marketed back in the late 1920s, when mass production of refrigerators began: an electric compression machine (similar to the one most Americans use now) and a gas absorption machine (which is currently available only in Europe). The gas absorption machine was an engineer's delight. It had no moving parts and was therefore both silent and almost maintenance free. The electric compression machine, however, was being manufactured by several of this country's largest companies—General Electric and Westinghouse—whose business, until then, had been largely to supply the equipment needed by electric utility companies. General Electric had considered bringing a gas absorption machine onto the market, but had eventually rejected the idea—and this is the salient point here—because the electric compression machine would be better for the economic interests of the utilities. Thus, when its electric compression refrigerator came on the market, late in the 1920s, the company had an overt objective, to replace the icebox and the icehouse, as well as an embedded objective, to generate income for electric utility companies (and also, since refrigerator motors ran twenty-four hours a day, to even out the "load" for the utility companies). In this particular case the embedded objectives also turned out to have some historical significance; since the companies manufacturing electric compression refrigerators had deeper pockets than the companies manufacturing gas absorption refrigerators, they were better able to struggle through the Depression, keeping the price of their product as low as possible, offering installment purchases, offering large discounts to jobbers and retailers. Eventually the price differential drove the gas absorption manufacturers out of business, or at least out of business in the United States.

Historians of technology have revealed the existence of many kinds of embedded objectives in technological systems: interservice rivalries, job security, egotism, bribery, market dominance, patent control. The wise consumer, consequently, is one—and this is a point to which I will have occasion to return shortly—who makes a point of knowing something about embedded objectives; in order to discover embedded objectives, the wise consumer must make an effort to learn something both about the social world in which the technology was produced and also something about the one through which it is being delivered.

The third characteristic of technological systems is that they all have *unintended consequences* once they have been diffused. People are ingenious; societies are complex; accidents will happen. The men who invented radio had no idea that someone was going to be able to sell air time to advertisers; the people who developed intrauterine devices did not suspect that they were going to cause pelvic inflammatory disease in some users; the engineers who widened highways did not anticipate that more drivers would be inclined to use them once widened; some of the men who worked on the atomic bomb imagined that the device would be used for deterrence, not destruction.

The existence of unintended consequences should not lead us to despair, however, or to a feeling that technology has a momentum which we mere mortals can do nothing to change.[4] First, we can learn to anticipate some of those consequences. Part of the "un" in "unintended" derives from ignorance and we can lessen our ignorance by learning more about the history, sociology, and economics of technology; we can also learn to tease out the embedded motives that were referred to above. Second, we can develop systems that will guard against even those consequences we cannot learn to anticipate, sociotechnical insurance, as it were. Such insurance surely will not be perfect, but just as we routinely sign up for life insurance and health insurance, we'd be foolish, modern life being what it is, not to carry it.

In order to explore the social and ethical implications of the genome project, we will need to read the history of prenatal diagnosis from the perspective of the history of technology: in terms of motives, embedded assumptions, and unanticipated consequences.

The first fetal condition that could be diagnosed prenatally was, ironically enough, sex. In 1949 two Canadian histologists, M. L. Barr and E. G. Bertram, described a particular morphologic feature of the nuclei of cat nerve cells that they could observe only in the cells of female cats.[5] Further investigation by Barr and others revealed that this distinction could be made in other mammals, including humans, and in other tissues, including desquamated cells—cells sloughed off—from mucous membranes. This morphologic feature, which derives from one of the two X chromosomes of the female, is now referred to, eponymously, as the Barr body, or as sex chromatin. When suitably stained it is visible—and this is crucial—during interphase and prophase, two steps in the process of cell division.

In 1955, within a few months of the publication of Barr's paper demonstrating that sex chromatin could be seen in desquamated mucous cells, four separate research groups (in New York, Jerusalem, Copenhagen, and Minneapolis) announced that the presence or absence of sex chromatin in desquamated cells of amniotic fluid could be used to determine the sex of a fetus; the cells did not even have to be cultured; sex determination could be made within a few hours after removing the amniotic fluid (by the withdrawal technique which was then, and still is, called amniocentesis).

In Copenhagen (and probably in other places as well) this piece of genomic information was immediately put to use within the context of a medical genetics practice.[6] Pregnant women whose presenting "symptom" was a family history of hemophilia were offered the opportunity to discover the sex of the fetus they were carrying and also the opportunity, if the "diagnosis" turned out "male," to abort their pregnancies (hemophilia is one of the sex-linked genetic diseases; the recessive gene is carried on the X chromosome; females can be carriers, but are rarely afflicted). Within a few years this very limited use of amniocentesis—in the diagnosis of sex-linked diseases for which there was a family history—had become routine practice in medical genetics units (which, in those days, were located only in major research hospitals) in Canada, the United States, Britain, and Western Europe—despite the fact that in several countries so-called therapeutic or eugenic abortions were still not easy to obtain.

Within the next decade, because of three different developments, amniocentesis became considerably more widespread. By 1966 media had been developed that allowed fetal cells from amniotic fluid to be successfully cultured, making karyotyping (examination of the number and morphology of the chromosomes) possible. In 1967 the first karyotypical diagnoses of fetal chromosomal disorders were reported, and in 1968 *Lancet* carried the first report of an abortion performed to prevent the birth of a fetus who had been previously diagnosed as suffering from Down's syndrome.[7] At this juncture the potential for widespread use of amniocentesis became considerably greater because the most common presenting symptom for chromosomal disorders is "advanced maternal age" and older pregnant women far outnumber pregnant women with a family history of sex-linked disease.

In 1972, in a separate development, D. J. H. Brock and his colleagues in Edinburgh discovered that several neural tube defects (such as spina bifida) could be diagnosed in a fetus by the presence of elevated levels of alphafetoprotein in the amniotic fluid. Once again the potential for widespread use of amniocentesis increased, since the incidence of these defects is fairly high in certain geographic areas (7.9 per 1,000 live births in Northern Ireland, for example).[8]

Finally, in November of 1975, at a meeting of the American Academy of Pediatrics, the U.S. government announced that the results of an NIH collaborative study indicated that amniocentesis for prenatal diagnosis was not only reasonably safe for mother and fetus but also very accurate.[9] Similarly favorable results were announced by Canadian and British researchers a few months later.

At this juncture amniocentesis began to move out of the domain of research institutions and into ordinary clinical practice. By the mid-1970s many thousands of women had had amniocentesis, and thousands more were having their amniotic fluid removed and analyzed every year. Laboratories began to develop routines for systematizing the analyses of fetal material and they also began developing routine documents for reporting the results of those analyses to physicians; the standard report form for karyotyping required that the technicians report the sex of the fetus, because all the chromosomes—autosomes and sex chromosomes—had to be described.

Clinical practice with regard to that information seems to have varied; some physicians automatically told parents the sex of the fetus; some asked whether the parents wanted to know and then provided the answer if it was requested; some—regarding the information as "not medically relevant"—withheld it.[10] Such a large number of women were having amniocentesis performed and, in the process, discovering the sex of their fetuses that the word eventually leaked out: a woman could discover the sex of her fetus in time to have a legal abortion if she was not happy with the sex of the fetus she was carrying. With the possible exception of a group of physicians in India, we have no record that any physician actually advertised amniocentesis as a way to preselect the sex of a child, but some physicians soon began to report that a few patients were requesting amniocentesis *solely* for the purpose of discovering the sex of their fetuses, most frequently women who had already had a large number of children of one sex.

Here was an outcome that no one had expected. In 1960, when

Fritz Fuchs and his colleagues had begun offering amniocentesis and therapeutic abortion to women with a family history of hemophilia, abortion was illegal in most Western countries and the women's movement was moribund. By 1975 the women's movement had come to life, nontherapeutic abortion was legal almost everywhere, and thousands of pregnant women had had hollow needles inserted in their abdomens. Apparently no one in the medical community had anticipated that patients would be interested in or able to use amniocentesis for nonmedical conditions, and no one had anticipated that women would be able, relatively easily, to abort pregnancies that were unwanted because the sex of the fetus was known. But among many things that the women's movement had insisted upon, and that had been given the force of law in the United States in the decision in Roe v. Wade, in 1973, was a woman's right to have an abortion performed by a physician during the first two trimesters of her pregnancy without any questions being asked about her motives: abortion, as it were, on demand.

Many physicians felt they were caught between a rock and a hard place. Some believed that abortion should be legal but should not be used in order to ensure that a newborn child is of the preferred sex. Others believed that scarce medical resources—the laboratory work involved in amniocentesis was (and is) both time-consuming and expensive, and in the 1970s there wasn't enough laboratory capacity to meet the demand—should not be expended on conditions that were not medically "relevant."[11] At the same time these same physicians realized that if they failed to provide information their patients wanted they were compromising the autonomy of their patients; yet others realized that they were also compromising the autonomy of their patients by refusing to provide abortion services for women whose motives they deemed suspect.

Two principles of medical ethics were clearly in conflict; the injunction to "do no harm" restrains physicians from performing surgical procedures for which there is no medical justification; the injunction to "respect patient autonomy" encourages them to provide all relevant information and to adhere to patient requests about treatment. In an effort to resolve the dilemma thus created, some genetics counselors tried to convince parents not to abort for sex preference—in direct contradiction to their professional ethic of nondirective counseling. Some laboratory directors instructed genetics counselors to tell parents who wanted information about sex that, at that very moment, their labs were too busy to take on the case. Some obstetricians would refer patients to medical geneticists for amniocentesis but would not then perform the requested abortion if "wrong sex" were the only reason the patient proffered. Others would refer a patient to another physician willing to do the abortion; still others (and the rumor mill in various cities soon revealed who they were) were willing to provide both the information and the surgery.

In an effort to resolve this ambiguous situation a consensus conference was held at the Hastings Center in 1979. The position taken, at that time, by the physicians and bioethicists who were present, was that, to paraphrase a very contorted paragraph: We think this is a terrible thing and parents shouldn't do it and we respect physicians who won't do it either; nonetheless we don't think any law should be passed preventing it.[12] Within a few months, however, one of the individuals signing that report, John C. Fletcher, a bioethicist who was then on the staff of NIH, published a reevaluation of his position, arguing, on the grounds of consistency, that physicians cannot refuse an abortion in one situation (when prenatal diagnosis for sex has been requested) while still being willing to provide it, no questions asked, in all others:

> . . . the woman's right to decide is the overriding consideration in the abortion issue. The rationale for the legal rule omitting a test of reasons is that a woman has the right to control her reproduction and the risks involved in a pregnancy. To employ public or medical tests of reasons provides opportunities to obstruct and defeat society's obligation to grant women the freedom to determine their own reproductive futures. To prevent obstruction of self-determination, it is better to have no public tests of reasons.[13]

Reflecting the difficulty of the issue and the extent to which deeply held ethical principles conflicted, Fletcher changed his mind again, four years later. By that time it had become clear that in China and India some physicians had offered patients prenatal diagnosis in order to facilitate the abortion of female fetuses. Reasoning that a substantive ethical principal (in this case nonmalfeasance, to do no harm) was more important than a procedural ethical principal (consistency) and calculating that the harm done to society from sex selection in general and sex selection of males in particular would outweigh the benefit derived by individuals from raising a child of a preferred sex, he concluded that physicians ought not to do it, and that the ought is strong enough that laws need to be passed to insure it; the duty of avoid harm, he reasoned, in this case outweighed the duty to respect individual autonomy.[14]

Clinical practitioners still have not reached consensus on the issue of abortion for sex selection, however. Fletcher and his colleague, Dorothy Wertz, recently undertook a multinational survey of the attitudes of genetics counselors.[15] Thirty-four percent of those who were surveyed asserted that they would provide prenatal diagnosis for patients who wished only to know the sex of their fetuses and 29 percent said that although they wouldn't do it themselves, they would refer the patient to someone who would. Most of these physicians had decided to act in this fashion on the grounds that patient autonomy ought to be respected; that a woman had a right to control her reproduction without interference from another party. Presumably this also means that 27 percent of physicians either would refuse the service if it was requested or would not inform their patients that it was available; these physicians were acting, essentially, on the grounds that the injunction to do no harm should take precedence over the injunction to respect patient autonomy.

Feminist ethics might be able to provide a path between the horns of this dilemma. Feminist ethics is a discipline much younger and less well developed than either molecular genetics or the history of technology. Some would date it from the rebirth of the women's movement in the early 1970s, others from the publication, in 1982, of Carol Gilligan's book, *In a Different Voice*. Either way, the field can be split into two slightly different branches which correspond, roughly, to two different branches—normative and nonnormative—of ethics itself. Normative ethicists are concerned with deducing ethical principles; a feminist normative ethicist might assert that feminist ethicists ought to deduce ethical principles that are unlike those that have previously dominated in patriarchal cultures. Nonnormative ethicists want to investigate what people actually do and how they explain what they do in specific moral situations. A feminist nonnormative ethicist might assert that some of women's life experiences are different from men's, particularly those that have to do with pregnancy. As a result women face moral dilemmas that are different from the ones men face and they have developed moral principles—still largely unarticulated—that are different from men's: a "different moral voice," a voice that ought to be heard, not only because it comes from half the population, but also because it may help us to deal with some of the moral dilemmas about life and death which we currently face and with which our older ethical systems have difficulty dealing. Nonnormative feminist ethics is the variety which seems to me most likely to shed helpful light on the dilemma that prenatal diagnosis for sex selection has created both for physicians and for patients.

Gilligan gives us a paradigm example of that "other" moral voice in the following example.[16] Two children have been asked how they would resolve "the Heinz dilemma," a standard case history used in investigations of nonnormative ethics: should a man steal a drug he cannot afford in order to save the life of his wife? One child answers in the affirmative, "constructing the dilemma," Gilligan tells us, "as a conflict between the values of property and life."

> For one thing a human life is worth more than money, and if the druggist only makes $1000, he is still going to live, but if Heinz doesn't steal the drug, his wife is going to die. (*Why is life worth more than money?*) Because the druggist can get a thousand dollars later from rich people with cancer, but Heinz can't get his wife again. (*Why not?*) Because people are all different and so you couldn't get Heinz's wife again.

This child goes on to say that if Heinz were caught the judge would probably release him, since the judge would see that stealing was, in this case, the right thing to do; furthermore, "the laws have mistakes and you can't go writing up a law for everything you can imagine."[17]

The other child's answer is in the negative; Heinz should not steal the drug.

> Well, I don't think so. I think there might be other ways besides stealing it, like if he could borrow the money or make a loan or

something, but he really shouldn't steal the drug—but his wife shouldn't die either.

Gilligan continues: asked why he should not steal the drug, this child "considers neither property nor law but rather the effect that theft could have on the relationship between Heinz and his wife."

> If he stole the drug, he might save his wife then, but if he did he might have to go to jail and then his wife might get sicker again, and he couldn't get more of the drug, and it might not be good. So, they should really just talk it out and find some other way to make the money.

This is the paradigm case. Gilligan asserts that this child, who happens to be a girl, analyzes moral problems in a fashion that is more common among females than among males, although it is not uniquely female by any means. This child, "seeing in the dilemma not a math problem with humans but a narrative of relationships that extends over time, envisions the wife's continuing need for her husband and the husband's continuing concern for his wife and seeks to respond to the druggist's need in a way that would sustain rather than sever connection."[18]

Much of Gilligan's book is concerned with investigations not of children who are presented with the Heinz dilemma but of women who have made a decision to abort a pregnancy; Gilligan noticed that they, like the second child quoted above, used the language of relationships, the need to sustain relationships, the need to provide good relationships, the need to nurture and to nurture well, in expressing their reasons for deciding to abort. Rayna Rapp, an anthropologist, and Barbara Katz Rothman, a sociologist, heard the same nurturing voice in their parallel investigations of couples who had made the painful decision to abort what had previously been a wanted pregnancy because, in the second trimester, information from amniocentesis had indicated that the fetus was, in some fashion, genetically afflicted. Both the women and men who were interviewed about their abortion decisions thought of themselves as ending the unwonted pain that an afflicted child would necessarily suffer; they worried that if they had an afflicted child they would not be able properly to nurture it, or that they would not be able to properly nurture their other children, or that with an afflicted child in the family, a child who would never become independent, they would not properly be able to nurture themselves.[19]

Is there an ethical principle that could be derived from these discourses about abortion? Several scholars, most particularly Rosalind Petchesky and Barbara Katz Rothman, have tried to abstract an ethical principle from the various things that people—both women and men—say about why, in their view, a particular abortion (not abortion in general, but this abortion in particular) was morally justifiable. Their conclusions can be briefly expressed in this formulation: *nurturance matters.*[20]

An embryo cannot become an infant unless it is nurtured; an infant cannot become an adult unless it is nurtured and—at the other end of the developmental spectrum—adults who are ill or disabled cannot continue to live unless they too are nurtured. Nurturance is a continuous, day-to-day, mundane process: feeding, sheltering, protecting, assisting. Its goal is, in the case of embryos, to create an individual who can have a relationship with other individuals, in the case of adults, to maintain and sustain the life of an individual who has relationships. Indeed no other human relationships are possible unless nurturance occurs, and thus no moral decisions can or ought to be made unless decisions relating to nurturance are made first. Hence, nurturance not only matters, it matters first.

If this principle were to be taken seriously it would follow that when individuals cannot, for whatever reasons, make decisions for themselves, the person or persons who have the right to make the decisions are those who are nurturing the individual in question. Whether or not we agree that a fetus is an individual we can still agree that it is not capable of making decisions about itself. This means that decisions about an embryo or a fetus which is *in utero* ought to be made by the person in whose uterus it is developing; this person may or may not be its biological mother or its intended social mother, but certainly won't be its father or a doctor or the governor of the state in which it happens to be located.[21]

The principle that *nurturance matters* also suggests the two bases on which nurturers should be making their decisions about what is morally correct behavior in any given instance. The goal of nurturing embryos, fetuses, and infants is to create independent, autonomous individuals, individuals who will no longer need nurturance. First, nurturers must weigh, in making their decisions, the chances that the embryos, fetuses, and infants will ever reach that goal: is this individual ever going to be able to function independently or make autonomous decisions? Second, since the resources with which people nurture are, unfortunately, not infinite (parents have just so much energy, and just so much money; societies provide just certain social services and not others), nurturers must also weigh the resources they will have to expend in nurturing one individual against the competing needs of others, including themselves, who are also dependent: If I have a child with Down's syndrome will I be able to care for my other children? Is there an institution that will be able to care decently for my spina bifida child if I die? If I raise a child with cystic fibrosis will I be able reach my own goals for myself?

Some will say that such considerations are essentially selfish and that the goal of an ethical system should be to encourage altruism. Feminist ethicists have replied to this objection with an observation that derives from their experiences as housewives, parents, nurses, and social workers: one cannot care properly for others unless one has taken proper care of oneself. Nurturance matters and self-nurturance is not the same thing as selfishness: selfishness is to "be for oneself and for nothing else"; self-nurturance is to "be for oneself so as better to be for others." Indeed those who call the decision to abort when the mother is pregnant with an afflicted fetus "selfish" are responding in the context of an ethical system in which both nurturers and nurturance have never been accorded primacy and in which the sacrifice of individual to group goals has always been favored—precisely the ethical system which feminist ethicists are hoping to supersede.

An abortion policy constructed in accordance with the principle that *nurturance matters* is clearly one in which the decision to abort should rest entirely in the hands of the woman who is pregnant. Physicians and others who will have to provide abortion services would, under such a policy, be morally obligated to abide by the patient's decision even if it has been made by a calculus, even a eugenic calculus, with which the provider did not not agree—unless the provider was willing and able to take over the continuing nurturance of the fetus.

Thus, a feminist ethic based on the principle of nurturance sanctions the behavior of those physicians who provide information about the sex of fetuses to their patients as well as those who provide abortion services to patients who request it for that reason—no matter how personally repugnant to the provider the decision of the woman may be. Their behavior would be consistent with Voltaire's oft-quoted remark, "I don't agree with a word you are saying, but I will defend, to the death if need be, your right to say it," as well as with the behavior of those who voted for universal suffrage despite their fear that "the uneducated and unwashed may lead us straight to despotism." Nurturance is, in short, a liberal ethical principle as well as a feminist one.

Many people worry—and in the waffling of the medical community about prenatal diagnosis for sex, we can see precisely the reasons why they worry—that with increasing genomic knowledge the number of abortions for "nonmedically relevant" reasons will increase and that people will use abortion to avoid having babies with diseases that may be only marginally disabling or IQs that may not get them into Harvard. Others worry, contrarily, that with increasing genomic knowledge there will be new and increasingly persuasive reasons to infringe individual rights. As an example they point to the proposal by the right-to-life movement in the United States that abortions be prohibited for sex selection (legislation to that effect has been drafted in more than a dozen states)—in a straightforward effort to draw feminist support for the right-to-life cause. Were genes for "intelligence" or "blue eyes" or "cowardice" to be identified tomorrow, these people fear that anti-eugenicists would line up with the right-to-lifers and the feminists—with the net result that abortion rights would be hopelessly compromised.

Which outcome is more to be feared? A future in which parents can freely choose the characteristics of fetuses that will be brought to term, or a future in which there are grounds on which a woman could be refused both information about her fetus and an abortion if she desired one?

We can begin answering those questions by exploring some of the overt motives and the hidden assumptions that determined the history of prenatal diagnosis. Some of the men who were involved in the discovery that sex could be diagnosed in fetuses are now unhappy with the uses to which their discovery has been put. Jerome Lejeune, the man who discovered the connection between Down's syndrome and trisomy 21, is distressed that his discovery has led to the termination of so many pregnancies; Ian MacDonald, who first developed obstetric ultrasound, was opposed to the use of his instrument to guide amniocentesis needles.[22] How might any of those men have prevented their discoveries from being used as means to ends of which they did not approve?

To start with they could have tried to keep their discoveries secret. LeJeune, for example, could have refrained from publishing his results until he (or others) had found a way to remove the disruptive third chromosome; he could have tried to keep this piece of genomic information internal to the scientific system, instead of letting it emerge into the medical world. Unfortunately, what we know about the scientific system suggests that such an effort would have run counter to the ethical and social norms of LeJeune's profession. In any event it would likely also have been unsuccessful, since the conditions under which modern science is done make simultaneous discovery very likely; in fact, trisomy 21 was recognized as the cause of Down's syndrome by British scientists independently of Lejeune and about the same time.[23] LeJeune's express motive was to find a cure for Down's syndrome, but his motive was—in this case—not determinate, because the embedded assumption of the scientific system was that results, even partial results, are better published than kept secret.

If the scientific results could not have been suppressed then possibly the reporting of them to patients could have been better controlled. Surely, clerical systems could easily have been developed—at the beginning—to insure that information about the sex of a fetus would be withheld, except at a physician's discretion. Patients, the sociologists tell us, tend to be passive in the medical system, and female patients more passive than male; they accept what information they are offered, but rarely demand to know more. In the early 1960s, several clerical systems already existed that prevented patients from acquiring specific information about their own condition: medical laboratories in the United States would not, for example, report the results of a pregnancy test directly to patients; similarly, patients could not find out (unless a physician told them) the results of tests done on their own blood and the radiological system prevented patients from looking at or handling their own X rays. Methods could have been found for closing the barn door before the horse had gotten out or, to put the matter another way, for keeping the information about the sex of fetuses privileged among physicians so that they could have divulged it only to patients for whom it was medically relevant. Such methods would not have been perfect—sociological methods rarely are—but far fewer people would have known that it was possible to diagnose the sex of a fetus during a pregnancy; the pressure on physicians to divulge that information would have been less intense—and therefore, given the power that physicians have in the medical system, possibly more controllable.

Had it been chosen, however, this latter option would have reinforced the paternalistic character of medical practice. Patients would be refused access to information about their own condition—and decisions that affect patient lives would have been made by physicians, essentially behind closed doors. The vast majority of patients in a medical genetics practice are not mentally incompetent; withholding information from them is a violation of their autonomy—and, as a result of a growing body of common law, of their rights—and is tantamount to treating them as if they were children. Patients' rights groups and women's health groups have been fighting such practices in the medical community— with some success—for several decades. Thus if the effort to suppress scientific information would have violated an established norm, a well-embedded assumption, in the scientific world, the effort to suppress medical information would have violated a developing norm, a just recently embedded assumption, in the medical world. From the perspective of both patients' rights and women's rights, it would have been akin to cutting off a nose in order to spite a face.

The history of prenatal diagnosis thus seems to suggest that in order to prevent a future in which parents will be able to choose the characteristics of fetuses that will be brought to term we will have to alter the norms of the scientific profession, *and* return medical practice to paternalistic modes of operation, *and* restrict women's rights to request and obtain abortions. As long as the ethics of scientific professionalism remain unchanged, scientists will not only publish their findings but will try to disseminate them broadly. Under these circumstances genomic information is bound to make its way, rapidly, into the medical system—just as information about sex chromatin did. Similarly, as long as the medical world continues to move away from paternalism and toward respect for patient autonomy, genetic information about fetuses cannot be treated as privileged to physicians alone. If patients have access to the information, some will undoubtedly request abortions for reasons with which their physicians will not agree—but as long as abortion on request remains legal they will be able to obtain abortions somewhere. The only way to prohibit such abortions would be to attach conditions to legality—an abortion is legal only if the pregnancy results from rape, for example—and these conditions would thus compromise reproductive rights. The only way, in short, to prevent a future in which mothers will be able to choose the characteristics of the fetuses they will bring to term will be to violate the norms of the scientific community, return the medical community to paternalism, and restrict women's access to abortion.

Is fear of such a future worth such a cost? The history of prenatal diagnosis certainly suggests that it isn't. And the investigations of feminist ethicists also suggest that we have, in fact, little to fear about this scenario as long as individual women are left in control of their own reproduction. For when left free to decide, most women decide to abort for reasons that have to do with their sense of good nurturance: for example, when they feel either that this is not a time when they can nurture a child properly or this is not the fetus that will grow into a child whom they can nurture properly. Why fear a future in which ever more children will be ever more wanted by their mothers?

What we have to fear, I believe, is government interference with any part of the process of prenatal diagnosis—from controls on scientific research, to controls on access to information, to controls on abortion. There are many who would assuage their fears about the future of the genome project by having governments intefere with it. Some people would, for example, like to withdraw public funding; others would like to prevent physicians from disclosing information to patients or prevent physicians from performing abortions for reasons that have to do either with the sex or the medical condition of the fetus. But those who would argue in this view ought carefully to consider the consequences. If governments prevent physicians from disclosing information to their patients, then the rights of patients will be diminished not just by medical but also by governmental authority. And, finally, if governments reobtain the right to prevent abortions they also reobtain the right to interfere in reproductive decisions. If nothing else, the history of the twentieth century ought to have taught us that individuals can sometimes behave badly, but they can never behave as badly, or as destructively, as governments can.

HENRY T. GREELY

Health Insurance, Employment Discrimination, and the Genetics Revolution

12

The ongoing revolution in genetics is eroding our ignorance about the genetic roots of specific diseases and about individuals' susceptibility to future illnesses. But ignorance has its advantages. As this revolution accelerates, the ignorance necessary for a broad-based health insurance market will begin to disappear. As many commentators have pointed out,[1] people at low risk for genetic diseases may pay less for their health insurance; people at high risk may pay more or, in many cases, find insurance completely unavailable. And, in the United States, where employers usually bear the financial burden of their employees' medical costs, those at high risk may also face discrimination in employment. The result has been an increasing call for limiting discrimination based on genotype. Some have argued that the genetics revolution is a good reason to move to national health insurance; a few have urged that these problems should make us reconsider our support for research in human genetics.

The discussion so far has identified important issues, but it has not put the nature of the problem or the merits of the proposed solutions in the full context of the complex American health care financing system. The root problem is not the genetic revolution but the increase in our ability to predict individual health, to which genetic research is only one contributor. The major consequence may not be discrimination but the further erosion of employer-provided health coverage. And limitations exist to most methods proposed for solving those problems, while other alternatives remain to be explored.[2] To flesh out these ideas, let me first describe the probable effects of the genetics revolution on the patchwork American health care financing system. I will then put those effects into context by describing some of the limitations of genetics for prediction and some of the strengths of other methods of predicting individual health. Finally, I will survey the solutions proposed for this problem and suggest some alternatives.

The United States has a patchwork system for financing health care. The large majority of our population of 250 million people is covered through a number of very different mechanisms, and about 34 million Americans are not covered at all. The largest group, about 150 million, have private group insurance, usually as an employee or as an employee's spouse or dependent. Another 10 to 15 million people, many of them self-employed, rely on insurance policies they have purchased themselves. About 33 million, mainly elderly, people have health coverage through Medicare and about 23 million of the poor are covered by Medicaid.[3]

In spite of its small share of the total population, individually purchased health insurance is what most people think of as health insurance.[4] Improvements in predicting individual health have enormous effects on this kind of insurance because it is based on medical underwriting. In medical underwriting, insurers take the applicants' health status into consideration in reviewing their applications. Although it is now uncommon for applicants to be required to visit an insurance company doctor, insurers usually require applicants to answer questions about their health or to provide statements from their own physicians. In addition, the companies share information on applicants through a computerized data base.

Insurers use this medical information in several ways. About 30 percent of all applications for individual health insurance are denied; small percentages are granted with higher premiums or with specific medical conditions excluded. If the insurer issues a policy but later discovers that the applicant gave false answers, it can try to rescind the policy. Finally, many individual policies exclude all of an applicant's "preexisting" conditions for some period of time, whether or not they were listed in the application.

The consequences of the genetics revolution for individual health insurance are straightforward: people who are known to be at higher risk for genetic illnesses will be denied insurance or sold insurance that excludes the conditions most important to them. This problem is neither new nor unique to genetic diseases. Millions of Americans have long been unable to buy useful individual health insurance because of a medical history of, for example, diabetes, uncontrolled high blood pressure, cancer, serious obesity, HIV infection, or any of a large number of other expensive ailments. The genetics revolution will not change the problem, but it will expand the number of people affected.

It is tempting to blame this state of affairs on insurance company greed, but health insurers are reacting to a real problem known as adverse selection.[5] Adverse selection can be explained simply: all other things being equal, people who know they are at high risk will be more likely to seek insurance. If screening for a particular genetic disease were widespread, an insurer that covered the disease and did not exclude people who knew they were at high risk would end up paying a disproportionately high number of claims for the condition.

Where applicants know their future risks and insurers do not, adverse selection can lead to the breakdown of the entire insurance market. Only customers who expect their future health costs to be higher than the premium will buy insurance, thereby forcing the insurers to raise the premium. At the new, higher premium, customers whose expected costs were higher than the first premium but lower than the new premium would stop buying insurance. Once again, the insurers would have to raise their premiums, starting the process again. The cycle will continue until the only insurance available would be priced for the highest risk group. Medical underwriting, rescission for misrepresentations, and exclusion of preexisting conditions—all of which will exclude people at known genetic risk—are ways companies selling individual insurance combat this real threat.

The only good news is that individual insurance is the primary source of health coverage for only about 5 percent of the population. Employment is the major source of coverage. Increased predictability will undercut this source of coverage as well, but in much more complicated ways. The background about employment-related health coverage that follows may seem very distant from genetics, but it is crucial to understanding these problems.

The United States is the only developed country that relies on the voluntary choices of employers to provide health coverage for most of its population. The source of this reliance lies both in the historically smaller role of government in American life and in a set of fortuitous, fifty-year-old legal decisions. During World War II, the War Labor Board ruled that medical benefits were not wages for purposes of binding wage controls. At the same time, the federal government, confronted for the first time with applying the vastly expanded income tax to ordinary workers, decided that health benefits were not taxable "income." Finally, the labor laws were interpreted to mean that employers had to bargain with their unions about health benefits.[6]

These decisions built the foundation for broad employment-related health coverage; the advantages of group insurance and huge tax benefits of employment-related coverage have expanded it. An insurer covering a preexisting group does not have to worry about adverse selection and has lower marketing and administrative costs. As a result, group coverage is both broader than individual coverage, in that it is offered without medical underwriting and excluded conditions, and cheaper. The tax benefits arise because an employer's payments for employee health costs are not income for purposes of the federal and state income and payroll taxes. In 1990, if an employer spent $1,000 to increase the salary of an unmarried California employee earning $30,000 per year, the employee took home an additional $511.[7] On the other hand, the employer could have paid $1,000 per year toward health insurance for her and her family. In that event, the employee would receive health coverage worth nearly twice as much as her posttax gains from an equivalent raise.

The roots of employment-related coverage are important in understanding the reaction of the system to improved prediction, but how employers pay for the coverage they buy is even more important. Three different methods have been used: community-rating, experience-rating, and self-insurance.[8]

In community-rating, the insurer charges each employer an amount per employee that corresponds to the insurer's average costs in that region. The insurers use the premiums to pay the employees' claims. The insurers bear the risk that the employees' claims will be higher than average, but gain the benefits if claims are lower than average.

Experience-rating insurers, on the other hand, offer different premiums to different employers. Rates are based on each employer's claims experience of the previous year or on a rolling average of the claims of past years. The more precisely experience-rating adjusts an employer's premiums for its employees' claims, the less risk the insurer takes.

As employers and insurers learn more about the costs of providing health benefits to particular groups of employees, the competitive nature of health insurance inevitably leads to experience-rating. Companies with healthier-than-average employees are offered lower premiums by insurers using experience-rating than by insurers using community-rating, who charge everyone the same. The community-rating insurers are therefore left with the more expensive groups, forcing them to raise their rates (because their "community" of insured people in the region had become more expensive). Just as individual insurance can be trapped in the adverse-selection cycle, community-rating firms under competition from experience-rating insurers must offer higher premiums, to a smaller group of companies, each year. Left to itself, this competitive spiral leads to experience-rating for all employment groups.

Self-insurance is an alternative to either community-rating or experience-rating. In self-insured plans, employers (and occasionally unions) agree to pay for the specified health benefits directly. In essence, the employer's health insurance "premium" will be exactly the same during each period as the covered costs of its employees and their dependents, plus the employer's administrative costs. In such a plan, the employer takes the risk the insurer would otherwise take: that the covered employees will have unusually high medical costs.

Self-insurance received an enormous, if unplanned, boost from the Employee Retirement Income Security Act of 1974 (ERISA). ERISA is primarily about pensions, but employer plans for both pensions and other employee benefits, including health coverage, are exempted from most state laws. The interaction between these preemption provisions of ERISA and state insurance laws are complex, but their result is simple: an employer purchasing health insurance is subject to state insurance laws, an employer that provides self-insured health benefits is not.[9]

In the 1980s, experience-rating met ERISA. As experience-rating became more thorough, employers began to bear all the risk of their employees' ill health. At that point, shifting from experience-rated insurance to self-insurance was a minor administrative matter. The employer had borne the risk before; now the employer continued to bear that risk but received the benefits of self-insurance: freedom from the ubiquitous state taxes on insurance premiums, exemption from state laws requiring insurers to provide certain mandated benefits, and the control of and interest on any unspent "premiums." There are no comprehensive statistics about employment-related health plans, but the existing survey evidence shows self-insurance is now the leading method of providing employment-related health coverage. The best recent survey estimated that about half the people covered by employment-related health insurance were in wholly or partly self-insured plans. Another survey found that 63 percent of the companies it examined used some kind of self-insurance, 23 percent bought experience-rated insurance, and only 14 percent had community-rated insurance.[10]

Thus, the economics of a competitive group health insurance market and the unintended consequences of the preemption provisions of a statute concerned mainly with pensions combined to push employers to self-insure. As a result, every covered dollar of medical care employees receive means one dollar less profit for the employer.[11] When employers can select their own workers and design their own health benefit plans, their incentive to save money on health costs makes added information about the expected future health of current or potential employees dangerous. By offering such information, the revolution in genetics could undermine the entire American health care financing system.

A specific example might help. Let's create a company called Gene Sequencers, Inc., or GSI. Its owners, intending to be socially responsible as well as successful in the labor market, provide comprehensive health benefits to their employees and their families. GSI spends the national average of about $3,000 per year on health benefits for each employee. To save money, it does so through its own self-insured plan, administered for it by a health insurance company.

Now assume GSI learns that one of its middle-aged employees carries the gene for Huntington's disease. Sometime within the next few years, he will begin to show the symptoms of this terrible malady and his health care costs will skyrocket. The exact amount of those costs is, of course, unknown, but GSI's best estimate is that they will be about $10,000 per year higher than the average employee's. The company had been paying him and other people doing his job about $30,000 per year, which it felt was their fair value in the labor market. The added $10,000 per year means the employee will cost GSI far more than it thinks his labor is worth. What can GSI do?

It could do nothing and bear these costs. Or it might try four actions to avoid them: (1) cut the worker's salary by $10,000 but maintain his health benefits, (2) fire him, (3) change its benefits package to exclude his disease, or (4) drop all health coverage. GSI wants to absorb those costs, but it discovers that its major competitors are either not covering genetic diseases or are not offering health benefits at all. Because it is still providing broad-based health coverage, it fears that it will attract employees at high risk for genetic disease and that its health care costs will eventually far exceed the national average. Ultimately, GSI decides it must choose between employee health benefits and its own survival.

By providing health benefits on a self-insured basis, GSI has essentially become an insurer offering individually underwritten health insurance. Instead of choosing whom to insure by passing on applications for health insurance, it chooses whom it will insure through its hiring and firing decisions. Like the individually underwritten insurers discussed earlier, it faces adverse selection. Each of its four alternatives would avoid the adverse selection problem by, one way or the other, making this employee bear his own predicted future medical costs.

How realistic are GSI's alternatives? As discussed in more detail below, the first two, paying lower salaries to those at high risk for illness or firing them, are probably, but not certainly, illegal, although hard to prove. The third is probably legal for employers who self-insure. The last alternative is clearly legal.[12]

Are employers actually using these strategies? Hiring and firing discrimination has been reported and occasionally litigated—for genetic risks, for AIDS, for multiple sclerosis, and for health risks such as smoking. There have also been scattered cases of employers or unions changing their health plans to limit or eliminate their coverage of AIDS. It seems that most employers are still covering high-cost employees, but we are only beginning to understand genetic diseases and to predict their onset. As long as employers continue to have the incentive to avoid health costs and the ability to act on that incentive, the threat is real.

Genetic information can improve predictions about an individual's future medical costs in three ways. Some diseases, like Huntington's disease, are entirely and definitively determined by genetics. A person's chances of contracting many other diseases appear to be affected by his or her genes.[13] And finally, in some diseases, an individual's prognosis, and thus expected medical expenses, may be predicted by the genotype of the affected cells. As we learn more about the human genome, screening will increasingly be able to distinguish among people at low, medium, and high risk for incurring great medical costs in the future. If employers or insurers can obtain such information and use it, discrimination and the decline of health coverage are likely. But this dire prospect must be viewed in its proper context. Two facts are crucial: genetic knowledge probably will not often lead to very powerful predictions, at least in the near future, whereas other kinds of research currently *are* leading to immediate and powerful predictions.

Genetic screening offers dramatic information about some conditions, but, fortunately, terribly debilitating genetic diseases are not very common. Some more common diseases, like colon cancer, breast cancer, heart disease, and diabetes, appear to have genetic links, but those links do not always lead to strong predictions. Genetic analysis can tell whether someone is at a slightly higher or slightly lower risk of colon cancer: it cannot say whether the person will get the disease, at what age it will strike, how quickly it will be diagnosed, and whether treatment will be successful. Indeed, even when genetic analysis can predict a physiological problem perfectly, medical costs may depend totally on environmental factors. Phenylketonuria, for example, is a genetic disease that causes severe brain damage, but the damage can be prevented if those affected modify their diets to avoid a chemical called phenylalanine.

While advances in genetics have been grabbing the headlines, research in other areas has been quietly improving the predictability of individual health in several different, and potentially more important, ways. Some of that research has been classic biomedical research into infection. AIDS is probably the most important example.[14] Infection with HIV causes AIDS, but from infection to a diagnosis of AIDS takes, on average, eleven years. During those years, evidence of infection can be used very accurately to predict that the person will incur the high medical costs of AIDS—and so can be used to discriminate against an HIV-infected person in apparent excellent health. HIV infection is just one of many nongenetic markers of future health costs; some other examples are the presence of some forms of human papillomavirus, the Lyme disease spirochete, or antibodies to the insulin-producing cells of the pancreas.[15] These markers have limited immediate health effects, but each helps predict later, expensive conditions.

Research of a very different sort is likely to be far more significant for health prediction, at least in the near term. The vast lode of data about people's health costs has been mined to find ways of predicting future health from factors that are easily ascertained: personal characteristics, personal activities, or past use of the health care system. These statistical studies have reached new levels of sophistication over the past decade, thanks in large part to the consequences of opening up Medicare to health maintenance organizations (HMOs).

In 1982, Congress decided to encourage HMOs to enroll Medicare patients by offering to pay the HMOs an annual fee per patient. The government was concerned that HMOs would be able to predict which Medicare patients would be expensive and which would be inexpensive. If they then enrolled only the healthier Medicare patients, the HMO would get an unearned additional profit (at the federal government's expense) and the high-risk patient would find it hard to locate an HMO that would accept her. The problem led the government to pay HMOs different amounts for different patients on the basis of age, sex, welfare status, and residence, through a calculation called "adjusted average per capita cost" (AAPCC). The use of AAPCC sparked research into its accuracy, with clear results: AAPCC does not predict very well. More important for our purposes, the research uncovered other ways of adjusting costs that worked much better: adjusting by measures of health status, by diagnostic health groups for particular illnesses, and by prior or current actual use of medical services.[16]

Let us focus on a study done by Joseph Newhouse and his colleagues.[17] This study tried to find equations to predict an individual's health costs for the next year by using the AAPCC variables, information from a medical history and physical examination, the patient's subjective assessment of her health, and the amount she spent on doctors and hospitals in the previous year.

The study found that while the AAPCC variables explained only 2.2 percent of the total variance between people in costs, the AAPCC variables plus prior use explained 6.4 percent, and all the variables combined to explain 9.0 percent. These may not seem like very useful results, but the annual variation in health charges is so vast that even slightly improved prediction is valuable. If an HMO could predict the annual health costs of Medicare enrollees better than the AAPCC by just one percentage point of total variance, it would save an estimated $630 per covered person by enrolling only those people whose predicted costs are below average. If it could improve on the AAPCC by 5.5 percentage points, the savings would be $1,170 per covered person; at 7.5 percentage points, it would save $1,320 per covered person.[18] If a self-insuring employer improved its selection of employees by 6.4 percent of the variance in annual medical costs—which it could do with only prior-year use and the AAPCC variables—it could save more than $1,200 per employee. For current employees, this data is easy for the employer to collect; as a self-insurer, it knows the amount of the previous year's bills because it paid them.

Improved prediction is coming, albeit in these two very different forms. In its medical form, through genetic or other medical research, it can identify individuals who will have some relatively uncommon diseases and a larger number of individuals who will have higher-than-average risks for various diseases. In its statistical form, it cannot identify individuals who definitely will have particular diseases, but it can sort a group—including a group of employees or job applicants—into those at higher or lower risk for future medical costs. Less powerful than medical information in specific cases, statistical prediction should prove much more powerful in the aggregate.

Increased predictability of individual health, whether from the genetics revolution, research in infectious disease, or statistical analysis, will threaten those at high risk for health costs with employment discrimination and catastrophic medical bills. Before seeking ways to solve those problems, we should ask whether a solution is needed. Life is full of inequalities, many of them inextricably linked to a person's innate characteristics, but our society makes no substantial effort to equalize most disparities between people. Why should predictable health risks be different?

Many will consider the answer obvious because they believe health care should be considered a fundamental right. For those who do not accept that position, I want to suggest two reasons why discrimination based on predicted health risks deserves attention. The first is utilitarian; the second is not.

The first argument is simply that the benefits of allowing improved prediction to ruin the health care financing system fall far short of its costs. The benefits are monetary and fall exclusively to those people thought to be at low risk. They will get better jobs, earn more money, and pay less for their health insurance than their riskier counterparts, to the extent of perhaps a thousand dollars a year. The costs, on the other hand, are broadly spread and, though difficult to quantify, likely to be enormous. They will fall on high-risk employees, on low-risk employees, and on society in general.

High-risk employees (and their spouses and children) would be harmed in several ways. Discrimination may keep them from jobs they would otherwise have had. Changes in health coverage will keep them from employment-related health insurance they would otherwise have had. With individual insurance unlikely to be available to high-risk people,[19] they will be completely uninsured and may face both financial ruin and, according to numerous studies, poorer health care.[20] Even when they get health care, they will be seen no longer as valued patients but as beggars, a change in status that reduces the respect they receive from the health care system and, perhaps, their own self-respect.

Low-risk employees would also bear some costs from the response to adverse selection if, as seems most likely, employers responded by limiting or eliminating health coverage. They might be able to purchase individually underwritten health insurance, but at higher prices and with smaller tax advantages. They would also face uncertainty about their own future status—the progress of medical science could easily change this year's low risk into next year's high risk.

Finally, the entire society will feel the adverse effects on the health care system and on the overall economy. Fewer insured people will mean more uncompensated care, but the need to provide uncompensated care would then warp the health care delivery system. It would make hospital emergency rooms the only source of health care for many of the uninsured, even though that care could often be provided more efficiently in other settings. The costs of uncompensated emergency care would lead hospitals to close their emergency rooms entirely or to expand into areas without indigent patients. And the need of hospitals and other providers to recover their costs for treating the uninsured would make them to try to pass those costs on to insured patients, distorting the prices paid by everyone in the health care system and creating new incentives for insurers and others to avoid these higher costs.[21]

These costs appear to swamp the monetary benefits to low-risk employees. The problem with this kind of utilitarian analysis is that, strictly speaking, different people's utility cannot be compared.[22] A benefit for one person cannot be weighed dollar for dollar against a loss for someone else. Health insurance is not immune from this problem, but the nature of the benefits and costs here—a few dollars on one side; financial ruin, bad health, a distorted health care sector, and general insecurity on the other—make this balance seem easy to draw, if not to prove.

The second argument against discrimination draws from a deep sentiment in our culture: people should not be "punished" for things that are beyond their control. The importance of criminal intent and mental competency to the criminal system is some evidence of the strength of this sentiment; so is the consensus against discrimination clearly based solely on race or sex, characteristics that an individual does not choose and, for the most part, cannot change. Some health risks are completely unavoidable and are in no way an individual's "fault," and, as a result, we should not "punish" him for them.

This argument is not entirely compelling. Many things are outside a person's control, but we do not compensate for all of them. One cannot choose to be born with talent or with rich parents. Even genetic health risks are often partially voluntary. A person may have a genetic susceptibility to lung cancer and also smoke. In spite of its flaws, though, this argument does speak to our deeply held beliefs. When the predicted risks are truly unavoidable, society will feel pressure to intervene.

Assume that, for whatever reasons, society decides to limit the effects of increased predictability. Three different strategies are available: banning discrimination, protecting patient confidentiality, and changing employer incentives.

The United States bans employment discrimination, including discrimination in fringe benefits, in many contexts. The Constitution's command of "equal protection of the laws" applies to the employment decisions of the federal, state, and local governments. Title VII of the Civil Rights Act of 1964 bans discrimination by most public or private employers on the basis of race, sex, religion, or national origin, and the Age Discrimination in Employment Act protects workers forty years old or older. The law most relevant to people at high risk for medical costs is the newly passed Americans with Disabilities Act (the ADA).[23]

The ADA, which expands the scope and extends the reach of the Federal Rehabilitation Act of 1973, prohibits employers from discriminating on the basis of disability. Following the earlier Federal Rehabilitation Act, it defines a disability as "(A) a physical or mental impairment that substantially limits one or more of the major life activities . . . ; (B) a record of such an impairment; or (C) being regarded as having such an impairment." An employer can take negative action based on a disability only if the individual cannot perform the essential functions of the job, with or without reasonable accommodations by the employer, or if the necessary accommodations would cause the employer "undue hardship."

Many high-risk people clearly will be protected under the ADA. Some will be covered because they have current conditions that "substantially limit" their "major life activities." Paraplegia, multiple sclerosis, AIDS, and insulin-dependent diabetes all fit that description while predicting higher-than-average future health costs. Others will be covered because they have what is clearly "a record of such an impairment," which may turn out to predict future illness. For example, if an employee had previously had a case of cancer, the cancer history would be a record of a past impairment and thus a disability, and in some cases it would also predict higher future medical costs.

But what about a person with the gene for Huntington's disease? He has no current or past impairment but will have serious and expensive future health problems. The ADA addresses past and present impairments in its definition, but it does not talk of future impairments. It does, however, include people who are "regarded as" being impaired. The Equal Employment Opportunity Commission has promulgated draft regulations that include people at heightened risk for future disabilities in the "regarded as" category, but those regulations could be changed or challenged. Given the limited legislative history and some cases under similar laws, the person carrying the Huntington's gene is probably covered by the Act, but that result cannot be guaranteed.[24]

Of course, even if the ADA does not reach these conditions, Congress might well amend it to deal with discrimination based on predicted health. Even a clear and complete ban on such discrimination suffers from two problems. First, it will not touch employers' most important reaction to increased prediction. Federal law, including both the ADA and ERISA, expressly permits employers to limit coverage or even to eliminate health benefits.[25] Second, even in the areas where their coverage is strongest— firing employees who have high health risks—the statutes may not work well. These statutes are not self-enforcing; plaintiffs still must bring (and win) lawsuits. Employment discrimination on the basis of race or sex has been illegal since 1964 under Title VII of the Civil Rights Act, but few would argue that it has been eliminated. Legislation to ban discrimination based on health risks, genetic or otherwise, is likely to be even harder to enforce than Title VII, in light of the many and subtle ways employers could discriminate against those at higher risk.

A second strategy is to keep employers from learning enough about a person's individual health risks to allow them to discriminate. Existing law protects information about an individual's health prospects in two ways. The first is the common-law confidentiality of the doctor-patient relationship. The fact that a doctor cannot reveal medical information without the patient's permission is of little value, however, if the employer can make an applicant answer questions about her medical condition or require medical tests. The second, the ADA and other statutes barring discrimination against the disabled, is more helpful.[26] The ADA prohibits employer-required medical disclosure or examinations except under certain conditions and for specific and limited purposes.

The ADA's provisions on medical privacy are helpful, but this strategy has the same problems as banning discrimination. First, it does nothing to prevent employers from reacting to the threat of adverse selection by limiting or eliminating their health coverage. And second, a ban on information cannot be perfectly enforced. Even under the ADA's limitations, employers will get useful information about some health risks. Some risks, like obesity, will be immediately visible. Others, like past use of health benefits, will be available from a self-insuring employer's own records. And some information about an employee's health will inevitably become known to fellow workers and supervisors.

The third strategy is more fundamental and, I believe, more promising. Rather than trying to keep employers from discriminating—directly or, through limiting their information, indirectly—the government could remove their incentive to discriminate. This incentive comes from the fact that the employer bears the financial risks of the future medical costs of its employees (and often of their families). Removing that risk removes the employer's main incentive to discriminate. How could the risk be eliminated? At least three different plausible reforms could eliminate or greatly reduce this kind of discrimination.

One answer would be a tax-financed system of mandatory national health insurance, perhaps similar to the Canadian system.[27] This kind of system would solve the problem of discrimination by ending the employer's responsibility for medical costs. Since the employer would not pay the medical costs of its employees, it would not care whether those costs were high or low.

A second solution would be to expand the existing framework of employment-related health coverage by requiring employers to provide health coverage to all their employees, but provide government subsidies for employees at high risk.[28] These "insurance vouchers" would make employers (and individual insurers) indifferent between people at high and low risk.

A third solution would also require employers to provide coverage for all their employees, but would require that the employers do so through community-rated insurance.[29] If the employer's coverage were community-rated, the higher-than-average medical costs of higher-risk employees would not fall on either the employee or the employer. They would fall on the insurer and, through community-rating, on all the companies it insured. As a result, an employer would have no incentive to discriminate against high-risk people and no reason to fear adverse selection. Suggesting a return to community-rating may seem ironic in light of the market's rejection of community-rating over the past few decades, but the rejection of community-rating was a result of competition from experience-rating and self-insurance. If experience-rating and self-insurance were banned, community-rating should be stable.

Each of these solutions has its own limitations. A national health insurance system, even if politically feasible, would sacrifice any of the benefits from competition and responsiveness that might exist in the current pluralistic system.[30] Subsidies could well be cumbersome to administer and would have to be based on the very best risk assessment—otherwise, they would become another AAPCC, a target to be surpassed by employers. Mandatory coverage with community-rating would give insurers an incentive to discriminate between companies by accepting or attracting only companies with low-risk employees. Both government regulation and employer vigilance would be necessary to prevent the cycle of discrimination from starting again at that level.

Reasonable people can differ about which approach may be best. It should be clear, however, that each holds out the promise of a broad solution to the problem of increased prediction, along with the potential for other valuable changes in America's overall health care financing system.

The erosion of the current health care financing system by the increased predictability of individuual health is a serious problem. The genetics revolution is not the source of the problem and, quite likely, will play a smaller role in it than other methods of predicting individual health. The true problem is a health care financing system that leaves the cost, terms, and the very existence of health coverage for most Americans at the discretion of employers and insurers. We can solve this problem, but to do so we must first understand it and then muster the political will to do something about it. Because of the unique visibility and political sensitivity of the revolution in genetics, it may contribute more to the problem's solution than it will to the problem's cause.

EVELYN FOX KELLER

Nature, Nurture, and the Human Genome Project

13

Genes became big business in the 1980s, and they are likely to become even bigger business in the decades to come. Plant genes, mouse genes, bacterial genes, and human genes are all in the news, but over the last couple of years it is human genes that have become the focus of particular interest. Daily, we are told—by Barbara Walters, by newspaper journalists, and above all, by proponents of the human genome project—that it is our genes that make us "what we are," that make some of us musical geniuses, Olympic athletes, or theoretical physicists and others alcoholics, manic-depressives, schizophrenics—even homeless. The Office of Technology Assessment concludes that "one of the strongest arguments for supporting human genome projects is that they will provide knowledge about the determinants of the human condition"; that, especially, the human genome project promises to illuminate the determinants of human disease, even of those diseases "that are at the root of many current societal problems."[1]

Some may worry about the "desirability of using genetic information to control and shape the future of human society," but others worry, perhaps equally, about a possible failure of courage.[2] To withhold support for this ambitious and expensive undertaking, writes Daniel Koshland, the editor of *Science* magazine, is to incur "the immorality of omission—the failure to apply a great

new technology to aid the poor, the infirm, and the underprivileged."[3*]

Thanks largely to the remarkable progress of molecular biology,
it is claimed that the controversy between nature and nurture that
has plagued us for so long has finally been resolved. To quote
Koshland again, we now know what "may seem obvious to a
scientist, but our judges, journalists, legislators, and philosophers
have been slow to learn"—namely, that if we want to induce
children to behave, to rehabilitate prisoners, to prevent suicides,
we must recognize that

> we are dealing with a very complex problem in which the structure
> of society and chemical therapy will [both] play roles. Better
> schools, a better environment, better counseling, and better rehabil-
> itation will help some individuals, but not all. Better drugs and
> genetic engineering will help others, but not all. It is not be going
> to be easy for those without scientific training to cope with these
> complicated relationships even when all the factors are well under-
> stood.[4]

In the same vein, Robert Weinberg, a prominent molecular biologist at MIT, says,

> Over the next decade, one may begin to stumble across genes that
> are surprisingly strong determinants of cognition, affect, and other
> aspects of human function and appearance. [To deny this would
> be] hiding one's head in the sand.[5]

Most responsible advocates are of course careful to acknowledge the role of *both* nature and nurture, but rhetorically, as well
as in scientific practice, it is "nature" that emerges as the decisive
victor. Like others, Koshland takes his object of advocacy to be
research—not on environmental influences, but on genetic determinants; similarly, he does not cite the importance of social, psychological, or political training, but only of scientific training. Also
like other commentators on the nature-nurture controversy, Kosh-

---

*In the address on which this editorial was based, delivered at the First Human
Genome Conference in October 1989, Koshland was even more explicit. In response
to the oft-raised question, "Why not give this money to the homeless?" he said, "What
these people don't realize is that the homeless are impaired . . . Indeed, no group will
benefit more from the application of human genetics." Just how the human genome
project will aid "the poor, the infirm, and the underprivileged," Koshland did not
say.

land is referring not to our physical development but to our emotional, intellectual, and behavioral development. Confidence in the genetic basis of our physiology has long been established; what appears to be new here is our confidence in the genetic basis of behavior. As Robert Plomin writes, "Just 15 years ago, the idea of genetic influence on complex human behavior was anathema to many behavioral scientists. Now, however, the role of inheritance in behavior has become widely accepted, even for sensitive domains as IQ."[6]

The shifts that Plomin, Koshland, and others note are real, and the usual assumption is that they are a direct consequence of developments in our scientific understanding of genetics. It is important to note, however, that our beliefs in nature and nurture have a cultural as well as a scientific history. There is indeed something new in the current configuration of our beliefs, and if we are to understand that novelty properly we must examine both histories, their mutual entwinement and their interdependence. In an effort to set the stage for asking what exactly is new in the current resurgence of genetic determinism, I will begin with a brief review of the nature-nurture debate from the early part of this century to the period following World War II. I then turn to an examination of the changing terms of this debate in the context of the rise of molecular biology.

Although it is commonly recognized that beliefs in genetic determinism ran strong among early eugenicists, it is perhaps less well known how prevalent such beliefs were among geneticists in the earlier part of this century. For example, the historian Diane Paul writes that a consensus "among geneticists concerning the role of heredity in the determination of intellectual, psychological, and moral traits"—a consensus that was to endure until the mid-1940s—was, by the 1920s, already "so complete that virtually no one . . . is to be found outside it." For many geneticists during these years, the desirability of at least some of the eugenic possibilities that genetic determinism implied seemed obvious. In 1939, for example, a "Geneticists' Manifesto," authored by H. J. Muller and cosigned by twenty-two distinguished geneticists, was issued at the Seventh International Congress of Genetics. It read:

The most important genetic objectives, from a social point of view, are the improvement of those genetic characteristics which make (a) for health, (b) for the complex called intelligence, and (c) for those temperamental qualities which favour fellow-feeling and social behavior . . . A more widespread understanding of biological principles will bring with it the realization that much more than the prevention of genetic deterioration is to be sought for, and that the raising of the level of the average of the population nearly to that of the highest now existing in isolated individuals, in regard to physical well being, intelligence and temperamental qualities, is an achievement that would—so far as purely genetic considerations are concerned—be physically possible within a comparatively small number of generations. Thus everyone might look upon "genius," combined of course with stability, as his birthright.[7]

What is important to add to Paul's argument is that belief in the power of genes to mold the character of human beings—along with the confidence that belief engendered in the uses to which an understanding of genetics could be put in shaping the future course of evolution—provided a major impetus behind the development of both classical and molecular genetics. For example, Warren Weaver, the head of the Natural Science Division of the Rockefeller Foundation during the 1930s, was explicit about his motivation for diverting the resources under his command to the study of biological problems:

[It] was perfectly clear that man wasn't going to suffer for lack of power to control the physical universe . . . What he really needed to know was more about himself . . . Increased knowledge of man is . . . the ultimate aim of the biological sciences.[8]

In his 1934 progress report, he wrote:

The challenge . . . is obvious. Can man gain an intelligent control over his own power? Can we develop so sound and extensive a genetics that we can hope to breed, in the future, superior men? . . . Can man acquire enough knowledge of his own vital processes so that we can hope to rationalize human behavior? Can we, in short, create a new science of Man?[9]

After World War II, however, as Paul (and others) go on to note, confidence in the power of genes over human behavior, and the concomitant eugenic potential of genetics, suffered a rapid decline. In large part, Paul attributes this decline to the "momentous social forces" unleashed by Hitler's rise to power in Germany. Especially, she observes, "biological explanations of nonphysical human differences rapidly lost favor in the general revulsion towards the uses to which they had been put by the Nazis."[10]

I would suggest that a more nuanced account of this decline may be in order. In particular, I suggest that in the revulsion against Nazi eugenics in Germany (as well as, it should be added, the revulsion against racist eugenics in the United States and "classist" eugenics in England), the direct link between genetics and its eugenic implications that had earlier been so visible, and so powerfully motivating, was no longer politically tolerable. For geneticists, this link was severed by demarcating the knowledge of genetics from its uses or, when that failed, distinguishing human from nonhuman genetics. In both cases, all reference to the human uses to which a knowledge of genetics might be put was abandoned. For others, the link between genetics and eugenics was avoided by a more general demarcation between biology and culture; the force of genetics was confined to purely physiological attributes, while behavior came increasingly to be seen as belonging to the domain of culture. Either way, attempts of earlier eugenicists to base a social program on genetic principles—in Germany, England, or the United States—could be discredited on the grounds that they were based on a misuse of genetics, on untenable (or at the very least, on excessively simplistic) assumptions about the genetic basis of behavior, without simultaneously discrediting the science of genetics per se.[11]

With these demarcations ready at hand, human behavior could be claimed as a free zone while at the same time the science of genetics—and with it, confidence in the genetic determination of everything but human behavior—could prosper. Because nurture (or culture) had always been seen as a more commodious force than nature, it was perhaps inevitable that in the aftermath of the war, it would be to nurture that the development of human behavior would be attributed. Simply put, it was nurture, not nature, that was seen as conducive to the kind of unfettered development imagined possible by a victorious and "free" republic. And in the absence of a strong public stance to the contrary by geneticists, the general optimism of the time inclined both popular and academic assumptions about the relative importance of nature and nurture (at least in the realm of human behavior) to undergo a decisive shift. In the mood that came to prevail, anything seemed possible, given the right environment and the right kind of nurturing.[12] The period of the 1950s and 1960s can, in retrospect, readily be described as the age of psychology. For psychoanalytically minded people, acceptable human behavior was attributed to "good-enough mothering"; for the behaviorists, to proper reinforcement and conditioning. With the possible exception of a small group of geneticists, almost no one, either in popular or scientific culture, was then looking to genetics.

At the same time, however, during the very period when behavior appeared to be so securely located in the domain of nurture, the science of genetics began to make unprecedented strides. With these strides came a dramatic change in the place of genetics in biology and, ultimately, on our larger cultural map. In the earlier part of the century, genetics had been a small, relatively recondite field of biology (to many, of dubious importance for an understanding of physiology and development), but by the late 1960s genetics had moved to center place in the life sciences. Furthermore, the postwar period had witnessed enormous expansion of the sciences as a whole, both in cultural influence and in actual size, and in that general expansion the life sciences grew apace. By the 1960s, geneticists thus constituted a far larger group (measured in absolute numbers) than they had in the period before World War II, with vastly more influence in both scientific and popular culture. Between 1950 and 1970, membership in the Genetics Society of America alone increased from 882 to 3,043, according to the *Federation of American Societies Directory*. Additional recruits to genetics would also be found among molecular biologists.

The simplest account for the growth in both the number and influence of geneticists is surely to be found in the extraordinary successes of molecular biology. Before 1953, genes had been abstract hypothetical units; in that year they became concrete, knowable entities. The achievement of James D. Watson and Francis Crick permitted the identification of genes as sequences of DNA and offered a solution to the mystery of genetic replication. In the years that followed, molecular biologists learned an enormous amount about the mechanisms by which genes—or, now, sequences of DNA—are said to regulate and control the essential processes of living organisms. DNA became the "Master Molecule" of life.

Marking their sense of accomplishment, the National Academy of Sciences conducted a comprehensive survey of the life sciences in 1968, entitling it *Biology and the Future of Man*. The bulk of this eminently sober and carefully edited report on the "state of the art" and future expectations of the biological disciplines had little direct reference to "man," and those parts that did focused almost entirely on questions of physiology. Of the first 900 pages of this report, the question of behavioral genetics occupies only five, with most of these referring to nonhuman behavior. In the last (and title) chapter, however, the question of "man" emerges with force, and with it the exemption of human behavior that had heretofore been so carefully maintained was dissolved. In this closing chapter, we reencounter all the themes familiar from the first half of the century, in a form marked only faintly by the traces of their recent controversiality. The authors write,

> We could breed for obesity or leanness, blue eyes or black, wavy or wiry hair, and any one of the obvious physical attributes in which human beings vary. Presumably, we could also breed for mental performance, for special properties like spatial perception or verbal capacity, perhaps even for cooperativeness or disruptive behavior, even, conceivably for high scores in intelligence tests . . .
>
> Man, although potentially able to select his own genetic constitution has not yet made use of this power. Selection is a harsh process. To make speedy progress, reproduction should be limited primarily to those who possess genotypes for the desired traits. But who will decide what is desirable? . . . Who would dare to prohibit procreation to a majority of men and women? And to whom would society entrust such decisions? May we expect changes in attitudes of whole societies so that they would accept the self-control of human evolution at the cost of forgoing the private decisions of most people to propagate themselves in their own children? It is extremely unlikely that such changes in attitude will come soon. The future of man, however, may well extend . . . long enough not only to ponder these possibilities but also to explore them in actuality.[13]

The final and closing paragraph of the volume reads:

> Man's view of himself has undergone many changes. From a unique position in the universe, the Copernican revolution reduced him to an inhabitant of one of many planets. From a unique position among organisms, the Darwinian revolution assigned him a place among the millions of other species which evolved from one an-

other. Yet, *Homo sapiens* has overcome the limitations of his origin. He controls the vast energies of the atomic nucleus, moves across his planet at speeds barely below escape velocity, and can escape when he so wills. He communicates with his fellows at the speed of light, extends the powers of his brain with those of the digital computer, and influences the numbers and genetic constitution of virtually all other living species. Now he can guide his own evolution. In him, Nature has reached beyond the hard regularities of physical phenomena. *Homo sapiens,* the creature of Nature, has transcended her. From a product of circumstances, he has risen to responsibility. At last, he is Man. May he behave so![14]

From these closing remarks of the NAS report, written in 1968, it would be tempting to conclude that, among geneticists, not much had changed since the 1920s—that, whatever changes in popular sentiment had occurred over the intervening years, the confidence of (at least some) geneticists in genetic determinism (and their concomitant interest in shaping the future course of evolution) was as strong as ever. Despite some visible signs of continuity across the decades, however, these remarks reflect not simply a resurgence of genetic determinism but the beginnings of its transfiguration. They may be read as a preview of a new era in our thinking about nature and nurture, an era in which the forces of nature and nurture are radically reconceived. This reconception is indeed intimately tied to developments in molecular biology, but in rather more complex ways than is usually imagined.

Of signal importance in the transfiguration of genetic determinism is the fact that, in the late 1960s, molecular biologists began to develop techniques by which they themselves could manipulate the "Master Molecule." They learned how to sequence it, how to synthesize it, and how to alter it. Out of molecular biology emerged a technological know-how that decisively altered our historical sense of the immutability of "nature." Where the traditional view had been that "nature" spelled destiny and "nurture" freedom, now the roles appeared to be reversed. The technological innovations of molecular biology invited a vastly extended discursive prowess, encouraging the notion that we could more readily control the former than the latter—not simply as a long-term goal but as an immediate prospect. This notion, though far in excess of the actual capabilities of molecular biology of that time, transformed the very terms of the nature-nurture debate; eventually, it would transform the terms of molecular biology as well.

For the first twenty years of molecular biology, research focused on organisms at the opposite end of the phylogenetic scale from humans, and to most people the implications for human beings seemed remote. For some, however, the distance from *Escherichia coli* to *Homo sapiens* had never seemed very large, and certainly by the late 1960s, with the development of new techniques for working with eukaryotic genes and mammalian viruses, that gap began to close. It was perhaps inevitable that the prospects of control invited by the new research would soon extend into the reaches of human nature. The first explicit formulations of such ambitions by molecular biologists began to appear around 1969. Even then, however, when molecular biology was just beginning to move into the domain of higher organisms, the kinds of control envisioned were already presented as crucially distinct from those of the older eugenics.

Whereas the eugenics programs of the earlier part of the century had had to rely on massive social programs, and hence were subject to social control, molecular genetics seemed to enable what Robert Sinsheimer called "a new eugenics"—a eugenics that "could, at least in principle, be implemented on a quite individual basis."[15] Sinsheimer added,

> The old eugenics was limited to a numerical enhancement of the best of our existing gene pool. The new eugenics would permit in principle the conversion of all the unfit to the highest genetic level.[16]

In short, in the vision inspired by the successes of molecular biology, "nature" became newly malleable, perhaps infinitely so; certainly it was vastly more malleable than anyone had ever imagined "nurture" to be. Sinsheimer continued:

> It is a new horizon in the history of man. Some may smile and may feel that this is but a new version of the old dream, of the perfection of man. It is that, but it is something more. The old dreams of the cultural perfection of man were always sharply constrained by his inherent, inherited imperfections and limitations . . . To foster his better traits and to curb his worse by cultural means alone has

always been, while clearly not impossible, in many instances most difficult . . . We now glimpse another route—the chance to ease the internal strains and heal the internal flaws directly, to carry on and consciously perfect far beyond our present vision this remarkable product of two billion years of evolution.[17]

In 1969, the molecular biology of human genetics was in its infancy. Just the year before, the first non-sex-linked gene had been mapped to a particular chromosome, and no human gene of any kind had yet been precisely located. And among those human genes of which molecular biologists could claim any knowledge at all, none had the remotest relation to human behavior. Yet Sinsheimer could already anticipate a genetic route to "the cultural perfection of man." From what might such extraordinary confidence have derived?

True, we have come to expect—perhaps even demand—gestures of clairvoyant prognostication from the scientific community, especially from those who could speak from the forefront of scientific research. But it would be a serious mistake to regard the statements that were made either as conclusions drawn from scientific experience or as mere speculation or fantasy. The hopes and anticipations that Sinsheimer expressed in 1969 (like the concluding remarks of the NAS report) must be understood to have been drawn at least as much from a culturally as from a scientifically inspired vision. In turn, we must understand the particular power that anticipations of this sort, posed in the form of scientific predictions, have to influence (directly or indirectly) the future course of scientific progress. Given the status and authority molecular biology had acquired since 1953, predictions from the world of science would inevitably influence popular beliefs, attitudes, and expectations considerably. Similarly, even though remote from current scientific reality, to the extent that they reflected the hopes and ambitions of working scientists (and of the agencies that funded them), they would also have the power to influence the direction that scientific research would take in the future. In this way, the hopeful statements I've quoted can be seen as expressions of a kind of intentionality; as such, they actively contributed to the construction of future scientific reality.

The themes in these scientific/utopian scenarios that had particular influence on popular belief are (1) the newly acclaimed malleability of "nature"; (2) the reach across the divide between biology and culture that had been at least tacitly in place since World War II; and (3) the emphasis on the role of individual choice in the kinds of interventions the new genetics would make possible. In turn, of course, the influence of such arguments on popular belief would prove critical for making available the resources and support required for these aspirations to exert a practical influence over the future course of research in molecular biology.

The last twenty years have exhibited just such a convergence of cultural attitudes and technical focus. After 1970, both the development of techniques permitting direct intervention in the structure of DNA sequences and the use of these techniques in the study of human genetics took off exponentially. In 1970, there was the first synthesis of a gene, by H. Gobind Khorana; in 1973, the first successful genetic modification effected through the splicing of a well-defined fragment of DNA from one organism into the genome of another (recombinant DNA). By the end of the decade, in large part through the use of recombinant techniques, the number of human genes identified and mapped to specific chromosomes had increased more than sixfold, or, if one counts only genes found on autosomal chromosomes, approximately 300-fold.[18]

Without doubt, the 1970s was a decade of extraordinary expansion for molecular biology: technically, institutionally, culturally, and economically. My aim is not to question that expansion per se, but rather to question the conventional understanding that the institutional, cultural, and economic expansion of molecular biology proceeded directly, and as a matter of course, from its technical successes. In particular, I want to focus on the ideological expansion of molecular biology into both popular culture and medicine, and at least to raise a question about the effect of this ideological expansion on subsequent technical developments. To this end, the historian Edward Yoxen's exploration of the construction of the idea of "genetic disease" provides an absolutely essential starting point, for it is this concept which both has provided the ground for the cultural and medical expansion of molecular genetics and, at the same time, distinguishes current formulations of genetic determinism from those of the earlier part of the century.[19]

As Yoxen points out, one need not dispute the fact that "many of the phenomena of genetic disease are grounded in material reality" in order to ask "why we isolate or delineate certain phenomena for analysis, why we say that they constitute diseases, and why we seek to explain their nature and cause in genetical terms."[20] Although an earlier generation of geneticists may not have doubted the power of genes to determine (and thus ultimately to transform) human well-being, they did not (except in isolated instances) link their claims to a concept of genetic disease, and their medical colleagues, failing to see any direct relation between genes and treatment even for those diseases that were understood to be genetic, regarded genetics as being of little relevance to medical practice. Today, however, the relation between genetics and the medical sciences has dramatically changed. Even though, in actuality, genetics remains of quite limited practical relevance to the healing arts, the concept of disease—now extended throughout the domain of human behavior—has increasingly come to be understood by health scientists in terms of genetics. Indeed, the volume of medical literature on genetic disease has increased exponentially over the past decade,* and much of this literature suggests a conceptual shift that one commentator describes as follows:

> [In the past,] most physicians and investigators have perceived that deleterious influences on human health are of two kinds: either a deficiency of a basic resource such as food or vitamins, or exposure to hazards that may be either natural . . . or man-made . . . Genetics is now showing that this view of the determinants of health as being external is too simplistic. It neglects a major determinant of disease—an internal one. Far from being a rare cause of disease, genetic factors are a very important determinant of health or illness in developed countries.[21]

But as Yoxen points out, in the course of this conceptual shift "genetic disease" has become an extremely large category, encompassing not only genetic disorders that are thought of as diseases but also genetic abnormalities associated with no known disorder as well as disorders that may be neither genetic nor diseases.[22]

---

*A count of review articles on genetic disease listed in Medline reveals a more than sevenfold increase over the years 1986 to 1989 alone. Fifty-one articles are listed for 1986, 152 for 1987, 288 for 1988, and 366 in 1989.

Many factors (both technical and cultural) have contributed to the expansion of the concept of genetic disease and, with it, the domain of clinical genetics. Among these one might note: increasingly general acceptance of the explanatory framework of molecular biology; the postwar diminution of the burden of acute disease; intensification of scientific training for medical practice; changing expectations for health in the general public; and patterns of resource distribution for scientific research. For example, Yoxen notes that, in the early 1970s, the National Institute of General Medical Sciences (a subdivision of the NIH) sought

> to mobilize support for its programs by representing genetic disorders as a significant cause of ill health. Here, genetics offers a strategy of territorial expansion through the redefinition of the causes of disease to a relatively low status institution.[23]

Yoxen's main point, however, is to indicate the many social, economic, political, and technical issues that must be taken into account if we are to understand how the "basic explanatory form of a 'genetic disease' has been constructed to fit the contemporary context."[24]

My point is an even more general one. It is to note that the concept of genetic disease, enthusiastically appropriated by the medical sciences for complex institutional and economic reasons, represents an ideological expansion of molecular biology far beyond its technical successes. I also want to argue that the general acceptance of this concept has, in turn, proved critical for the direction that subsequent technical developments in molecular biology have now begun to take. Without question, it was the technical prowess that molecular biology had achieved by the early 1980s that made it possible even to imagine a task as formidable as that of sequencing what has come to be called "the human genome." But it was the concept of genetic disease that created the climate in which such a project could appear both reasonable and desirable.

I want to focus on two arguments that surfaced early in the advocacy of the human genome project. First is the startling promise that the full sequence of the human genome will teach us, finally, "what it means to be human"; it will enable us to "decipher the mysteries" of our own existence. In spite of the fact that the actual genomes of any two individuals will differ by as much as three million bases, from a molecular biological point of view, the "essential underlying definition" of the human being is a single entity.[25] Advocates for the human genome project continue by arguing that the characterization of this entity (namely, its genetic sequence) therefore constitutes a critical question for medicine. But what is sometimes presented as a sequitur is more commonly presented as an independent appeal to the "major [or 'revolutionary'] impact" such a data base will have "on health care and disease prevention." In the official report issued in 1988 by the National Research Council Committee on Mapping and Sequencing the Human Genome, the value of this information for the "diagnosis, treatment, and prevention" of human disease is repeatedly emphasized. It is argued:

> Encoded in the DNA sequence are fundamental determinants of those mental capacities—learning, language, memory—essential to human culture. Encoded there as well are the mutations and variations that cause or increase susceptibility to many diseases responsible for much human suffering.[26]

The committee concludes "that a project to map and sequence the human genome should be undertaken" in order to "allow rapid progress to occur in the diagnosis and ultimate control of many human diseases."[27] James Watson makes the point even more strongly. For him, the human genome project is "our best go at diseases." Indeed, he goes further. Referring to manic depression as an instance of the kind of disease we seek to control, he argues that we must find the gene because without it "we are lost."[28]

The two central images of the rhetoric employed here—on the one hand the idea of a base-line norm, indicated by "the human genome," and on the other the specter of a panoply of genetic diseases (currently estimated at well over 3,000)—definitively distinguish this discourse from its precursors. The emphasis now is not so much on the "cultural perfection of man" or on the "conscious" and "direct" employment of genetic technology to engineer our "transition to a whole new pitch of evolution,"[29] or even on improving the quality of our genetic pool, but rather on the use of genetics—through diagnosis, treatment, and prevention—to guarantee to all human beings an individual and natural right, the right to health.* In its 1988 report on the human genome project, the Office of Technology Assessment concluded that "new technologies for identifying traits and altering genes make it possible for eugenic goals to be achieved through technological as opposed to social control."[30] But even more important, the report sets the project's eugenic implications apart from earlier precedents by distinguishing a "eugenics of normalcy": that is, "the use of genetic information . . . to ensure that . . . each individual has at least a modicum of normal genes." The report cites an argument that "individuals have a paramount right to be born with a normal, adequate hereditary endowment."[31]

Just as Sinsheimer predicted twenty years ago, the nineties version of the "new eugenics" (though the word *eugenics* is not now used) is no longer construed as a matter of social policy, the good of the species, or the quality of our collective gene pool; the current concern is the problem (as Watson puts it) of the "disease-causing genes" that "some of us *as individuals* have inherited [my italics]." Accordingly, it is presented in terms of the choices that "they as individuals" will have to make.[32] Genetics merely provides the information enabling the individual to realize an inalienable right to health, where "health" is defined in reference to a tacit norm, signified by "*the* human genome," and in contradistinction to a state of unhealth (or abnormality), indicated by an ever growing list of conditions characterized as "genetic disease."

A number of fairly obvious questions come to mind at this point about the concepts of both "individual" and "choice" that are invoked in this discourse, but first, some basic points stand in need of clarification. The first is that, despite the repeated emphasis on health care, on the diagnosis, treatment, and prevention of genetic disease, it is in fact primarily the possibility of diagnosis that is considered of practical relevance for the near future by even the most enthusiastic proponents of the human genome project; estimates of arrival times for therapeutic benefits run, optimisti-

---

*This point was brought home with particular clarity in a recent Barbara Walters special (aired on July 18, 1990). Although entitled "The Perfect Baby," the real point of the program was that "perfection" was in fact *not* the goal of the new human genetics. Joan Marks, director of human genetics at Sarah Lawrence College, made the point explicitly at the end of the program: "What we're talking about here is not perfect babies or perfect human beings. We're talking about healthy people. And I think it's wonderful to imagine a day when we can do a lot more than we can do now to assure that babies coming into this world will be healthy."

cally, as long as fifty years hence. Thus, "treatment" is at best a long-term goal, and "prevention" means preventing the births of individuals diagnosed as genetically aberrant—in a word, it means abortion. The choices "individuals" are asked to make are therefore choices not on behalf of their own health but on behalf of the health of their offspring and, implicitly, on behalf of the nation's health costs. Pointing to schizophrenia, which he claimed currently accounts for one-half of all hospital beds, Charles Cantor, the former head of the Human Genome Center at the Lawrence Berkeley Laboratory, recently argued in a lecture that the project would more than pay for itself by preventing the occurrence of just this one disease. When asked how such a saving could be effected he could only say: "by preventing the birth" of schizophrenics.[33]

Which brings us to the second point requiring clarification: namely, that these newly available choices, though ostensibly made by individuals, are in fairly obvious ways preconstructed by the categories of disease already presented to the decisionmaker, often on the basis of rather dubious evidence. Psychiatric disorders are a good case in point. In 1987, reports of a genetic locus for manic depression received extensive publicity, as did a similar report for a genetic locus for schizophrenia published in 1988. Less well publicized was the retraction of both these claims in 1989. Three months before Cantor's lecture, *Nature* had reported that the retraction "leaves us with no persuasive evidence linking any psychiatric disease to a single locus." As David Baltimore, then Director of the Whitehead Institute at MIT, said, "Setting myself up as an average reader of *Nature,* what am I to believe?"[34] Even more pressing for my point is the question of what the average reader of *Time* and *Newsweek* is to believe. If the scientific community were in closer agreement on genetic definitions of disease, an individual's choices might be clearer, but they would not be any more "autonomous."

The current disarray surrounding attempts to define "genetic disease" bears in part on a third point that I briefly indicated earlier—namely, the elusiveness of a norm against which the concept of abnormality is implicitly defined. Molecular analysis of human DNA indicates that the genomes of any two individuals will, on average, differ in approximately three million bases. In an attempt to bypass the enormous diversity among even "normal" human beings, a composite genome, with different chromosomes obtained from different individuals, has been adopted as the standard for genomic analysis. This "solution" does nothing, however, to address either the de facto variability in nucleotide sequence within individual chromosomes or the consequent difficulty in deciding what a "normal" sequence would be.

A fourth and final point that needs at least to be mentioned is that many of the categories of genetic disease—especially those referring to mental competence—put into question the very capacity of those individuals who carry the purported "disease-causing genes" to make choices. Such individuals might well be expected, in Watson's own words, to be "genetically incapable of being responsible."[35]

Forty years ago, when the specter of eugenics aroused such intense anxiety, the aims of genetics were made safe by a clear demarcation between biology and culture. The province of genetics, particularly of molecular genetics, was biology—primarily, the biology of lower organisms. To most people in or out of genetics, molecular biology seemed to have little if any bearing on human behavior. At that time, it was culture, not biology, that "made us human"; culture was simultaneously the source and the object of our special, human, freedom to make choices. Today we are being told—and judging from media accounts, we are apparently coming to believe—that what makes us human is our genes. Indeed, the very notion of "culture" as distinct from "biology" seems to have vanished; in the terms that increasingly dominate contemporary discourse, "culture" has become subsumed under biology.

But if culture is to be subsumed under biology, and if it is our biological or genetic future that we now seek to shape, where are we to locate the domain of freedom by which this future can be charted? The disarming suggestion that is put forth is that this domain of freedom is to be found in the elusive realm of "individual choice"—a suggestion that invokes a democratic and egalitarian ideal somewhere beyond biology. But since there is in this discourse no domain "beyond biology," since it is our genes that "make us what we are," and since they do so with a definitive inequality that compromises even those choices some of us can make, we are obliged to look elsewhere for the implied realm of freedom. I suggest that the locus of freedom on which this discourse tacitly depends is to be found not in the domain of "individual choice," comforting as such a notion might be, but rather in a domain protected by the ambiguous designation of "normality." More generally, I suggest that the distinction that had earlier been made by the demarcation between culture and biology (or between nurture and nature) is now made by a demarcation between the normal and the abnormal; the force of destiny is no longer attached to culture, or even to biology in general, but rather more specifically to the biology (or genetics) of disease. Far from teaching us "what it means to be human," in actual practice, the burden of the new human genetics turns on the elucidation not of human order but of human disorder. Our genes may make us "what we are," but, it would appear, they do so more forcefully for some of us than for others. By general consensus, molecular geneticists do not seek genetic loci for traits that they—and we—accept as normal. Indeed, they, like us, do not even seek to define the meaning of "normal."

It is perhaps inevitable that the appeal to the desire for health translates into a search for the genetic basis of unhealth, but the net effect of this translation is that the nature of normality is allowed silently to elude the gaze of genetic scrutiny—and thereby tacitly to evade its determinist grip. The freedom molecular biology promises to bring is the freedom to rout the domain of destiny inhering in "disease-causing genes" in the name of an unspecified standard of normality—a standard that remains unexamined not simply by oversight but by the internal logic of the endeavor. The "normal" state can be specified in this endeavor only by negation—by the absence of those alleles said to cause disease.

More problematic still is the insistent ambiguity inhering in the very term *normal*, an ambiguity that the philosopher and historian of science Ian Hacking traces to Auguste Comte:

> Comte . . . expressed and to some extent invented a fundamental tension in the idea of the normal—the normal as existing average, and the normal as figure of perfection to which we may progress. This is an even richer source of hidden power than the fact/value ambiguity that had always been present in the idea of the normal . . . On the one hand there is the thought that the normal is what is right, so that talk of the normal is a splendid way of preserving or returning to the status quo . . . On the other hand is the idea

that the normal is only average, and so is something to be improved upon.[36]

This ambiguity permits all of us a certain latitude in our hopes and expectations for a "eugenics of normalcy." It also clears a large field for the operation of distinctly nongenetic, ideological forces.

Both the definition and the routing of genetic disease express human choices, and even if "individual choice" is an inadequate model for describing the process by which choices actually get made, the very possibility of choice depends on a residual domain of agency that can remain free only to the extent that it remains unexamined. The question, of course, is where, and how, this residual domain of agency gets constructed and articulated, how the authority for prescribing the meaning of "normal" is distributed. The notion of culture (like that of nurture) may have vanished from contemporary biological discourse, but it is here, hidden from view, that the facts of culture continue to exert their undeniable force.

There is no question that eugenics has become a vastly more realizable prospect than it was in the earlier part of the century, and it must be granted that, in many ways, the very notion remains as disturbing as it was in 1945. As Watson has written,

> We have only to look at how the Nazis used leading members of the German human genetics and psychiatry communities to justify their genocide programs, first against the mentally ill and then the Jews and the Gypsies. We need no more vivid reminders that science in the wrong hands can do incalculable harm.[37]

It is of course true that, in 1990, we have no Nazi conspiracy to fear. All we have to fear today is our own complacency that there are some "right hands" in which to invest this responsibility—above all, the responsibility for arbitrating normality.

DANIEL J. KEVLES
AND LEROY HOOD

Reflections

14

In February 1990, Martin Rechsteiner, a professor in the biochemistry department at the University of Utah, sent a letter to colleagues around the United States contending that the human genome project is "a waste of national resources" and urging like-minded scientists to raise a protest against the project to key government officials, including the president's science advisor. In April, six biologists from across the country posted a joint "Dear Colleague" letter on the electronic mail system BioNet, which links the country's molecular biology laboratories, declaring, "The human genome project can be stopped. Please join our effort." In July, Bernard Davis, with the support of twenty-two—virtually all—of his colleagues in the department of microbiology and molecular genetics at Harvard University Medical School, published a letter in *Science* magazine urging a reassessment of the government's commitment to the project. That month, Rechsteiner and Davis made their cases in a hearing on Capitol Hill. By early 1991, a U.S. Senate aide was remarking of the human genome effort, "There's no groundswell of support. In fact, there's a groundswell against it."[1]

Thematically, the renewed criticism of the project echoed, in part, the main animadversion cast upon it in 1987—that it repre-

We are grateful to Rebecca Ullrich for assistance in research for this chapter.

sented the subjugation of biology to the directed, hierarchical mode of Big Science. Dissidents called into evidence Watson's decision, as director of the NIH's genome project, to fund genome centers, while Davis and his colleagues stressed that the project had begun "illogically, as a means of expanding the biological activities" of one of the country's principal agents of Big Science, the Department of Energy. Rechsteiner, undoubtedly pointing to Senator Pete Domenici, of New Mexico, averred that "the project owes its existence to a powerful U.S. Senator who desired funds for a national laboratory located in his state."[2]

Adding a new twist to the familiar charge that it would be a waste of money to sequence junk DNA, Rechsteiner held that ascertaining the sequence even of coding regions would not necessarily advance biological science: the effort would obtain DNA data for the sake of acquiring the information, independent of hypotheses that the data might address and with insufficient attention to the physiological or biochemical environment in which genes function. Nor, he insisted, would it necessarily foster medical progress: the revelations concerning the genetics of cancer or of cholesterol metabolism had not, after all, required human genome analysis; and detailed knowledge of genetic mutations had not led to therapies or cures. Rechsteiner told a *New York Times* reporter, "The human genome project is bad science, it's unthought-out science, it's hyped science." Rechsteiner had a sharp idea of what the genome project would do—divert funds from most other areas of basic biological research and training, limiting opportunities for substantively creative research while producing, in its genome centers, "armies of technicians" skilled only at DNA sequencing and data entry.[3]

The new band of dissidents were troubled, even angered, by the fact that, while the genome project had been prospering, general basic research in the biological sciences had been financially squeezed. A key indicator of crisis was the amount of money available for grants to investigators outside the National Institutes of Health from a key part of the agency, its National Institute for General Medical Sciences (NIGMS). Between 1988 and 1990, when the genome budget had risen from roughly $17 million to $88 million, the NIGMS budget exclusive of AIDS research had risen from $613 million to only $667 million, an increase that failed even to offset inflation in the cost of biomedical research. In its external grants programs, NIH considers two kinds of competitive applications—those for new projects and those for renewals of previous awards whose term has expired. In 1988, NIGMS had been able to award 981 new and competing renewal grants for research projects unrelated to the genome; in 1990, it awarded only 555, a 43 percent decrease and about 150 fewer grants than the nadir of the preceding decade.

During the same period, across all sections of NIH, the combined number of competitive grants had fallen from 6,000 a year to 4,600, smaller than the number funded in 1981. Of applications deemed worthy of support, the fraction actually funded had fallen from 40 percent to less than 25 percent; in some areas of research, the portion was as low as 12 percent. Money was so scarce as to prompt Davis to speak of "famine" in these sciences.[4] In a letter to *Science* magazine, John C. Lucchesi, chairman of the Genetics Study Section in the Division of Research Grants at NIH, predicted that "a few rounds of funding at the present award rates will very quickly result in a reduction in the number of active laboratories to less than half their current number," adding, "Arguments are made that the human genome project will give birth to a new generation of technologies. What good will that do in the absence of individuals trained and capable of applying these technologies . . . ?" Rechsteiner's resentment of the trend was typical: "During these hard times we saw the disbursal of previously unheard of amounts of money to a handful of genomists."[5]

In the view of the dissidents, the $200 million a year that the project was eventually slated to get would be much better spent easing the straitened circumstances of basic biomedical research. Rechsteiner noted that although this sum might not be large by, say, defense department standards, it might appear to be "all the money in the world" to a "struggling young assistant professor." The critics commonly pointed out how many basic biomedical research grants—of the current average size of $212,000—might be carved from the genome project budget. The total estimated 1991 genome budget—$154 million for NIH and DOE combined—would support 385 such grants, which Bernard Davis declared would provide "a substantial amount of famine relief for untargeted research."[6]

However, the pinch in general biomedical research, deplorable as it is, cannot be blamed entirely—or even significantly—on the genome project. The difficulty has been rooted partly in the general politics and administration of NIH research monies. Pork-barrel legislation has moved some percentage of NIH support into science projects that fall outside the competitive process. More important, in the mid-1980s NIH extended the average grant period from 3.3 to 4.3 years, mainly to provide greater stability for individual research projects and to relieve investigators of the burden of frequent reapplications for support. However admirable in intent, the shift established a pulse of fixed, downstream claims on the agency's budget that, in the absence of a sufficient compensatory increase in appropriations, reduced by roughly 25 percent the monies available for new or renewed awards. The reduction about equaled the decline that had occurred in the number of competitive awards and was, hence, enough by itself to account for the drop.[7]

The pinch is symptomatic also of deeper trends pertaining to growth and limits in biomedical research. Between 1977 and 1987, the number of new biomedical investigators, measured by the number of Ph.D.s awarded in biology, swelled by some 48,000, about twice the total of doctorates added to the life sciences during the 1960s. Compared with the physical sciences, which awarded about the same number of doctorates in the 1980s as in the 1960s, the life sciences have been doing a booming business.[8] In 1977, some 70,000 life-science doctorates were employed; in 1987, slightly more than 107,000—a rise of more than 50 percent. Between 1981 and 1990, in constant dollars, the NIH budget also rose about 50 percent[9]—two-thirds more than the constant-dollar increase in total federal outlays, which came to about 30 percent. The biomedical research community, having been multiply cloning itself every year, may be bumping up against the fact that funds for biomedical research will not—because they cannot—increase indefinitely at the disproportionately high rate necessary to accommodate all the new Ph.D.s. Then, too, the overall shortage of funds available per investigator is compounded for younger scientists by the high start-up costs for research—between several hundred thousand and half a million dollars typically to equip a beginning assistant professor's molecular biological research laboratory.

Even under the most affluent circumstances, the allocation of public resources in science, as in other areas, involves political decisions—political in the best sense that politics is the process by which democratic government determines how much it will spend to meet various public needs. In recent years, the annual NIH budget for research in the manifest public health problem of AIDS has skyrocketed, reaching some $800 million in 1991, almost ten times that of the NIH genome budget. Many biologists doubt the scientific merits of some AIDS research, but they are reluctant to question the magnitude of the investment because the battle against the disease, commanding powerful public support, is sacrosanct. The only sharply identified NIH program left to attack is the genome project.[10]

The attacks seem undeserved from several perspectives. In 1991, NIH expenditures on the project accounted for only 1 percent of the agency's total budget of $8 billion. Should the project come to be funded at the $200 million a year recommended by the National Academy of Sciences, the NIH share would amount to just 1.5 percent of the agency's total 1991 budget, or roughly 3 percent of its resources for external grants. A case can be made that the genome project has brought appropriations to biomedical research that it would not otherwise have received. The exact amount of the surplus may be debatable, but, like any other NIH program, the project has no obligation to defend itself primarily on grounds that it assists the biomedical research budget. Its principal justification is scientific—that the technologies, data, methods, and trained personnel that it is fostering will substantially strengthen the infrastructure of the biomedical enterprise. It is for this reason that the project has a legitimate claim on biomedical research funds (it is perhaps the only major institute or center in the National Institutes of Health that derived from a report of the National Academy of Sciences instead of from a congressional initiative to combat a particular disease).[11] And it is for this reason that the political system—the Congress, the president, and NIH—has resolved to allocate to the project the public resources that it requires.

In its deliberate emphasis on technological and methodological innovation, the genome project flies in the face of tradition and preference in the biomedical research community. Some of the rhetoric raised against the project seems to suggest that technology is no more than an auxiliary to genuine biological research, that it is even somehow alien to the enterprise, and that progress in the biomedical sciences is best achieved by lone investigators using simple methods and simple tools. The rhetoric has frequently surfaced before in the annals of twentieth-century experimental biology, as though experimental biologists had not gotten over the turn-of-the-century charges by natural historians that the study of life could not be conducted in the environment of Petri dishes. It achieved a sufficiently commonplace status to make its way, for example, into the 1954 Nobel Prize ceremonies, when a member of the Royal Caroline Institute remarked, "The electronics, radioactive isotopes, and complicated biochemistry of our age has threatened to turn medical science into something dangerously resembling technology. Now and again we need to be reminded of its fundamental biological elements."[12]

The fact of the matter, however, is that progress in the biomedical sciences has been empowered and accelerated to a considerable extent by sophisticated tools and technologies—notably the ultracentrifuge, radioactive isotopes, X-ray diffraction, chromatography, electrophoresis, and electron microscopy. None of these technologies was indigenous to biology. All were originally the products of the physical sciences or physical scientists at the fringes of biology, and many were developed partly with philanthropic or commercial support for use in biological research. For example, the builders of cyclotrons, the first abundant source of radioactive isotopes, found much of their initial financial patronage in medically oriented philanthropies eager to foster the construction of machines that would supply the isotopes cheaply and in profusion. (Since World War II, the principal source has been the atomic piles of the Atomic Energy Commission and its successor agencies.)[13]

The indispensability of sophisticated technology is obvious—even critics of the genome project concede it—but perhaps not so obvious, though also important, are the ways that various technologies have affected activity in the scientific workplace. The issue is pertinent to the charge that gene sequencing will necessarily move biology in the direction of gene drudgery. The indictment rests on a romantic premise—that conventional biology is without drudgery, that every young laboratory investigator constantly confronts demanding intellectual challenges at the workbench. A moment's reflection about what people have actually done in the laboratory will reveal that the premise is generally false. Part of what people have done in molecular biology is commonplace biochemistry—for example, identifying restriction enzymes, determining protein, DNA, and RNA sequences, synthesizing genes and cloning them. Recall that H. Ghobind Khorana required a band of collaborators and roughly five years to synthesize a small gene, succeeding finally in 1970, and that much bigger genes can now be synthesized with table-top instruments in a day. Commercial firms that supply off-the-shelf materials, such as clones and restriction enzymes, have relieved molecular biologists of some of the laboratory tedium. Technologies—automatic sequencers and synthesizers, for example—that were innovated in academic environments have emancipated them from a good deal else.

The genome project's technological emphasis has undoubtedly contributed to its Big Science image—and helped fuel the celebration of conventional, small-scale research in the biomedical sciences as a preferable alternative. The portrayal of the individual investigatory enterprise as small science seems to lack informed perspective: the capital and operating costs, number of graduate students, postdoctoral fellows, and technicians, laboratory space and equipment surely combine to make the enterprise large-scale, if not altogether big, science when compared with, say, Thomas Hunt Morgan's microscopes, jars of fruit flies, supply of rotting bananas, and handful of graduate students, or Watson and Crick's tinker-toy and paper-cutout modeling of hypotheses for the structure of DNA.

Unfortunately, the way that the genome project has been identified with Big Science in scientific journals and the press has tended to cloud matters. The discussion has been selective—the genome project is Big Science but the AIDS program, which spends far more annually, is somehow not—and it has been less discriminating than one would wish. The project has been lumped together with the Superconducting Supercollider and the space station—efforts that involve not only big money but big machines and big organizations.[14] The fact of the matter is that the genome project is a type of Big Science, but it is not the type that its critics deplore.

Big Science has come in significantly different versions. A simple taxonomy of the genre might break it down, in the American context, into three different forms: centralized, federal, and mixed, each suitable to its function. The centralized form has been characteristic of big technological missions—for example, the Manhattan Project to develop atomic bombs; the Apollo Program to land men on the Moon; or the current program to build, launch, and operate a space station. The features of big-mission science include centralized control of widespread efforts to produce and operate a major technological system.

The federal form has been typical of research aimed at the acquisition of knowledge concerning big subjects—for example, a physical or geological map of the continent, a catalogue of stars and galaxies, or the intricacies of major diseases such as cancer. Programs seeking this sort of knowledge have been marked by coordinated encouragement of local initiative, by pluralist, decentralized efforts to develop as necessary the tools necessary to the task, and by integration of the information obtained into some systematically organized data base.

The mixed form has been a standard feature of big-facility science—research programs that depend on major technological instruments such as high-energy particle accelerators, planetary probes, or arrays of radio telescopes. The creation, maintenance, and operation of the facility fall under the centralized control and direction of a large cadre of scientists and engineers (in the case of the Superconducting Supercollider, of an army of them, organized to design and develop the accelerator as well as its detector technologies). However, the uses of the facility for research are federally determined, the product of pluralist initiatives by research groups distributed in a variety of institutions.

The human genome project falls in neither the centralized nor the mixed categories but in the federal form of Big Science. As such, it is by no means new in the annals of scientific investigations sponsored by the American government. It finds precedent, for example, in the program of the United States Coast Survey, which was established, in 1807, to map the country's coasts and was extended eventually to include a geodetic map of the country, a task that came to involve investigations in many geographical regions. Similarly, in 1879, Congress established the United States Geological Survey under the directorship of John Wesley Powell, who organized it to draw a geological map of the western interior, inaugurating a research program conducted partly by agency em-

ployees and partly through a grant system that drew many different scientists—geologists, paleontologists, and mineralogists—
into the endeavor.[15]

The big subject of the genome project is, of course, the map
and sequence of the human genome. It has proceeded largely by
awarding grants for research to many small groups of scientists
scattered across the country for work initiated by them on problems and organisms relevant to the overall goal. In 1991, for example, NIH funded some 175 different genome projects, at an average grant size of $312,000 a year (about 1.5 times the average
NIH grant for basic research and about equal to the average AIDS
research grant). NIH has also, to be sure, established eight centers
to foster interdisciplinary work on special aspects of technological
development and large-scale mapping and sequencing, but—like
the individual research grants—the centers have been supported
on a competitive, peer-reviewed basis and are, in any case, mod-
est in size. The largest 1991 center budget amounted to $4 million
and was spread over several individual research groups.[16] It
should be clear from the numbers and mode of operation that
what characterizes the human genome project is not central direction, hierarchy, and concentration but loose coordination, local
freedom, and programmatic as well as institutional pluralism.

It should also be evident that the purposes of the genome project fall within a historical tradition of technological and methodological innovation that has paid high dividends to the biomedical
sciences. It is not just that technologies such as electrophoresis or
chromatography became useful in biological research. It is that
several of them started out big, expensive, and, hence, comparatively exclusive—and that, unlike particle accelerators, they became small, cheap, widely obtainable and dispersed.[17] In the genome project, the agreement to use the method of sequence
tagged sites (STS) for identifying and locating genomic clones has
already done away with the need for establishing clone libraries,
which were originally estimated to cost $60 million over the
fifteen-year life of the genome project. Once the STS of a particular
clone is recorded in a data base, anyone can promptly re-create
the clone in his or her laboratory, whether the laboratory is big or
small. As David Botstein early remarked, "It gives the individual
investigator the power to map things. He doesn't have to join up
with Los Alamos."[18]

The genome project's commitment to devising superior sequencing technologies means, in part, attempting to increase the sequencing throughput—the number of base pairs identified per unit time. The goal is decidedly ambitious: rates a hundred- to a thousand-fold higher than the steady pace of perhaps 5,000 base pairs a day of which the automated fluorescent DNA sequencer is currently capable. The escalation in throughput will reduce the current cost of DNA sequencing—estimates range from $2 to $15 per base pair—by at least a hundred-fold, thus placing large-scale sequencing of interesting portions of any organism's DNA within the financial reach of small laboratories. A multitude of cheap sequencers would further decentralize genome studies, provide still more room for independent projects and approaches, allow easier corroboration of results, and promote still further technological innovation.

Enthusiasts of the genome project are not blind to the magnitude of its technical challenges. They recognize the difficulties and uncertainties in producing the necessary genetic maps and sequencing technologies. They are well aware that straightforward sequencing of the entire human genome cannot, as a matter of financial practicality, be accomplished unless and until the cost of sequencing a base pair is drastically reduced. Still, as uncertain outcomes in science and technology go, the genome project is a decidedly good bet—a better one, it might be argued, than many other technologically dependent enterprises. A giant accelerator that fails to work or is abandoned before completion will likely produce little if anything of scientific value. In contrast, obtaining just a fraction of the human genome sequence, particularly the fraction containing the genes for disease, will pay high scientific and medical dividends.[19]

Completely or partially successful, the genome project will yield what one observer has termed "an orgy of information," a harvest that, in the judgment of Francis S. Collins, one of the co-discoverers of the gene for cystic fibrosis, will "drive the research enterprise for at least the next 100 years." Victor McKusick has likened expanding human map and sequence information to creating a latter-day version of Vesalius' human anatomy, a compendium of comparably fundamental knowledge that will serve as a basis for the medicine of future decades. The quick location of the gene for Huntington's disease was lucky, but tracking down the gene for cystic fibrosis was highly demanding—and still more so will be finding the genetic sources for diseases that, like cystic fibrosis, arise from subtle changes in DNA but occur much less frequently than cystic fibrosis in the population. In such cases, which will probably be numerous, meticulous map and sequence data will be indispensable.[20]

Sequence information, allowing the comparison of sequence patterns across species, will open a dramatic new chapter in the study of organic evolution. It will also permit comprehensive assessment of whether most of mammalian DNA is in fact junk—a view that the Nobel laureate Paul Berg describes as expressing "a prejudiced definition of genes." Berg adds that perhaps 50 percent of the genome's sequence is genetically active, with many introns including important regulatory signals. He asks: "Shall we foreclose on the likelihood that the so-called noncoding regions within and surrounding genes contain signals that we have not yet recognized or learned to assay? Are we prepared to dismiss the likelihood of surprises that could emerge from viewing sequence arrangements over megabase rather than kilobase distances?"[21]

New techniques will be necessary for managing, storing, analyzing, and distributing the avalanche of information that mapping and sequencing will yield. Employment of the techniques will, in turn, require new kinds of biologists, men and women capable of applying sophisticated technologies and methods of data analysis to fundamental and interesting problems in biology. As Victor McKusick has noted, "The genomics laboratories will be superb settings for training a new breed of scientist—one who is prepared to capitalize on both the molecular genetics revolution and the computation revolution. These will be the leaders in biology in the 21st century."[22]

The Big Science indictment merely distracts attention from the thorny issues that do surround the mapping and sequencing of human genes. As the chapters in this book indicate, the issues are primarily economic and social in nature, and they are numerous. Although the human genome project has not created them, it has certainly exacerbated them, magnifying tensions inherent in the genome's political economy.

One of the leading sources of tension is the matter of data sharing. In general, the project's record in cooperation has been substantial nationally and, as the creation of HUGO indicates, internationally, too. Laboratories submit their genomic information to data bases, responding to the type of incentives offered by the Centre d'Etude du Polymorphisme Humain (or of mild coercion imposed by a growing number of journals, which will not publish genomic articles without proof that the authors have submitted their data electronically to GenBank, in Los Alamos). In Europe, some 35 laboratories participate in a joint network devoted to the sequencing of yeast chromosomes, and a network centered on CEPH shares clones as well as linkage-map data.[23]

However, the eagerness for scientific priority has lured some laboratories to maintain a tight hold on map and sequence data, which they will release to common data bases only after they have analyzed the information themselves.[24] When in 1989 the Japanese appeared to be foot-dragging in their support for the genome project, James Watson threatened to cut off their scientists from access to American genomic know-how, explaining that countries that did not share in the cost of the work ought not to participate in its benefits. He set the buy-in price for the Japanese at $300,000 a year. "We'll swap it, but we won't give it," Watson reportedly said of the genome data developed in the United States.[25]

Probably with Watson's blast in mind, Walter Bodmer remarked that "some Americans have a chauvinistic attitude—they think it is going to be their project." Yet what was at stake was not so much national pride as economic interest. Indeed, the human genome project seeks to serve two frequently conflicting purposes—international cooperation, which is a proxy for the ideals of open science, and national competitiveness, which turns on the acquisition and protection of self-interested advantages.[26]

To understand how this cross-purpose dynamic is likely to work, one can compare the genome initiative with another Big Science project involving technology and information—high-energy physics. The technology—the cyclotron and its derivative accelerators—originated in the 1930s. Although the cyclotron was patented in the hope that profits might come from licensing it to produce radioactive isotopes, it proved to have little commercial value either before the war or afterwards.[27] The early accelerator scientists and engineers, knowing nothing and caring less about the cyclotron patent, worked in a commercially unconstrained environment. This openness much accelerated the development of accelerators before the war, and so did a similar policy pursued by the Atomic Energy Commission after it. Both law and policy have tended to vest in the AEC and its successor agencies, including the Department of Energy, ownership of patentable inventions made in its laboratories or under its contracts and to make freely available the technologies of particle physics, including particle detectors, to scientists engaged in basic research.[28]

A similar freedom has characterized the exchange of information among high-energy physicists. Particle physicists have achieved an astonishing level of integration, at least in respect to creating, evaluating, and banking data about the properties of elementary particles. All participating physics centers, in continental Europe, the United Kingdom, and the Soviet Union, report using a data base management program and a computer language developed at Berkeley. The system works well, partly because the data base is relatively small and its users are all experts in the field. If particle physicists had a wish list, it would consist only of bigger accelerators. Whence this exemplary cooperation and consensus? The answer, according to a member of the British group: "Particle physics data have no economic or strategic worth."[29] Neither do particle physics technologies, which is why they have been internationally disseminated with such exemplary freedom.[30]

In contrast, new gene-sequencing technologies will have considerable commercial value. Indeed, a key dissimilarity to accelerator physics up to the present day is the degree to which technological innovation in the genome area has resulted from commercial activity. Several years ago, at a conference at Cold Spring Harbor on human molecular biology, some 25 percent of the papers came from the corporate sector. To be sure, patenting of inventions makes information about them public; but expectation of profits discourages open discussion of technical detail during the critical R&D phase before patent filing. It has been said that commercial considerations have interfered with the free exchange of results and ideas about genome research.[31]

Similar problems could well invade technologically oriented academic work in the project, since federal policy now encourages investigators in nonprofit institutions to collaborate with commercial firms and permits nonprofit recipients of federal research grants to patent inventions made in their laboratories. Two reviewers of progress and prospects in DNA sequencing have noted that "a disproportionate amount of the available literature is in the form of patent documentation."[32]

The genome project is breaking new ground in that, unlike the information of particle physics, the data it is generating also have high commercial potential. Recall, for example, that the sequences of genes reveal the existence and composition of particular proteins: some of the proteins so identified may have enormous therapeutic and, hence, market value. The raw sequence as such is a product of nature, which makes it unpatentable under American and most western European law. What is patentable are products and processes devised by human beings. American courts have interpreted that standard to mean that naturally occurring substances such as vitamins can be patented if they are isolated and purified. Thus a natural protein genetically engineered from the sequence and purified could be patentable—a fact that could encourage scientists to keep sequence information secret long enough to manufacture and claim a property right in the protein.[33]

As with proteins, so even with genes themselves. Genes are products of nature, at least as they occur with their introns and exons in the cellular chromosomes. However, the so-called copy DNA (cDNA) version of the gene, with the introns edited out, does not occur naturally. It is coded into messenger RNA by the process that reads the raw cellular DNA, but it is not itself physically realized in the cell. Since it can be physically realized by a devising of human beings, using the enzyme reverse transcriptase, it is patentable. Several cDNA genes—for example, the gene for human insulin—have been incorporated into bacterial plasmids by genetic engineers and patented in that combined form.[34]

The methods and technologies of the genome project will undoubtedly accelerate the patenting of cDNA sequences. Since the typical human cDNA ranges in size from 1,000 to 8,000 base pairs, advanced sequencing machines will permit a single laboratory to specify hundreds, and perhaps thousands, of them in a year. A cDNA can also be identified by the method of expressed sequence tags, which identifies it by a sequence of only 400 to 500 base pairs. One can obtain such short sequences very quickly; indeed, a single automated sequencer could sequence more than 5,000 cDNAs annually. Britain, France, and Japan have resolved to concentrate their sequencing efforts on cDNAs—and, where possible, to patent every cDNA that is completely sequenced. Some lawyers have speculated that obtaining the expressed sequence tags of cDNAs might be enough to identify them for a patent.[35]

The lawyers may well be mistaken about the patentability of cDNAs defined merely by expression site tags; such specification for a particular cDNA would not prevent other researchers from obtaining the same cDNA from the genome by another means—for example, by using a different expression site tag. Still, since cDNAs specified by their full sequences can be patented, the technological acceleration of sequencing rates that is already under way could lead to rampant patenting—to a biological gold rush resulting in the staking of patent claims on the cDNA form of virtually every gene in the human genome.

The prospect is, to say the least, troubling. At first glance, the reasons might seem obvious: if anything is literally a common birthright of human beings, it is the human genome. It would thus seem that if anything should be avoided in the genomic political economy, it is a war of patents and commerce over the operational elements of that birthright. Indeed, the European Community has decided not to allow its genome-project contractors to exploit on an exclusive basis any property rights in human DNA.[36] Yet the obvious reasons do not in and of themselves constitute a sufficient moral or economic argument against the patenting of cDNAs.

The primary purpose of the patent system is to encourage technological innovation. To that end, the criteria for patentability include the requirements that an invention must be nonobvious to practitioners in the field and must lend itself to some utility—that is, have a practical use. It would seem defensible to patent a cDNA in connection with a determination of its function and its manipulation to some practical end. Indeed, the absence of such patentability would chill the investment of time, money, and expertise necessary to develop cDNA biotechnology. However, it would be a perversion of the patent system to allow the patenting of cDNAs as such, independent of any utility except the trivial and obvious one of obtaining the gene that it encodes. Awarding such patents would be tantamount to granting gold-mining claims on parcels of land that are unworked, a practice prohibited by both custom and public policy. It would also amount, de facto, to granting patents merely on genomic information, which would pervert the fundamental purpose of the patent system because it would place obstacles to the use of the information for technological development.

Genomic information—of human beings or any other organism—is what is in principle common property. It should be maintained as such as a matter of practical equity, since the mapping and sequencing of genomes will be—is already—the product of the ingenuity of a multinational community of scientists and of investments by many countries. Hard and imaginative thought needs to be given to means of preserving what is rightly common property while providing incentives for private development of research results for human benefit. For example, an international corporation might be established that would hold patents on human cDNA as such—if such patents are allowed—license the patents on an auction basis to those who would develop them, and plow the proceeds back into basic research. Whatever the particular means, one of the fundamental challenges to the political economy of the genome is how to achieve and maintain international cooperation in the face of the high commercial stakes in genomic information and technologies.[37]

In April 1991, an exposition opened in the hall atop the great arch of La Defense, in Paris, under the title *La Vie en Kit: Éthique et Biologie*. This exhibit concerning "life in a test tube" included displays about molecular genetics and the human genome project. The ethical worries were manifest in a statement by the writer Monette Vaquin that was printed in the catalogue and was also prominently placarded at the genome display:

> Today, astounding paradox, the generation following Nazism is giving the world the tools of eugenics beyond the wildest Hitlerian dreams. It is as if the preposterous ideas of the fathers' generation haunted the discoveries of the sons. Scientists of tomorrow will have a power that exceeds all the powers known to mankind: that of manipulating the genome. Who can say for sure that it will be used only to avoid hereditary illnesses?[38]

Vaquin's apprehensions, echoed frequently by scientists and social analysts alike, indicate that the shadow of eugenics continues to hang over the genome project. Commentators have suggested that the project may stimulate state attempts at positive eugenics, the use of genetic engineering to foster or enhance characteristics such as scholastic, scientific, and mathematical intelligence, musical ability, or athletic prowess. The ultimate goal will be the creation of new Einsteins, Mozarts, or Kareem Abdul-Jabbars (curiously, brilliantly talented women—such as Marie Curie or Nadia Boulanger or Martina Navratilova—are rarely if ever mentioned in the pantheon of superpeople). Other commentators have warned that the project will more likely spark a revival of negative eugenics—state programs of intervention in reproductive behavior so as to discourage the transmission of "bad" genes in the population.

Negative-eugenic programs could well be prompted by economic incentives. Concern for financial costs played a role in the eugenics movement of the early twentieth century, when social pathologies were said to be increasing at a terrible rate. At the Sesquicentennial Exposition in Philadelphia, in 1926, the American Eugenics Society exhibit included a board that, in the manner of the population counters of a later day, revealed with flashing lights that every fifteen seconds a hundred dollars of the observer's money went for the care of persons with "bad heredity" and that every forty-eight seconds a mentally deficient person was born in the United States. The display implied that restricting the reproduction of people with deleterious genes would not only benefit the gene pool but reduce state and local expenditures for "feeblemindedness" in public institutional settings—that is, state institutions and state hospitals for the mentally deficient and physically disabled or diseased. Perhaps indicative of this reasoning is that, in California and several other states, eugenic sterilization rates increased significantly during the 1930s, when state budgets for the mentally handicapped were squeezed.[39]

In our own day, the more that health care becomes a public responsibility, payable through the tax system, and the more expensive this care becomes, the greater the possibility that taxpayers will rebel against paying for the care of those whom genetics dooms to severe disease or disability. Public policy might feel pressure to encourage, or even to compel, people not to bring genetically disadvantaged children into the world—not for the sake of the gene pool but in the interest of keeping public health costs down.

Eugenic promptings might also come from scientists, who, having been lured by ideas of biological imperatives in the past, could find them equally seductive in the future. It is worth bearing in mind that eugenics was not an aberration, the commitment merely of a few oddball scientists and mean-spirited social theorists. It was embraced by leading biologists—not only of the political right but of the progressive left—and it was integral to the research programs of prominent, powerful institutions devoted to the study of human heredity. Indeed, eugenics remained a powerfully attractive idea even after the social prejudice of its early form was recognized and exposed. Objective, socially unprejudiced knowledge is not ipso facto inconsistent with eugenic goals of some type. Indeed, such knowledge may assist in seeking them. The enrichment of human genetics by molecular biology moved Robert Sinsheimer, in 1969, to raise with enthusiasm the possibility of a "new eugenics"—a eugenics that could be free of social bias and, as a result of DNA engineering, scientifically achievable. The more that is learned in the future about human genetics, the more might some biologists be tempted to reunite it with eugenic goals.

In recent years, crude eugenic policies have been promulgated by several governments. In Singapore in 1984, Prime Minister Lee Kwan Yew deplored the relatively low birth rate among educated women, resorting to the fallacy that their intelligence was higher than average and that they were thus allowing the quality of the country's gene pool to diminish. Since then, the government has adopted a variety of incentives—for example, preferential school enrollment for offspring—to increase the fecundity of educated women, and it has offered a similar incentive to their less-educated sisters who would have themselves sterilized after the birth of a first or second child. In 1988, China's Gansu Province adopted a eugenic law that would—so the authorities said— improve "population quality" by banning the marriages of mentally retarded people unless they first submit to sterilization. Since then, similar laws have been adopted in other provinces and have been endorsed by Prime Minister Li Peng. The official newspaper *Peasants Daily* explained, "Idiots give birth to idiots."[40]

Geneticists know that idiots do not necessarily give birth to

idiots and that mental retardation may arise from many nongenetic causes. Analysts of civil liberty also know that reproductive
freedom is much more easily curtailed in dictatorial governments
than in democratic ones. Eugenics profits from authoritarian-
ism—indeed, almost requires it. The institutions of political democracy may not have been robust enough to resist altogether
the violations of civil liberties characteristic of the early eugenics
movement, but they did contest them effectively in many places.
The British government refused to pass eugenic sterilization laws.
So did many American states, and where eugenic laws were enacted, they were often unenforced. It is far-fetched to expect a
Nazi-like eugenic program to develop in the contemporary United
States so long as political democracy and the Bill of Rights continue in force. If a Nazi-like eugenic program becomes a threatening reality, the country will have a good deal more to be worried
about politically than just eugenics.

What makes contemporary political democracies unlikely to em-
brace eugenics is that they contain powerful anti-eugenic constitu-
encies. Awareness of the barbarities and cruelties of state-
sponsored eugenics in the past has tended to set most geneticists
and the public at large against such programs. Geneticists today
know better than their early-twentieth-century predecessors that
ideas concerning what is "good for the gene pool" are highly
problematic. (We might add, however, that even though they
know better, they may not know enough and that, given the
human genome project, education in the social and ethical implications of genetic research and genetic claims should probably
become a required part of every biologist's professional training.)
Then, too, although prejudice continues against persons living
with a variety of disabilities and diseases, today such people are
politically empowered, as are minority groups, to a degree that
they were not in the early twentieth century. For example, in
1990 they obtained passage of the Americans with Disabilities Act,
which, among other things, prohibits discrimination against disabled people in employment, public services, and public accommodations. They may not be sufficiently empowered to counter
all quasi-eugenic threats to themselves, but they are politically
positioned, with allies in the media, the medical profession, and
elsewhere, to block or at least to hinder eugenic proposals that
might affect them.

The advance of human genetics and biotechnology has created the capacity for a kind of "homemade eugenics," to use the insightful term of the analyst Robert Wright—"individual families deciding what kinds of kids they want to have." At the moment, the kinds they can choose are those without certain disabilities or diseases, such as Down's syndrome or Tay-Sachs. Most parents would probably prefer a healthy baby. In the future, they might have the opportunity—for example, via genetic analysis of embryos—to have improved babies, children who are likely to be more intelligent or more athletic or better looking (whatever that might mean).[41]

Will people exploit such possibilities? Quite possibly, given the interest that some parents have shown in choosing the sex of their child or that others have pursued in the administration of growth hormone to offspring who they think will grow up too short. Benedikt Härlin's report to the European Parliament on the human genome project noted that the increasing availability of genetic tests was generating increasingly widespread pressure from families for "individual eugenic choice in order to give one's own child the best possible start in a society in which hereditary traits become a criterion of social hierarchy." A 1989 editorial in *Trends in Biotechnology* recognized a major source of the pressure: "'Human improvement' is a fact of life, not because of the state eugenics committee, but because of consumer demand. How can we expect to deal responsibly with human genetic information in such a culture?"[42]

However, genetic enhancement would inevitably involve the manipulation of human embryos, and, for better or for worse, human-embryo research faces governmental prohibitions in the United States and powerful opposition in virtually all the major western democracies, especially from Roman Catholics. The European Parliament did resolve in 1989 to allow for research on human embryos, but only under very restricted circumstances—for example, only if it would be "of direct and otherwise unattainable benefit in terms of the welfare of the child concerned and its mother." The Parliament's action was based on a report from its Committee on Legal Affairs and Citizens' Rights entitled *Ethical and Legal Problems of Genetic Engineering and Human Artificial Insemination*. The rapporteur for the section of the report concerned with genetic engineering was Willi Rothley, who is not only a Green but a Catholic, and the report itself argued against genetic manipulation of the embryo on several philosophical grounds, including the claim that "each generation must be allowed to wrestle with human nature as it is given to them, and not with the irreversible biological results of their forebears' actions."[43]

The idea of human genetic engineering as such offends many non-Catholics, too. A broad spectrum of lay and religious opinion on both sides of the Atlantic agrees with the European Parliament's 1989 declaration that genetic analysis "must on no account be used for the scientifically dubious and politically unacceptable purpose of 'positively improving' the population's gene pool" and its call for "an absolute ban on all experiments designed to reorganize on an arbitrary basis the genetic make-up of humans."[44] In any event, human genetic improvement is not likely to yield to human effort for some time to come. While the human genome project will undoubtedly accelerate the identification of genes for physical and medically related traits, it is unlikely to reveal with any speed how genes contribute to the formation of those qualities—particularly talent, creativity, behavior, appearance—that the world so much wants and admires. The idea that genetic knowledge will soon permit the engineering of Einsteins or even the enhancement of general intelligence is simply preposterous.[45] Equally important, the engineering of designer human genomes is not possible under current reproductive technologies and is not likely in the near future to become much easier technically.

The prospect and possibilities of human genetic engineering remain tantalizing, of course, even if they are still the stuff of science fiction, and they will continue to elicit both fearful condemnation and enthusiastic speculation. However, the near-term ethical challenges of the human genome project lie neither in private forays in human genetic improvement nor in some state-mandated program of eugenics. They lie, as several of the chapters in this book reveal, in the grit of what the project will produce in abundance: genetic information. They center on the control, diffusion, and use of that information within the context of a market economy, and they are deeply troubling.

Numerous individuals and families already seek genetic information. In 1990, genetic testing was commonplace enough to warrant assessment in an issue of *Consumer Reports*. Yet the acquisition of certain information may produce wrenching ripple effects. A test may reveal, for example, that collateral family members such as siblings possess a disease gene—that for Huntington's disease, say—for which no treatment, let alone cure, is known.[46] Genetic counseling may help a couple make important reproductive decisions, but post-conception genetic tests may just show that the fetus has lost the roll of the genetic dice. The pregnant couple is then confronted with the only therapeutic choice available at the moment—to abort or not abort a child that is usually wanted. The problem can be compounded by uncertainty: for example, the test for whether an individual is a carrier of the recessive cystic fibrosis gene is only 75 percent reliable—that is, it detects the gene in only three out of four of the people who carry it. In consequence, it reveals only 56 percent (75 percent times 75 percent) of the couples who are truly at risk for bearing a child with the disease, which is to say that it misses 44 percent of such couples. Even if couples are well counseled about the meaning of the tests—a considerable and costly task in and of itself—most will face anxiety about how to act on the results.[47]

The more disease genes that RFLP-mapping and other technologies pinpoint, the greater the number of people who will be drawn into the testing network. Many people may not wish to obtain their genetic profiles, particularly if they are at risk for an inheritable disease for which no treatment is known, but commercial and medical interests may pressure them to be tested anyway. It has been estimated that the potential market for genetic carrier screening and prenatal testing is enormous, including eventually 2.8 million people getting themselves tested each year to learn whether they are carriers of the recessive genes for cystic fibrosis, sickle cell anemia, hemophilia, and muscular dystrophy. The physicians Benjamin S. Wilfond and Norman Fost, noting that more than 8 million Americans may be carriers of the cystic fibrosis gene alone, have observed that "entrepreneurial interests can be expected to promote screening in what could become a billion-dollar industry."[48] Still, genetic testing, prenatal or otherwise, can be liberating if it reveals to individuals that either they or their newly conceived children are safe from some specific genetic doom. A young woman tested and found to be without the gene for Huntington's declared, "After 28 years of not knowing, it's like being released from prison. To have hope for the future . . . to be able to see my grandchildren."[49]

The torrent of new human genetic information will undoubtedly pose challenges across a broad spectrum of socioeconomic values and practices. It has been rightly emphasized that employers and medical or life insurers may seek to learn the genetic profiles of, respectively, prospective employees or clients. Employers might wish to identify workers likely to contract disorders that allegedly affect job performance or whose onset could be brought about by features of the workplace. Both employers and insurers might wish to identify people likely to fall victim to diseases that result in costly medical or disability payouts. The employers could use the information to assign susceptible people to risk-free duties or environments. They might also use it to deny them jobs, just as medical or life insurers might exploit it to exclude them from coverage. Whatever the purpose, genetic identification would brand people with what an American union official has called a lifelong "genetic scarlet letter" or what some Europeans term a "genetic passport."[50] Benedikt Härlin's report on the genome proposal to the European Parliament warned that parents, customers, and employees might be subject to pressure for genetic testing by "health authorities, insurance institutions, employers and other bodies," and declared that genetic knowledge so obtained would be in "every sense monstrous."[51]

A good deal of evidence suggests that apprehensions about the use of genetic information are not unfounded. Around 1970, it came to be feared that people with sickle cell trait—that is, who possess one of the recessive genes for the disease—might suffer the sickling of their red blood cells in the reduced-oxygen environment of high altitudes. They were prohibited from entering the Air Force Academy, restricted to ground jobs by several major commercial air carriers, and often charged higher premiums by insurance companies.[52] Recently, a couple whose first child suffers from cystic fibrosis became pregnant and sought to have their fetus diagnosed prenatally for the disease. Their medical insurer agreed to pay for the test so long as the mother would abort the second child if the results were positive. Otherwise, the company would cancel the family's health plan. (The company relented, but only under threat of a law suit.)

A good deal of the genetic discrimination so far appears to have been arbitrary, callous, and, especially in the employment area, a product of ignorance—for example, taking the presence of a single recessive disease gene as evidence that the applicant is vulnerable to the workplace environment. A current survey originated by several Harvard Medical School faculty turned up some thirty instances of genetic discrimination. People with inherited bio- chemical disorders were denied insurance even though they had been successfully treated and were not ill. An auto insurer refused to cover a man with a genetically based neuromuscular disorder who suffered no disability, and an employer declined to hire a woman after she revealed that she had the same disorder. Paul Billings, a medical geneticist and one of the survey authors, noted that the study was not methodologically of the kind to determine whether these agencies "have active policies of genetic discrimina- tion," but he added that the findings "suggest that such policies exist."[53]

A number of commentators have argued that employers and insurers ought to be prohibited from nosing into anybody's geno- mic passport. In 1991, the California state legislature passed (and the governor vetoed) a bill banning employers, health service agencies, and disability insurers from withholding jobs or protec- tion simply because a person is a carrier of a single gene associated with disability. About the same time, the possibility of a similar prohibition was raised in Britain by Baronness Warnock, who was earlier a leading figure in the shaping of a British policy for embryo research. However, insurers could sidestep such a prohibition by setting common rates high and offering discounts to clients with healthy genetic profiles, which such clients would, of course, sub- mit voluntarily.[54] Insurers have a natural interest in information that bears on risk. To them, rate discrimination based on genuine knowledge of risk is neither aribtrary nor illegitimate: it is sound actuarial and business practice.

The prevailing view in the industry, which has become increas- ingly well informed about genetic disease, is no doubt exemplifed by a June 1989 report entitled "The Potential Role of Genetic Test- ing in Risk Classification" that was largely prepared by Robert Pokorski and circulated by the American Council of Life Insur- ance. The report noted: "If insurers were unable to use genetic tests during the underwriting process because 'risks should only be classified on the basis of factors that people can control,' then equity would give way to equality (equal premiums regardless of risk) and private insurance as it is known today might well cease to exist."[55]

Industry representatives hold that equality would disadvantage not only insurance providers but also many of the people they cover. If a client has a high genetic medical risk that is not reflected in her premiums, then she would receive a high payout at low cost to herself but high cost to the company. The problem would be compounded if she knows the risk—while the company does not—and she purchases a large amount of insurance. In either case, the company would have to pass its increased costs along to other policy holders, which is to say that high-risk policyholders would be taxing low-risk ones.[56]

To keep to the principle of equity, insurance companies want to know about their clients at least what the clients know, genetically and otherwise, about themselves. They may also decide to pursue the principle of equity further and to require genetic testing of clients so that they can tailor rates to risk. The industry rightly, and somewhat ruefully, expects consumer resistance. Rob Bier, the managing director of communications for the American Council of Life Insurance, has remarked, "It seems unavoidable that there will be lots of legal battles as this technology unfolds. The insurance industry actually wishes genetic testing had never been developed."[57]

The legal battles could grow more heated in response to the accumulating data of the human genome project. More detailed understanding of the relationship between genetics and disease will increase the specification of an individual's risk to the point, perhaps, where risk becomes certainty and lifetime medical costs can be exactly calculated. In that case, medical insurance premiums would amount to payment for lifetime medical care on the layaway plan. Thus, the acquisition of human genetic information could not only accelerate the movement from community-based to experience-based ratings. It could also establish medical and, possibly, life insurance on a what-you-are base, with a policy specific to each genomic essence.[58]

Alternatively, the human genome project could help move medical insurance back to a community-based scheme. The more that is learned about the human genome, the more will it become obvious that everyone is susceptible to some kind of genetic disease or disability; everyone carries some genetic load and is likely to fall ill in one way or another. Of course, the cost and severity of the illnesses will vary, but everyone's being aware of his or her genetic jeopardy might well increase interest in a rating system that emphasizes equality rather than equity, that expresses what the Europeans call solidarity. According to G. W. de Wit, a professor of insurance economics at Erasmus University in the Netherlands, equality has long operated to a significant degree in the private insurance sector and will continue to prevail amid the availability of genetic information. He writes that if, for example, parents with a genetically diseased fetus choose to have the child, "all medical expenses for that child will be borne by the insurer," adding, "It seems fully justified to have the other policy holders contribute (solidarity), because otherwise the free choice of the parents is jeopardized." De Wit doubts that European medical insurers will demand genetic information from clients, though he notes: "If the private insurance industry is to be maintained in the present form, it will be necessary to balance continually and carefully the theoretial risk and social needs, both in private insurance and social insurance."[59]

Social insurance—that is, national health insurance—is the ultimate form of solidarity, and the human genome project, by revealing how everyone is in genetic jeopardy, might contribute to bringing about some form of it in the United States. All the same, genetic information could undermine national health insurance, too. If the cost of medical services continues to escalate, even national health systems might choose to ration the provision of care on the basis of genetic propensity for disease, especially to families at risk for bearing diseased children.[60]

To suggest that a genetically based disease or disorder is unacceptably costly is to cast a shadow over people who suffer from it. Already, the attitude that a newly conceived child with such an affliction merits abortion has been attacked as stigmatizing the living who have the ailment. Protests have come from individuals and families with diseases such as cystic fibrosis and sickle cell anemia, but especially from the handicapped and their advocates. Barbara Faye Waxman, an activist for the disabled who herself has a neuromuscular impairment, criticizes her fellow workers in a Los Angeles Planned Parenthood clinic for displaying "a strong eugenics mentality that exhibited disdain, discomfort and ignorance toward disabled babies." In the European Parliament, the Committee on Legal Affairs has warned against seeing the birth of handicapped children "only as an avoidable technical error," pointing out that selective abortion of the handicapped "not only undermines our ability to accept the disabled but also makes no significant impact on the problem of disability." In the United States, some advocates for the disabled have joined the anti-abortion movement.[61] It would seem to make little sense to seek to preserve the dignity of one group by limiting the reproductive freedom of another. What would make a good deal more sense is to recognize that values of social decency compel us to live in a state of conflicted practices—endorsing the use of genetic information in personal reproductive choices while upholding the rights and dignity of the diseased and disabled.

Genetic stigmatization has come in many varieties, but the most reckless has involved claims about the linkage of genes with behavior. The early eugenics movement stigmatized recent immigrant groups from eastern and southern Europe by declaring them biologically inferior in intelligence and inclined to criminality, alcoholism, prostitition, and the like. Eugenic science was obviously riddled with social prejudice, but even purified of social prejudice and vagueness of categories, behavioral genetics poses difficult problems—not only of distinguishing between nurture and nature but, equally important, of defining behavioral traits, measuring them, and recognizing spurious correlations. In its ongoing fascination with questions of behavior, human genetics will undoubtedly yield information that may be wrong, or socially volatile, or, if the history of eugenic science is any guide, both.

The search for genetic origins of human behavior continues, but while it is a scientifically legitimate goal, it also persists in demonstrating to a considerable degree that the quest is both scientifically and socially treacherous. For example, in recent years, several RFLP-family studies appeared to reveal specific genetic susceptibilities to manic depression and schizophrenia, but follow-up studies failed to confirm the initial results. At Washington University, St. Louis, the psychiatrist C. Robert Cloninger and his colleague Eric Devor, having scrutinized the families of alcoholics and the adopted sons of alcoholics, proposed a broad genetic theory of propensity to two types of alcoholism, each of which they tie to a chemically based set of personality traits. The traits include tendencies to novelty seeking and exploratory activity; fearfulness and shyness; reward dependency and social coolness.[62]

Cloninger and Devor's attentiveness to the alleged genetic foundation of such traits is by no means anomalous. In 1990, the Harvard psychologist Jerome Kagan reported that among a group of 379 students, those who suffered hay fever also scored high on a shyness index. "We think there is a small group of people who inherit a set of genes that predispose them to hay fever *and* shyness," Kagan said.[63] Whatever the merits of conclusions like these, one hears in such assessments echoes of earlier trait categories like love of the sea. The plasticity of these categories, not to mention of personality traits in general, would suggest the need for considerable caution in both the conduct of behavioral genetics and the promulgation of its claims, especially by the media.

It was front-page news in 1990 when researchers at UCLA and in Texas jointly announced that, having examined the brains of all of seventy deceased people—half of them severe alcoholics, half of them not—they had detected a gene for alcoholism. (It was page-ten news in *The New York Times* when, in December 1990, scientists at the National Institutes of Health reported that they could not confirm the UCLA/Texas results.)[64] Reporters often take as firm conclusions what scientists announce as tentative conclusions, yet scientists are complicitous in the process when they hold press conferences to proclaim attention-getting results in the behavioral area, however fragile they may be. These mutually misleading tendencies will no doubt be exacerbated as the flow of information from mapping and sequencing the human genome increases—which suggests that both scientists and the press could well attend to developing an ethics for handling socially charged genetic information, particularly of a racial, ethnic, or sexual type. Such an ethics might be founded on the proposition that the greater the social explosiveness of the conclusions, the more demanding should be the requirements of rigor and reliability for publishing and, especially, promoting them.

It is not only politically wise but in principle right to recognize that the human genome project has to be pursued in conjunction with some degree of ethical assessment and constraint. James

Watson's successful insistence that the project include a related program of ethical analysis—the first such venture in the annals of major scientific research initiatives in the United States—has been repeated in Europe. In response to arguments raised in the European Parliament, the European Community's genome project has established a group for the exploration of ethical issues.[65] In the last several years, the genome project has helped stimulate widespread consideration of bioethical questions at scientific meetings, international congresses, working groups, and in the press. The gatherings and the analyses have cultivated the reflections of "scientists, lawyers, physicians, theologians, philosophers, opinion leaders, writers, and journalists," to cite the observation of Claude Cheysson, the president of the Fondation l'Arche de la Fraternité, the arch at La Defense, who adds that the reflections represent not only "current average thinking but also the predictions of charlatans and makers of miracles."[66] The fears that the genome project will foster a drive for the production of superbabies or the callous elimination of the unfit are grossly exaggerated. They also divert attention from the scientific and social issues that the project actually raises—particularly how human genetic information should be used by geneticists, the media, insurers, employers, and government—and that are tangled enough to challenge any society's capacities for knowledgeable judgment and sympathetic tolerance.

Postscript to the Paperback Edition

Since the completion of this book, in September 1991, the social issues raised or exacerbated by the human genome project have received increasing attention in both the private and public sectors, though without alteration of their fundamental nature. In the same period, the technical pace of the genome project has been accelerated by advances in mapping and sequencing techniques.

In France, the biologist Daniel Cohen has devised an industrial approach to physical mapping that starts with the insertion of one-megabase segments of human DNA into yeast artificial chromosomes: the use of such huge lengths of DNA substantially reduces the difficulty of determining the order of the segments along the chromosome; and the large size lends itself to physical mapping with assembly-line methods. Cohen has already produced a complete map of chromosome 21. In the United States, Eric Lander has pioneered genetic mapping techniques that utilize simple repeat polymorphisms comprising, for example, two bases—say, AC—that are tandemly repeated from five to perhaps fifty times in different members of the population. Since these repeat sequences are easily identified and are variable (informative) within the human population, they will enable the rapid creation of a genetic map with the comparatively high density of two centimorgans.

The approaches of both Cohen and Lander benefit from economies of scale and thus require a certain concentration of effort. Similar advantages have been recognized in DNA sequencing, with the result that several institutes devoted to large-scale sequencing have been recently established. The Sanger Institute, in Cambridge, England, expects to achieve an overall sequencing rate of 30 to 40 megabases in its first year of operation, and its leaders believe that this throughput can be improved significantly in succeeding years.

The need for concentrating certain types of work at large centers with a wide range of interdisciplinary scientific skills does not imply that the genome project will become Big Biology in the manner that Fermilab is Big Physics. For example, the physical mapping capabilities of Cohen's center form a resource for a network of dozens of smaller groups on both sides of the Atlantic. What will probably continue to characterize the human genome project is not central direction, hierarchy, and concentration but loose coordination, local freedom, and programmatic as well as institutional pluralism.

Notes

1. Out of Eugenics

1. Francis Galton, *Inquiries into the Human Faculty* (London: Macmillan, 1883), pp. 24–25; Karl Pearson, *The Life, Letters, and Labours of Francis Galton*, 3 vols. in 4 (Cambridge: Cambridge University Press, 1914–1930), vol. IIIA, p. 348.

2. Historical accounts of eugenics, which itself produced a vast literature, have multiplied in recent years. For treatments of the subject in the United States and Britain, see Daniel J. Kevles, *In the Name of Eugenics: Genetics and the Uses of Human Heredity* (New York: Alfred A. Knopf, 1985); Kenneth M. Ludmerer, *Genetics and American Society: A Historical Appraisal* (Baltimore: Johns Hopkins University Press, 1972); and G. R. Searle, *Eugenics and Politics in Britain, 1900–1914* (Leyden: Noordhoff International Publishing, 1976). For Germany, see Benno Müller-Hill, *Murderous Science: Elimination by Scientific Selection of Jews, Gypsies, and Others, Germany, 1933–1945* (New York: Oxford University Press, 1988); Robert N. Proctor, *Racial Hygiene: Medicine under the Nazis* (Cambridge, Mass.: Harvard University Press, 1988); Sheila Faith Weiss, *Race Hygiene and National Efficiency: The Eugenics of Wilhelm Schallmayer* (Berkeley: University of California Press, 1987); and Paul Weindling, *Health, Race and German Politics between National Unification and Nazism, 1870–1945* (Cambridge: Cambridge University Press, 1990). For France, see William H. Schneider, *Quality and Quantity: The Quest for Biological Regeneration in Twentieth-Century France* (Cambridge: Cambridge University Press, 1990); and Gérard Lemaine and Benjamin Matalon, *Hommes supérieurs, hommes inférieurs* (Paris: Armand Colin, 1985). For Russia and elsewhere, see Mark B. Adams, ed., *The Wellborn Science: Eugenics in Germany, France,*

Brazil, and Russia (New York: Oxford University Press, 1990); Loren R. Graham, "Science and Values: The Eugenics Movement in Germany and Russia in the 1920s," *American Historical Review*, 83 (1978), 1135–1164; Nils Roll-Hansen, "Eugenics before World War II: The Case of Norway," *History and Philosophy of the Life Sciences*, 2 (1980), 269–298, and Nils Roll-Hansen, "The Progress of Eugenics: Growth of Knowledge and Change in Ideology," *History of Science*, 26 (1988), 295–331.

3. The establishment of these institutions and the research programs they pursued may be followed in Kevles, *In the Name of Eugenics;* Proctor, *Racial Hygiene;* and Sheila Faith Weiss, "The Race Hygiene Movement in Germany, 1904–1945," in Adams, ed., *The Wellborn Science*, pp. 8–68.

4. Charles B. Davenport, *Heredity in Relation to Eugenics* (New York: Henry Holt, 1911), pp. 66–67, 72–74, 77, 79–80, 93, 102, 157, 126; Charles B. Davenport, *Eugenics—the Science of Human Improvement by Better Breeding* (New York: Henry Holt, 1910), pp. 11–12; Charles B. Davenport and Morris Steggerda, *Race Crossing in Jamaica* (Washington, D.C.: Carnegie Institution, Publication No. 395, 1929), pp. 472–473, 468–469.

5. Henry H. Goddard, *Feeble-mindedness: Its Causes and Consequences* (New York: Macmillan, 1914), pp. 4, 7–9, 14, 17–19, 413, 504, 508–509, 514, 547. Davenport provided field-workers for Goddard's surveys of the mental characteristics of a local population in New Jersey and advice on genetic interpretation of the data. The surveys contributed essential material to this book and to Goddard's equally influential *The Kallikak Family: A Study in the Heredity of Feeblemindedness* (New York, 1912). Kevles, *In the Name of Eugenics*, p. 78.

6. Weiss, "The Race Hygiene Movement in Germany," pp. 44–46; Proctor, *Racial Hygiene*, esp. chaps. 4 and 7.

7. Ludmerer, *Genetics and American Society*, p. 60.

8. Davenport, *Heredity in Relation to Eugenics*, pp. 216, 218–219, 221–222.

9. Kevles, *In the Name of Eugenics*, pp. 61–62.

10. Ibid., pp. 107–112, 114–116; *Buck v. Bell*, 274 U.S. 201–207 (1927).

11. Proctor, *Racial Hygiene*, pp. 44, 292, 307

12. Kevles, *In the Name of Eugenics*, p. 199.

13. Lancelot Hogben, *Genetical Principles in Medicine and Social Science* (London: Williams and Norgate, 1931), pp. 82–84.

14. Julia Bell to Landsborough Thomson, June 10, 1936, quoted in Kevles, *In the Name of Eugenics*, p. 202.

15. Author's interview with James V. Neel, May 1982. Like other institutional opportunities in human genetics in that era, Neel's arose from a eugenic commitment. Initiative for the creation of his post seems to have come from the head of the laboratory, Lee R. Dice, a eugenically inclined ecologist who had persuaded the university to establish a small outpatient heredity clinic to help people learn whether they might have "bad" genes. Neel assumed responsibility for the clinic, and he spent part of his time exploring how the carriers of genetic disorders might be detected. Ibid.; James V. Neel, "The Detection of Genetic Carriers of Hereditary Disease," *American Journal of Human Genetics*, 26 (March 1949), 19–36.

16. James V. Neel, "The Inheritance of Sickle Cell Anemia," *Science*, 110 (1949), 64–66; Linus Pauling et al., "Sickle Cell Anemia: A Molecular Disease," Ibid., 543–548. Both papers are reprinted in Samuel H. Boyer IV, ed., *Papers on Human Genetics* (New York: Prentice-Hall, 1963), pp. 110–125.

17. T. C. Hsu, *Human and Mammalian Cytogenetics: An Historical Perspective* (New York: Springer-Verlag, 1979), pp. 27–29; Kevles, *In the Name of Eugenics*, pp. 238–250.

18. Kevles, *In the Name of Eugenics*, p. 258 and n. 21.

19. Lionel S. Penrose, "The Influence of the English Tradition in Human Genetics," in James F. Crow and James V. Neel, eds., *Proceedings of the Third International Congress of Human Genetics* (Baltimore: Johns Hopkins University Press, 1967), pp. 22–23.

20. Robert L. Sinsheimer, "The Prospect of Designed Genetic Change," *Engineering and Science*, 32 (April 1969), 8; Robert L. Sinsheimer, "The Santa Cruz Workshop—May 1985," *Genomics*, 5 (1989), 955; James D. Watson and Robert Mullen Cook-Deegan, "Origins of the Human Genome Project," *FASEB Journal*, 5 (January 1991), 8; Joel Davis, *Mapping the Code: The Human Genome Project and the Choices of Modern Science* (New York: John Wiley, 1990), pp. 88–91.

21. Robert Mullen Cook-Deegan, "The Alta Summit, December 1984," *Genomics*, 5 (1989), 661–662; Charles R. Cantor, "Orchestrating the Human Genome Project," *Science*, 248 (April 6, 1990), 50; Davis, *Mapping the Code*, pp. 95–96.

22. Davis, *Mapping the Code*, pp. 97–98; Joseph Bishop and Michael Waldholz, *Genome: The Story of the Most Astonishing Scientific Adventure of Our Time—The Attempt to Map All the Genes in the Human Body* (New York: Simon and Schuster, 1990), pp. 217–218.

23. Stanley N. Cohen, Annie C. Y. Chang, Herbert W. Boyer, and Robert B. Helling, "Construction of Biologically Functional Bacterial Plasmids *In Vitro*," *Proceedings of the National Academy of Sciences*, 70 (November 1973), 3240–44.

24. A. M. Maxam and W. Gilbert, "A New Method for Sequencing DNA," *Proceedings of the National Academy of Sciences*, 74 (1977), 560–564; F. Sanger and A. R. Coulson, "A Rapid Method of Determining Sequences in DNA by Primed Synthesis with DNA Polymerase," *Journal of Molecular Biology*, 94 (1975), 441–448; Lloyd M. Smith et al., "Fluorescence Detection in Automated DNA Sequence Analysis," *Nature*, 321 (June 12, 1986), 674–679.

25. D. C. Schwartz and C. R. Cantor, "Separation of Yeast Chromosome-Sized DNAs by Pulsed Field Gradient Gel Electrophoresis," *Cell*, 37 (1984), 67–75. Pulsed-field electrophoresis could separate lengths of DNA ranging in size from 10 to 10,000 kilobase pairs.

26. David Botstein et al., "Construction of a Genetic Linkage Map in Man Using Restriction Fragment Length Polymorphisms," *American Journal of Human Genetics*, 32 (1980), 314–331; Bishop and Waldholz, *Genome*, pp. 61–68.

27. Bishop and Waldholz, *Genome*, p. 214.

28. Ibid., pp. 100–101; James Gusella et al., "A Polymorphic DNA Marker Genetically Linked to Huntington's Disease," *Nature*, 306 (1983), 234–238.

29. Bishop and Waldholz, *Genome*, pp. 188–191; Renato Dulbecco, "A Turning Point in Cancer Research: Sequencing the Human Genome," *Science*, 231 (1986), 1055–56.

30. Robert Mullan Cook-Deegan, "The Human Genome Project: The Formation of Federal Policies in the United States, 1986–1990," in Kathi E. Hanna, ed., *Biomedical Politics* (Washington, D.C.: National Academy Press, 1991), pp. 112, 124, 129; Cantor, "Orchestrating the Human Genome Project," p. 49.

31. Cook-Deegan, "The Human Genome Project," p. 123.

32. Ibid., pp. 115–116; Tracy L. Friedman, "The Science and Politics of the Human Genome Project," Senior Thesis, Woodrow Wilson School of Public and International Affairs, Princeton University, April 1990, pp. 31–32.

33. Cook-Deegan, "The Human Genome Project," p. 118; Friedman, "Science and Politics of the Human Genome Project," pp. 31–32; Benjamin J. Barnhart, "The Human Genome Project: A DOE Perspective," BioDoc collection, European Economic Community, DG-XII, Brussels, file CB27.188, p. 2. The Brussels collection is referenced hereafter as BioDoc.

34. Cook-Deegan, "The Human Genome Project," p. 120; Friedman, "Science and Politics of the Human Genome Project," pp. 38–39; U.S. Congress, Senate, Subcommittee on Energy Research and Development, Committee on Energy and Natural Resources, *Department of Energy National Laboratory Cooperative Research Initiatives Act: Hearing before the Subcommittee on Energy Research and Development of the Committee on Energy and Natural Resources*, 100th Cong., 1st Sess., September 15 and 17, 1987.

35. Cook-Deegan, "The Human Genome Project," pp. 124–125; Watson and Cook-Deegan, "Origins of the Human Genome Project," p. 9.

36. Friedman, "Science and Politics of the Human Genome Project," pp. 32–35, 40–41, 56–59.

37. Ibid., pp. 38–42, 45; Cook-Deegan, "The Human Genome Project," pp. 134–135, 141; Watson and Cook-Deegan, "Origins of the Human Genome Project," p. 10.

38. Davis, *Mapping the Code*, pp. 126–128; Robert A. Weinberg, "The Case Against Gene Sequencing," *The Scientist*, November 16, 1987, p. 11. See also Francisco J. Ayala, "Two Frontiers of Human Biology: What the Sequence Won't Tell Us," *Issues in Science and Technology*, Spring 1987, p. 56; Christopher Joyce, "The Race to Map the Human Genome," *New Scientist*, March 5, 1987, p. 35.

39. Quoted in Friedman, "Science and Politics of the Human Genome Project," p. 95; Bishop and Waldholz, *Genome*, pp. 221–222; Joyce, "The Race to Map the Human Genome," p. 35.

40. Watson and Cook-Deegan, "Origins of the Human Genome Project," p. 9; Cantor, "Orchestrating the Human Genome Project," p. 50; Leslie Roberts, "Academy Backs Genome Project," *Science*, 239 (February 12,

1986), 725–726; Committee on Mapping and Sequencing the Human Genome, Board on Basic Biology, Commisson on Life Sciences, National Research Council, *Mapping and Sequencing the Human Genome* (Washington, D.C.: National Academy Press, 1988), pp. 1–10.

41. Cook-Deegan, "The Human Genome Project," p. 141; Friedman, "Science and Politics of the Human Genome Project," pp. 60–61, 63–64; John M. Barry, "Cracking the Code," *The Washingtonian*, February 1991, p. 182.

42. Friedman, "Science and Politics of the Human Genome Project," p. 45.

43. Walter Bodmer, "Genes and Atoms for Health and Prosperity," Presidential Address, 1988, British Association for the Advancement of Science, *Science and the Public*, September/October 1988, p. 7; European Parliament, Committee on Energy, Research, and Technology, *Report Drawn up on Behalf of the Committee on Energy, Research and Technology on the Proposal from the Commission to the Council (COM/88/424-C2–119/88) for a Decision Adopting a Specific Research Programme in the Field of Health: Predictive Medicine: Human Genome Analysis (1989–1991)*, Rapporteur Benedikt Härlin, European Parliament Session Documents, 1988–89, 30.01.1989, Series A, Doc. A2–0370/88 SYN 146, p. 21 (hereafter cited as Härlin Report); "U.S. Dominates Publishing of Genome Mapping Articles," *The Scientist*, June 13, 1988, p. 20.

44. "Frontiers Open Up," *Nature*, 347 (October 4, 1990), 413. In addition to Japan, the Group of Seven includes the United States, Canada, Britain, France, Italy, and Germany.

45. Davis, *Mapping the Code*, pp. 178–181; Akiyoshi Wada and Eiichi Soeda, "Strategy for Building an Automatic and High Speed DNA-Sequencing System," manuscript, 1986, p. 10; Leslie Roberts, "Human Genome: Questions of Cost," *Science*, 237 (September 18, 1987), 1411–12. As it turned out, Wada's expectations were greatly exaggerated. See the close of this chapter.

46. David Dickson, "Focus on the Genome," *Science*, 240 (May 6, 1988), 711; Watson and Cook-Deegan, "Origins of the Human Genome Project," p. 9; Härlin Report, p. 21; *European Technology Newsletter*, May 13, 1988, p. 5, copy in BioDoc; M. A. Ferguson-Smith, "European Approach to the Human Genome Project," *FASEB Journal*, 5 (January 1991), 63. Greece and Portugal had programs to analyze specific regional diseases; Härlin Report, p. 21.

47. Jean-Paul Gaudillere, "French Strategies in Molecular Biology," paper delivered at Harvard conference on the Human Genome Project, June 15, 1990, pp. 3, 8–9, 10–11.

48. Jean Dausset, Howard Cann, and Daniel Cohen, "Collaborative Mapping of the Human Genome," unpublished manuscript, CEPH, April 3, 1987, pp. 3–4.

49. Victor McKusick, "Mapping and Sequencing the Human Genome," *New England Journal of Medicine*, 320 (April 6, 1989), pp. 913–914; Davis, *Mapping the Code*, pp. 162–164; Walter F. Bodmer, "HUGO: The Human Genome Organization," *FASEB Journal*, 5 (January 1991), 73. The organi-

zation is also sometimes recognized informally as "Victor's Hugo," after Victor McKusick, the editor of the standard reference *Mendelian Inheritance in Man* and one of HUGO's principal founders. "Genentechnologie (IX)," *Basler Zeitung,* September 9, 1989, p. 38, BioDoc.

50. "Origines de la vie: le difficile de décryptage," *Le Figaro,* June 20, 1988; "Débat," *Biofutur,* February 1988, p. 21. The participants in the Paris debate included Professor François Gros, of the Collège de France, Daniel Cohen, vice-president of the Centre d'Etude du Polymorphisme Humain, and Jean Weissenbach, Director of Research at the Centre National de la Recherche Scientifique, Paris.

51. Ferguson-Smith, "European Approach," p. 63; Gaudillere, "French Strategies in Molecular Biology," pp. 2–3, 5–7.

52. "Bioactualité," *Biofutur,* September 1989, pp. 6–7; Sharon Kingman, "Buried Treasure in Human Genes," *New Scientist,* July 8, 1989, p. 36.

53. "Débat," *Biofutur,* February 1988, p. 20.

54. Lennart Philipson and John Tooze, "The Human Genome Project," *Biofutur,* June 1987, 101; David Dickson, "Europe Seeks Strategy for Biology," *Science,* 240 (May 6, 1988), p. 710.

55. Dickson, "Focus on the Genome," p. 711; "Débat," *Biofutur,* February 1988, p. 21.

56. The Commission called its proposal "a European response to the international challenges presented by the large-scale biological research projects in the United States . . . and Japan (Human Frontier Science Programme)," adding, "Although it is a programme of basic precompetitive research, both new information and new materials of potential commercial value will result; new technological processes will also be developed. These will all contribute to the development of Europe's biotechnology industry—often based in small and medium-sized enterprises." Commission of the European Communities, COM (88) 424 final—SYN 146, Brussels, 20 July 1988, *Proposal for a Council Decision Adopting a Specific Research Programme in the Field of Health; Predictive Medicine: Human Genome Analysis (1989–1991),* p. 1.

57. Ibid., p. 3.

58. Ibid., pp. 10, 12, 20, 30. The Commission argued formally that the genome proposal built on the Framework Program for Community Activities in the Field of Research and Technological Development (1987–1991), which had been adopted in 1986. Ibid., p. 10.

59. *The Institutions of the European Community* (Brussels: Information Services of the European Parliament and the Commission of the European Communities, European file no. 16/89, 1989), pp. 6–10; *Le Parlement européen* (Brussels: Bureau of Information of the European Parliament, July 1989).

60. Härlin Report, p. 3. Auxiliary opinions were also requested of the Committee on Budgets and the Committee on the Environment, Public Health, and Consumer Protection. Ibid.

61. London *Financial Times,* May 10, 1989, p. 18; Davis, *Mapping the Code,* p. 175; Michael Specter, "Petunias Survive German Debate over Biotechnol-

ogy," *International Herald Tribune,* April 12, 1990. The German fear of genetics and eugenics would intensify, leading some activist groups on a number of occasions to intimidate and even suppress debate on biomedical subjects in universities using methods reminiscent of the Nazis. Peter Singer, "On Being Silenced in Germany," *New York Review of Books,* August 15, 1991, pp. 36–42. In 1988, the human genome project provoked stormy debate in the Bundestag. Lecture by Hans Sass, "II Workshop on International Cooperation for the Human Genome Project: Ethics," Valencia, Spain, November 12, 1990.

62. Härlin Report, pp. 23–28.
63. Specter, "Petunias Survive German Debate over Biotechnology."
64. Härlin Report, pp. 3, 5–7, 10–11, 14. Härlin's committee was strongly supported in its position by the Committee on the Environment, Public Health, and Consumer Protection, which recommended modification of the genome project proposal to the end that the medical, ethical, legal, and social implications of such research be investigated before any specific technical projects were promoted or continued. To this committee's members, it was "quite clear that ethical problems will arise, particularly concerning eugenic problems and access to information by individuals, States, employers, insurance companies (etc.), if the programme is successful in its long term ambitions." Committee on the Environment, Public Health, and Consumer Protection, *Opinion for the Committee on Energy, Research and Technology on the Proposal from the Commission of the European Communities for a Council Decision Adopting a Specific Research Programme in the Field of Health: Predictive Medicine: Human Genome Analysis (1989–1991) (COM (88)424 final—SYN 146—Doc. C2–119/88)* (Draftsman: Mrs. Lentz-Cornette), pp. 3, 5–8.
65. Commission of the European Communities, *Modified Proposal for a Council Decision, Adopting a Specific Research and Technological Development Programme in the Field of Health: Human Genome Analysis (1990–1991)* (COM 89) 532 final—SYN 146, Brussels, November 13, 1989), p. 2.
66. London *Financial Times,* April 5, 1989, BioDoc; Dirk Stemerding, "Political Decision-Making on Human Genome Research in Europe," paper delivered at Harvard workshop on the Human Genome Project, June 15, 1990, p. 2.
67. Commission of the European Communities, *Modified Proposal,* pp. 2–4, 11–17; *Scrip,* December 8, 1989, p. 5, copy in BioDoc.
68. European Community, *Common Position Adopted by the Council on 15 December 1989 . . . Programme in the Field of Health: Human Genome Analysis (1990–1991),* Brussels, December 14, 1989 [sic], 10619/89; *Official Journal of the European Communities,* No. L 196/8, 26/7/90, Council Decision of 29 June 1990, adopting a specific research and technological development programme in the field of health, human genome analysis (1990–1991), (90/395/EEC).
69. Ferguson-Smith, "European Approach," p. 62; Watson and Cook-Deegan, "Origins of the Human Genome Project," pp. 9–10; "M. Curien

lance un programme national de recherche sur le génome humain," *Le Monde*, October 19, 1990; Tabitha M. Powledge, "Toward the Year 2005," *AAAS Observer*, November 3, 1989, pp. P1–P6; *Scrip*, September 26, 1990, p. 6; telephone conversation with Gordon Lake, of the European Parliament staff, April 24, 1991; Gaudillere, "French Strategies in Molecular Biology," p. 4; "List of Genome Research Projects Under CEC Programmes," *Technology Transfer Newsletter*, 11 (October/November 1989), p. 19. The Japanese had originally expected the Human Frontiers Scientific Programme to consist primarily of an international biological research institute that would be located in Japan, but no western nations would agree to participate. Though the Japanese government resisted establishing the program outside Japan, it agreed to do so in 1987 because of the widespread criticism of its all-take and little-give relationship to western science and technology. In 1991, the Japanese government's contribution to the Human Frontiers Programme is about $40 million, roughly 90 percent of the program's total annual budget. Conversation with Tateo Arimoto, a Japanese official who headed the effort to create the program, September 20, 1991.

70. Robert Wright, "Achilles' Helix," *The New Republic*, 203 (July 9 & 16, 1990), 27; Cook-Deegan, "The Human Genome Project," p. 149; *Genetic Technology News*, May 1988, pp. 1–2; S. E. Luria, letter to the editor, *Science*, 246 (November 17, 1989), 873.

71. Stephen S. Hall, "James Watson and the Search for Biology's 'Holy Grail,'" *Smithsonian*, February 1990, pp. 47–48; James D. Watson, "Moving Toward Clonal Man: Is This What We Want?" *The Atlantic*, 227 (May 1971), p. 53; Cook-Deegan, "The Human Genome Project," p. 149.

72. Hall, "James Watson and the Search for Biology's 'Holy Grail,'" pp. 47–48; Davis, *Mapping the Code*, p. 262.

73. Cook-Deegan, "The Human Genome Project," p. 127; Friedman, "Science and Politics of the Human Genome Project," pp. 45, 57–58, 61.

74. Cook-Deegan, "The Human Genome Project," p. 152; Watson and Cook-Deegan, *Origins of the Human Genome Project*, p. 10; Friedman, "Science and Politics of the Human Genome Project," pp. 61, 107. Leslie Roberts, "Genome Center Grants Chosen," *Science*, 249 (September 28, 1990), 1497. Early in 1990, NIH and DOE managed to submit a five-year joint genome research plan to Congress.

75. Cook-Deegan, "The Human Genome Project," p. 152; Watson and Cook-Deegan, *Origins of the Human Genome Project*, p. 10; Friedman, "Science and Politics of the Human Genome Project," pp. 61, 107. In fiscal 1991, $2.6 million of the NIH genome appropriation was budgeted for ethics. Talk by Eric Juengst of the National Center for Genome Research, Salk Institute, December 17, 1990.

76. Roberts, "Genome Center Grants Chosen," p. 1497. Baylor College of Medicine was later added to the six institutions listed in Roberts' article.

77. Davis, *Mapping the Code*, pp. 178–181; Friedman, "Science and Politics of the Human Genome Project," pp. 50–51.

78. Victor A. McKusick, "Current Trends in Mapping Human Genes," *FA-SEB Journal*, 5 (January 1991), 15–16; J. Claiborne Stephens et al., "Mapping the Human Genome: Current Status," *Science*, 250 (October 12, 1990), 239.

2. A History of the Science and Technology Behind Gene Mapping and Sequencing

1. Gregor Mendel, "Versuche über Pflanzen-Hybriden," *Verhandlungen des Naturforschenden Vereines in Brunn*, 4 (1865), 3–47. Best English translation is by Eva R. Sherwood, in Curt Stern and Eva R. Sherwood, eds., *The Origin of Genetics* (San Francisco: W. H. Freeman, 1966), pp. 1–48.
2. August Weismann, *The Germ-Plasm: A Theory of Heredity*, trans. W. Newton Parker and Harriet Rönnfeldt (London: Walter Scott, 1893), especially pp. 1–35; E. B. Wilson, *The Cell in Development and Inheritance* (New York, 1896), pp. 182–185.
3. Archibald E. Garrod, "The Incidence of Alkaptonuria: A Study in Chemical Individuality," *Lancet*, 2 (13 December 1902), 1617–1620; Garrod, *Inborn Errors of Metabolism* (London: Frowde & Hodder, 1909).
4. Victor A. McKusick, C. A. Francomano, and S. E. Antonarakis, *Mendelian Inheritance in Man*, 9th ed. (Baltimore: Johns Hopkins University Press, 1990). For access, contact GDB/OMIM User Support, Welch Medical Library, Johns Hopkins University, 1830 East Monument Street, Baltimore MD 21205; telephone (301) 955-7058, FAX (301) 955-0054.
5. Walter S. Sutton, "The Chromosomes in Heredity," *Biological Bulletin*, 4 (1903), 231–251.
6. William Bateson and R. C. Punnett, "Experimental Studies in the Physiology of Heredity," *Reports to the Evolution Committee of the Royal Society*, Reports 2, 3, and 4 (1905–1908), reprinted in part in James A. Peters, ed., *Classic Papers in Genetics* (Englewood Cliffs, N.J.: Prentice-Hall, 1959), pp. 42–60.
7. Edward M. East, "A Mendelian Interpretation of Variation That Is Apparently Continuous," *American Naturalist*, 44 (February 1910), 42–60.
8. T. H. Morgan, "Sex Limited Inheritance in Drosophila," *Science*, 32 (1910), 120–122.
9. E. B. Wilson, "The Sex Chromosomes," *Archiv für Mikroskopische Anatomie und Entwicklungsmechanik*, 77 (1911), 249; cited in Victor McKusick, "The Gene Map of *Homo sapiens*: Status and Prospectus," *Cold Spring Harbor Symposia on Quantitative Biology*, 51 (1986), 15, 24.
10. McKusick, *Mendelian Inheritance in Man*, passim.
11. A. H. Sturtevant, "The Linear Arrangement of Six Sex-linked Factors in Drosophila, as Shown by Their Mode of Association," *Journal of Experimental Zoology*, 14 (1913), 43–59.
12. See, for example, T. H. Morgan, A. H. Sturtevant, H. J. Muller, and C. B. Bridges, *The Mechanism of Mendelian Heredity*, revised ed. (New York: Henry Holt, 1922), frontispiece, p. ii.

13. McKusick, "Gene Map of *Homo Sapiens*," p. 19.

14. H. J. Muller, "Variation Due to Change in the Individual Gene," *American Naturalist*, 56 (1922), 32–50.

15. H. J. Muller, "Artificial Transmutation of the Gene," *Science*, 66 (1927), 84–87.

16. T. S. Painter, "A New Method for the Study of Chromosome Rearrangements and Plotting of Chromosome Maps," *Science*, 78 (1933), 585–586.

17. Conversation, Max Delbrück and Horace Freeland Judson, 9 July 1972, quoted and cited in Horace Freeland Judson, *The Eighth Day of Creation: Makers of the Revolution in Biology* (New York: Simon & Schuster, 1979), pp. 58–60.

18. Sewall Wright, "Color Inheritance in Mammals," *Journal of Heredity*, 8 (1917), 224–235; Wright, "The Physiology of the Gene," *Physiological Reviews*, 21 (1941), 487–527.

19. William B. Provine, *Sewall Wright and Evolutionary Biology* (Chicago: University of Chicago Press, 1986).

20. Judson, *The Eighth Day of Creation*, pp. 214, 356, 368.

21. Discussed at greater length with citations, in Horace Freeland Judson, "Reflections on the Historiography of Molecular Biology," *Minerva*, 18 (Autumn 1980), 369–421.

22. Frederick Sanger, "The Terminal Peptides of Insulin," *Biochemical Journal*, 45 (1949), 563–574, quote at 573; Sanger, "Some Chemical Investigations on the Structure of Insulin," *Cold Spring Harbor Symposia on Quantitative Biology, 14: Amino Acids and Proteins* (meeting 8–16 June 1949), 153–160.

23. Erwin Chargaff, "Chemical Specificity of Nucleic Acids and Mechanism of Their Enzymatic Degradation," *Experientia*, 6 (1950), 201–209; for other papers and discussion, see Judson, "Historiography."

24. S. E. Luria and M. Delbrück, "Mutations of Bacteria from Virus Sensitivity to Virus Resistance," *Genetics*, 28 (1943), 498–511; Jacques Monod and Alice Audureau, "Mutation et adaptation enzymatique chez Escherichia coli-mutable," *Annales de l'Institut Pasteur, Paris*, 72 (1946), 868–878; published late because of wartime restrictions.

25. Judson, *The Eighth Day of Creation*, pp. 19–195.

26. François Jacob and Elie L. Wollman, "Induction of Phage Development in Lysogenic Bacteria," *Cold Spring Harbor Symposia on Quantitative Biology*, 18 (1953), 101–121; William Hayes, "The Mechanism of Genetic Recombination in *Escherichia coli*," ibid., pp. 75–93; Elie L. Wollman and François Jacob, "Sur le mécanisme du transfert de matériel génétique au cours de la recombinaison chez *Escherichia coli* K12," *Comptes rendues Académie des Sciences*, 240 (séance du 20 Juin 1955), 2449–2451; interviews with François Jacob, 3 December 1975 and 28 May 1977.

27. The best contemporary summary of the early work is François Jacob and Elie Wollman, "Genetic and Physical Determination of Chromosomal Segments in *Escherichia coli*," *Symposium of the Society for Experimental Biology*, 12 (1958; held in 1957), 75–92.

28. Seymour Benzer, "The Elementary Units of Heredity," in W. D. McElroy and Bentley Glass, eds., *The Chemical Basis of Heredity* (Baltimore: Johns Hopkins University Press, 1957), pp. 70–93.

29. Barbara McClintock, "The Origin and Behavior of Mutable Loci in Maize," *Proceedings of the National Academy of Sciences*, 36 (1950), 344–355.

30. The discovery and naming of the operon and the repressor were accomplished in an intricate series of small experimental steps; for thorough discussion and citations, Judson, *The Eighth Day of Creation*, pp. 400–435.

31. François Jacob and Jacques Monod, "Genetic Regulatory Mechanisms in the Synthesis of Proteins," *Journal of Molecular Biology*, 3 (1961), 318–356.

32. M. W. Nirenberg and J. H. Matthaei, "The Dependence of Cell-Free Protein Synthesis in *E. coli* upon Naturally Occurring or Synthetic Polynucleotides," *Proceedings of the National Academy of Sciences*, 47 (1961), 1588–1602.

33. F. H. C. Crick, conversation, 15 September 1970, cited in Judson, *Eighth Day of Creation*, pp. 204, 640.

34. For the history of discovery, with full citations, see Howard M. Temin, Nobel lecture, 12 December 1975, published as "The DNA Provirus Hypothesis," *Science*, 192 (1976), 1075–1079.

35. Frederick Sanger, "The Croonian Lecture, 1975: Nucleotide Sequences in DNA," *Proceedings of the Royal Society of London, B*, 191 (1975), 317–333; quotes at 326.

36. F. Sanger, S. Nicklen, and A. R. Coulson, "DNA Sequencing with Chain Terminating Inhibitors," *Proceedings of the National Academy of Sciences*, 74 (1977), 5463–5467.

37. A. M. Maxam and W. Gilbert, "A New Method for Sequencing DNA," *Proceedings of the National Academy of Sciences*, 74 (1977), 1258.

38. Review of all this work with extensive citations: McKusick, "Gene Map of *Homo sapiens*."

39. Interview, Victor McKusick, 6 June 1987.

40. McKusick, "Gene Map of *Homo sapiens*"; interview, McKusick, 6 June 1987.

41. David Botstein, R. L. White, M. Skolnick, and R. W. Davis, "Construction of a Genetic Linkage Map in Man Using Restriction Fragment Length Polymorphisms," *American Journal of Human Genetics*, 32 (1980), 314–331.

42. An early review is: J. F. Gusella et al., "Molecular Genetics of Huntington's Disease," *Cold Spring Harbor Symposia on Quantitative Biology*, 51 (1986), 359–364.

43. Interview, McKusick, 6 June 1987.

44. D. C. Schwartz and C. R. Cantor, "Separation of Yeast Chromosome-sized DNAs by Pulsed Field Gel Electrophoresis," *Cell*, 37 (1984), 67–75.

45. D. T. Burke, G. F. Carle, and M. V. Olson, "Cloning of Large Segments of Exogenous DNA into Yeast by Means of Artificial Chromosome Vectors," *Science*, 236 (1987), 806–812.

46. K. F. Mullis, F. Faloona, S. Scharf, R. Saiki, G. Horn, and H. Ehrlich, "Specific Enzymatic Amplification of DNA *in Vitro:* The Polymerase

Chain Reaction," *Cold Spring Harbor Symposia on Quantitative Biology*, 51 (1986), 263–273; Elise A. Rose, "Applications of the Polymerase Chain Reaction to Genome Analysis," *FASEB Journal*, 5 (1991), 46–53.

47. L. M. Smith, J. Z. Sanders, R. J. Kaiser, P. Hughes, C. Dodd, C. R. Connell, C. Heiner, S. B. H. Kent, and L. E. Hood, "Fluorescence Detection in Automated DNA Sequencing Analysis," *Nature*, 321 (1986), 674–679; R. K. Wilson, A. S. Yuen, S. M. Clark, C. Spence, P. Arakelian, and L. E. Hood, "Automation of Dideoxy Nucleotide DNA Sequencing Reactions Using a Robotic Workstation," *BioTechniques*, 6 (1988), 776–787; interview, Leroy Hood, 12 June 1991. For a skeptical view, see Bart Barrell, "DNA Sequencing: Present Limitations and Prospects for the Future," *FASEB Journal*, 5 (1991), 40–45.

5. DNA-Based Medicine

1. James D. Watson, "The Human Genome Project: Past, Present, and Future," *Science*, 248 (1990), 44–49.

2. V. A. McKusick, C. A. Francomano, and S. E. Antonarakis, *Mendelian Inheritance in Man: Catalogs of Autosomal Dominant, Autosomal Recessive, and X-linked Phenotypes*, 9th ed. (Baltimore: Johns Hopkins University Press, 1990); J. C. Stephens et al., "Mapping the Human Genome: Current Status," *Science*, 250 (1990), 237–244.

3. D. L. Nelson et al., "*Alu* Polymerase Chain Reaction: A Method for Rapid Isolation of Human-specific Sequences from Complex DNA Sources," *Proceedings of the National Academy of Sciences*, 86 (1989), 6686–6690.

4. R. Guthrie, "Blood Screening for Phenylketonuria," *Journal of the American Medical Association*, 178 (1961), 863.

5. R. K. Saiki, P. S. Walsh, C. H. Levenson, and H. A. Erlich, "Genetic Analysis of Amplified DNA with Immobilized Sequence-specific Oligonucleotide Probes," *Proceedings of the National Academy of Sciences*, 86 (1989), 6230–6234.

6. A. Cao, "Results of Programmes for Antenatal Detection of Thalassemia in Reducing the Incidence of the Disorder," *Blood Review*, 1 (1987), 169–176.

7. E. Arpaia et al., "Identification of an Altered Splice Site in Ashkenazi Tay-Sachs Disease," *Nature*, 333 (1988), 85–86; R. Myerowitz, "Splice Junction Mutation in Some Ashkenazi Jews with Tay-Sachs Disease: Evidence against a Single Defect within This Ethnic Group," *Proceedings of the National Academy of Sciences*, 85 (1988), 3955–3959; R. Myerowitz and F. C. Costigan, "The Major Defect in Ashkenazi Jews with Tay-Sachs Disease Is an Insertion in the Gene for the α-Chain of β-Hexosaminidase," *Journal of Biological Chemistry*, 263 (1988), 18587–18589; K. Ohno and K. Suzuki, "A Splicing Defect Due to an Exon-Intron Junctional Mutation Results in Abnormal β-Hexosaminidase α-Chain mRNAs in Ashkenazi Jewish Patients with Tay-Sachs Disease," *Biochemical and Biophysical Research Communications*, 153 (1988), 463–469.

8. B. S. Kerem et al., "Identification of the Cystic Fibrosis Gene: Genetic Analysis," *Science*, 245 (1989), 1073–1080.
9. J. S. Chamberlain et al., "Multiplex PCR for the Diagnosis of Duchenne Muscular Dystrophy," in M. Innis et al., eds., *PCR Protocols: A Guide to Methods and Applications* (Orlando: Academic Press, 1990), pp. 272–281; J. S. Chamberlain, personal communication.
10. R. A. Gibbs et al., "Multiplex DNA Deletion Detection and Exon Sequencing of the Hypoxanthine Phosphoribosyltransferase Gene in Lesch-Nyhan Families," *Genomics*, 7 (1990), 235–244.
11. M. Grompe, D. M. Muzny, and C. T. Caskey, "Scanning Detection of Mutations in Human Ornithine Transcarbamylase by Chemical Mismatch Cleavage," *Proceedings of the National Academy of Sciences*, 86 (1989), 5888–5892.
12. D. M. Zaller et al., "Prevention and Treatment of Murine Experimental Allergic Encephalomyelitis with T Cell Receptor Vβ-specific Antibodies," *Journal of Experimental Medicine*, 171 (1990), 1943–1955.
13. J. W. Belmont et al., "Expression of Human Adenosine Deaminase in Murine Hematopoietic Cells," *Molecular Cell Biology*, 8 (1988), 5116–5125; E. A. Dzierzak, T. Papayannopoulou, and R. C. Mulligan, "Lineage-specific Expression of a Human β-Globin Gene in Murine Bone Marrow Transplant Recipients Reconstituted with Retrovirus-transduced Stem Cells," *Nature*, 331 (1988), 35–41.
14. S. A. Rosenberg, "Gene Transfer into Humans—Immunotherapy of Patients with Advanced Melanoma, Using Tumor-infiltrating Lymphocytes Modified by Retroviral Gene Transduction," *New England Journal of Medicine*, 323 (1990), 570–578.
15. J. A. Wolff et al., "Direct Gene Transfer into Mouse Muscle in Vivo," *Science*, 247 (1990), 1465–1468.
16. S. Thompson, "Germ Line Transmission and Expression of a Corrected HPRT Gene Produced by Gene Targeting in Embryonic Stem Cells," *Cell*, 56 (1989), 313–321.
17. S. Hartung, R. Jaenisch, and M. Breindl, "Retrovirus Insertion Inactivates Mouse α1(I) Collagen Gene by Blocking Initiation of Transcription," *Nature*, 320 (1986), 365–367.

6. Biology and Medicine in the Twenty-First Century

1. S. J. Horvath, J. R. Firca, T. Hunkapiller, M. W. Hunkapiller, and L. Hood, "An Automated DNA Synthesizer Employing Deoxynucleoside 3' Phosphoramidites," *Methods in Enzymology*, 154 (1987), 314–326.
2. M. Olson, L. Hood, C. Cantor, and D. Botstein, "A Common Language for Physical Mapping of the Human Genome," *Science*, 245 (1989), 1434–1435.
3. D. A. Nickerson, R. Kaiser, S. Lappin, J. Stewart, L. Hood, and U. Landegren, "Automated DNA Diagnostics Using an ELISA-based Oligonucleotide Ligation Assay," *Proceedings of the National Academy of Sciences*, 87 (1990), 8923–8927.

4. L. M. Smith, J. Z. Sanders, R. J. Kaiser, P. Hughes, C. Dodd, C. R. Connell, C. Heiner, S. B. H. Kent, and L. E. Hood, "Fluorescence Detection in Automated DNA Sequence Analysis," *Nature*, 321 (1986), 674–679.

5. T. Hunkapiller, R. J. Kaiser, B. F. Koop, and L. Hood, "Large-scale and Automated DNA Sequence Determination," *Science*, 254 (October 4, 1991), 59–67.

6. M. D. Adams et al., "cDNA Sequencing: 'Expressed Sequence Tags' and the Human Genome Project," *Science*, 252 (1991), 1651–1656.

8. The Social Power of Genetic Information

1. Dana Fradon, *The New Yorker*, June 29, 1987. The material in this chapter has been further developed in Dorothy Nelkin and Laurence Tancredi, *Dangerous Diagnostics: The Social Power of Biological Information* (New York: Basic Books, 1989; paperback, 1991).

2. National Institute of Mental Health, *Approaching the Twenty-First Century: Opportunities for NIMH Neurosciences Research*, Report to Congress on the Decade of the Brain (Washington, D.C.: U.S. Department of Health and Human Services, January 1988).

3. Charles R. Scriver, "Presidential Address," *American Journal of Human Genetics*, 40 (1987), 199–211.

4. Nelkin and Tancredi, *Dangerous Diagnostics*.

5. Michel Foucault, *Discipline and Punish: The Birth of the Prison* (New York: Vintage Books, 1979), pp. 183–184, 304.

6. Walter Reich, "Diagnostic Ethics: The Uses and Limits of Psychiatric Explanation," in Laurence Tancredi, ed., *Ethical Issues in Epidemiologic Research* (New Brunswick: Rutgers University Press, 1986), pp. 37–69.

7. Mary Douglas, *How Institutions Think* (Syracuse: Syracuse University Press, 1986), pp. 63, 92.

8. Dorothy Nelkin, *Selling Science: How the Press Covers Science and Technology* (New York: W. H. Freeman, 1988).

9. E. O. Wilson, *Sociobiology: The New Synthesis* (Cambridge: Harvard University Press, Belknap Press, 1975).

10. Daniel J. Kevles, *In the Name of Eugenics: Genetics and the Uses of Human Heredity* (New York: Alfred A. Knopf, 1985); Garland Allen, "The Misuse of Biological Hierarchies: The American Eugenics Movement," *History and Philosophy of the Life Sciences*, 5 (May 2, 1983), 105–128.

11. Marjorie Shaw, "Conditional Prospective Rights of the Fetus," *Journal of Legal Medicine*, 63 (1984), 63–116.

12. Daniel Koshland, Jr., "Nature, Nurture, and Behavior," *Science*, 235 (March 20, 1987), 1445.

13. See discussion of this language in the President's Commission for the Study of Ethical Problems in Medicine, Report #83–600502 (Washington, D.C.: U.S. Government Printing Office, 1983), and in Howard Kaye, *The Social Meaning of Biology* (New Haven: Yale University Press, 1986).

14. Irving Harris, *Emotional Blocks to Learning: A Study of the Reasons for Failure in School* (Glencoe, Ill.: The Free Press, 1962).

15. American Psychiatric Association, *Diagnostic and Statistical Manual*, III (1980) and IIIR (1987).
16. Gerald Coles, *The Learning Mystique* (New York: Pantheon, 1987).
17. U.S. Congress, *Genetic Screening in the Workplace* (Washington, D.C.: U.S. Government Printing Office, 1990).
18. Mark Rothstein, *Medical Screening of Workers* (Washington, D.C.: Bureau of National Affairs, 1984), p. 10; Elaine Draper, "High Risk Workers or High Risk Work," *International Journal of Sociology and Social Policy*, 6 (1986), 12–28; and Dorothy Nelkin and Michael S. Brown, *Workers at Risk* (Chicago: University of Chicago Press, 1984).
19. Paul Starr, *The Social Transformation of American Medicine* (New York: Basic Books, 1982).
20. U.S. Congress, Office of Technology Assessment, *Medical Testing and Health Insurance* (Washington, D.C.: U.S. Government Printing Office, 1988), pp. 3–4.
21. Society of Actuaries, *Record*, Annual Meeting, October 5–7, 1986, pp. 2943–2962.
22. Rochelle C. Dreyfuss and Dorothy Nelkin, "The Jurisprudence of Genetics" (forthcoming).
23. P. Low, J. Jeffries, and R. Bonnie, *The Trial of John W. Hinckley* (New York: Foundation News Press, 1986), p. 119; and M. Moore, "Causation and Its Excuses," *California Law Review*, 73 (1985), 1091–1149.
24. N. Volkow and L. Tancredi, "Neural Substrates of Violent Behavior: A Preliminary Study with Positron Emission Tomography," *British Journal of Psychiatry*, 151 (1987), 668–673. See also B. White, "Biological Causes for Violent Behavior: Research Could Affect Legal Decisions," *Texas Bar Journal*, 50 (1987), 446.
25. C. Ray Jeffrey in collaboration with R. V. Del Carmen and J. D. White, *Attacks on the Insanity Defense: Biological Psychiatry and New Perspectives on Criminal Behavior* (Springfield, Ill.: Charles C. Thomas, 1985), p. 82.
26. C. Ray Jeffrey, *Criminology: An Interdisciplinary Approach* (Englewood Cliffs, N.J.: Prentice Hall, 1990).
27. Neil A. Holtzman, *Proceed with Caution: Predicting Genetic Risks in the Recombinant DNA Era* (Baltimore: Johns Hopkins University Press, 1989).
28. Jack Ballantyne, ed., *DNA Technology and Forensic Science*, Banbury Report 32 (Cold Spring Harbor, N.Y.: Cold Spring Harbor Laboratory Press, 1989).
29. Peter Rowley, "Genetic Discrimination," *American Journal of Human Genetics*, 43 (July 1988), 105–106. Also Paul Billings, personal communication.
30. Diane Chapman Walsh, *Corporate Physicians: Between Medical Care and Management* (New Haven: Yale University Press, 1987).

9. DNA Fingerprinting

1. *Florida v. Andrews*, Case No. 87–2166 (1988).
2. *Frye v. U.S.*, 293 F. 1030 (1923).

3. See, e.g., *New York v. Wesley*, 140 Misc. 2d 306 (1989).
4. G. Sensabaugh and J. Witkowski, eds., *DNA Technology and Forensic Science*, Banbury Report 34 (Cold Spring Harbor, N.Y.: Cold Spring Harbor Press, 1989).
5. *New York v. Castro*, 144 Misc. 2d 956, 545 N.Y.S. 2d 985 (1989).
6. For a detailed discussion of the problems, see Eric S. Lander, "DNA Fingerprinting on Trial," *Nature*, 339 (1989), 501–505.
7. *Pennsylvania v. Shorter*, Case No. 7565–88. See also W. C. Thompson and S. Ford, "The Meaning of a Match: Sources of Ambiguity in the Interpretation of DNA Prints," in M. A. Farley and J. J. Harrington, eds., *Forensic DNA Technology* (Chelsea, Mich.: Lewis Publishers, 1991).
8. "Maine Case Deals Blow to DNA Fingerprinting," *Science*, 246 (1989), 1556–1558.
9. Ibid.
10. *California v. Collins*, 68 Cal. 2d 319 (1968).
11. See, e.g., Eric S. Lander, "Population Genetic Considerations in the Forensic Use of DNA Fingerprinting," in Sensabaugh and Witkowski, eds., *DNA Technology and Forensic Science*.
12. J. Gardner, *Saturday Evening Post*, December 1, 1962.
13. Such a situation occurred de facto in the famous Colin Pitchfork case described in Joseph Wambaugh's book *The Blooding* (New York: Morrow, 1989). In this rape case, an entire village submitted to DNA typing at the request of the police. The DNA testing was voluntary, but the peer pressure to submit was enormous.

10. Clairvoyance and Caution

1. James Gusella, Nancy Wexler, P. Michael Conneally, et al., "A Polymorphic DNA Marker Genetically-Linked to Huntington's Disease," *Nature*, 306 (1983), 234–238.
2. Johanna M. Rommens et al., "Identification of the Cystic Fibrosis Gene: Chromosome Walking and Jumping," *Science*, 245 (September 8, 1989), 1059–1065; John R. Riordan et al., "Identification of the Cystic Fibrosis Gene: Cloning and Characterization of Complementary DNA," ibid., pp. 1066–1073; Bat-Sheva Kerem et al., "Identification of the Cystic Fibrosis Gene: Genetic Analysis," ibid., pp. 1073–1080.
3. C. T. Caskey et al., "The American Society of Human Genetics Statement on Cystic Fibrosis Screening," *American Journal of Human Genetics*, 46 (1990), 393; B. S. Wilfond and N. Fost, "The Cystic Fibrosis Gene: Medical and Social Implications for Heterozygote Detection," *Journal of the American Medical Association*, 263 (May 23/30, 1990), 2777–2783; Workshop on Population Screening for the Cystic Fibrosis Gene, "Statement from the National Institutes of Health Workshop on Population Screening for the Cystic Fibrosis Gene," *New England Journal of Medicine*, 323 (July 5, 1990), 70–71.
4. Lori B. Andrews, "Legal Aspects of Genetic Information," *Yale Journal of Biology and Medicine*, 64 (1991), 29–40; N. S. Wexler, "Will the Circle Be

Unbroken? Sterilizing the Genetically Impaired," in A. Milunsky, ed., *Genetics and the Law* (New York: Plenum Press, 1980); Nancy S. Wexler, "The Oracle of DNA," in L. P. Rowland, ed., *Molecular Genetics in Diseases of Brain, Nerve, and Muscle* (New York: Oxford University Press, 1989).

5. Symposium, "A Legal Research Agenda for the Human Genome Initiative," Arizona State University, Center for the Study of Law, Science and Technology, May 17–18, 1991.

6. Reported in Marc Lappe, *Genetic Politics* (New York: Simon and Schuster, 1977).

7. G. Stamatoyannopoulos, "Problems of Screening and Counseling in the Hemoglobinopathies," A. G. Motulsky and W. Lenz, eds., *Birth Defects* (Amsterdam: Excerpta Medica, 1974), pp. 268–276.

8. A. Cao et al., "Prevention of Homozygous Betathalassemia by Carrier Screening and Prenatal Diagnosis in Sardinia," *American Journal of Human Genetics,* 33 (1981), 593–605.

9. Personal communication, Dr. Jason Brandt, Johns Hopkins Hospital, Baltimore, Md.

10. Amos Tversky and Daniel Kahneman, "The Framing of Decisions and the Psychology of Choice," *Science,* 211 (1981), 453–458.

11. Melissa A. Rosenfeld et al., "Adenovirus-Mediated Transfer of a Recombinant α_1-Antitrypsin Gene to the Lung Epithelium in Vivo," *Science,* 252 (April 19, 1991), 431–434.

12. Mitchell L. Drumm et al., "Correction of the Cystic Fibrosis Defect in Vitro by Retrovirus-Mediated Gene Transfer," *Cell,* 62 (September 21, 1990), 1227–1233; D. P. Rich et al., "Expression of Cystic Fibrosis Transmembrane Conductance Regulator Corrects Defective Chloride Channel Regulation in Cystic Fibrosis Airway Epithelial Cells," *Nature,* 347 (September 27, 1990), pp. 358–363.

13. Natalie Angier, "Team Cures Cells in Cystic Fibrosis by Gene Insertion," *New York Time,* September 21, 1990, p. A1.

11. Genetic Technology and Reproductive Choice

1. This example has been used to make this point in Langdon Winner, *The Whale and the Reactor: A Search for Limits in an Age of High Technology* (Chicago: University of Chicago Press, 1986), pp. 19–39.

2. Robert Caro, *The Power Broker: Robert Moses and the Fall of New York* (New York: Knopf, 1974), pp. 318, 481, 514, 546, 951–958.

3. Ruth Schwartz Cowan, *More Work for Mother: The Ironies of Household Technology from the Open Hearth to the Microwave* (New York: Basic Books, 1983), chap. 5.

4. Langdon Winner, *Autonomous Technology: Technics-Out-of-Control as a Theme in Political Thought* (Cambridge: MIT Press, 1977).

5. For a review of the literature on this subject, see Lars L. Cederqvist and Fritz Fuchs, "Antenatal Sex Determination: An Historical Review," *Clinical Obstetrics and Gynecology,* 13 (March 1970), 159–177.

6. Some of the information which follows is based on the author's interview with Fritz Fuchs, M.D., conducted in New York in the summer of 1989.

7. There have been many reviews of this history in the medical literature: e.g., Merton Sandler, *Amniotic Fluid and Its Clinical Significance* (New York: M. Dekker, 1981), chap. 1. The abortion was reported in Carlo Valenti, Edward J. Schutta, Tehila Kehaty, "Prenatal Diagnosis of Down's Syndrome," *Lancet*, 27 July 1968, 220.

8. See D. J. H. Brock, *Early Diagnosis of Fetal Defects* (Edinburgh and London: Churchill Livingstone, 1982), chap. 4.

9. "Amniocentesis: HEW Backs Test for Prenatal Diagnosis of Disease," *Science*, 190 (5 November 1975), 537–540. The report of the NICHD National Registry for Amniocentesis Study Group did not appear in print for another year: "Midtrimester Amniocentesis for Prenatal Diagnosis," *Journal of the American Medical Association*, 236 (27 September 1976), 1471–1476.

10. The information in this paragraph is based on several interviews the author has conducted with men and women who directed medical genetics services in the late 1970s and early 1980s in the United States.

11. J. C. Fletcher, "Prenatal Diagnosis: Ethical Issues," in Warren T. Reich, ed., *The Encylopedia of Bioethics* (New York: The Free Press, 1978), pp. 1336–1346.

12. T. M. Powledge and J. Fletcher, "Guidelines for the Ethical, Social and Legal Issues in Prenatal Diagnosis: A Report from the Genetics Research Group of the Hastings Center, Institute of Society, Ethics and the Life Sciences," *New England Journal of Medicine*, 300 (1979), 168–172.

13. John C. Fletcher, "Sounding Board: Ethics and Amniocentesis for Fetal Sex Identification," *New England Journal of Medicine*, 301 (1979), 551.

14. John C. Fletcher, "Is Sex Selection Ethical?" in K. Berg and K. E. Tranoy, *Research Ethics* (New York: Alan R. Liss, 1983), p. 333.

15. Dorothy Wertz and John C. Fletcher, "Attitudes of Genetics Counselors: A Multinational Survey," *American Journal of Human Genetics*, 42 (April 1988); also Dorothy Wertz and John C. Fletcher, "Ethics and Human Genetics: A Cross-Cultural Study in 17 Nations," in F. Vogel and K. Sperling, eds., *Human Genetics: Proceedings of the 7th International Congress of Human Genetics* (Berlin and New York: Springer-Verlag, 1987), pp. 655–672.

16. Carol Gilligan, *In a Different Voice: Psychological Theory and Women's Development* (Cambridge: Harvard University Press, 1982).

17. Ibid., p. 26.

18. Ibid., p. 28.

19. Rayna Rapp, "Constructing Amniocentesis: Maternal and Medical Discourses" (typescript, 5 December 1986); Barbara Katz Rothman, *The Tentative Pregnancy: Prenatal Diagnosis and the Future of Motherhood* (New York: Viking Penguin, 1986).

20. Rosalind Pollack Petchesky, *Abortion and Women's Choice: The State, Sexuality and Reproductive Freedom* (New York: Longman, 1984), esp. chaps.

1, 9, 10; Barbara Katz Rothman, *Recreating Motherhood: Ideology and Technology in a Patriarchal Society* (New York: W. W. Norton, 1989).

21. The principle that nurturance matters also means that decisions about a patient who is comatose ought to be made by the person who is providing day-to-day care of the patient, even if that person is not related by blood or marriage to the patient—but that is an argument which ought to be pursued in an essay other than this one.

22. Information on the reaction of some innovators in prenatal diagnosis for sex comes from the author's interview with Kurt Hirschhorn, summer 1989; on Lejeune, see Daniel J. Kevles, *In the Name of Eugenics: Genetics and the Uses of Human Heredity* (New York: Knopf, 1985), p. 287; on MacDonald, see Robert Reid, *My Children, My Children* (London: British Broadcasting Corp., 1977), chap. 6.

23. Kevles, *In the Name of Eugenics*, pp. 247–249.

12. Health Insurance, Employment Discrimination, and the Genetics Revolution

1. The possibility of employment discrimination based on genetic analysis has long been discussed. See, e.g., Bruce Hilton et al., *Ethical Issues in Human Genetics: Genetic Counseling and the Use of Genetic Knowledge* (New York: Plenum Press, 1973); National Academy of Sciences, *Genetic Screening: Programs, Principles and Research* (Washington, D.C.: National Academy of Sciences, 1975); Philip Reilly, *Genetics, Law, and Social Policy* (Cambridge: Harvard University Press, 1977); President's Commission for the Study of Ethical Problems in Medicine and Biomedical and Behavioral Research, *Screening and Counseling for Genetic Conditions: A Report on the Ethical, Social and Legal Implications of Genetic Screening, Counseling and Education Programs* (Washington, D.C.: Government Printing Office, 1983); Arno G. Motulsky, "Impact of Genetic Manipulation on Society and Medicine," *Science*, 219 (1983), 135–140; Nancy S. Wexler, "Genetic Jeopardy and the New Clairvoyance," *Progress in Medical Genetics*, 6 (1985), 277–304.

In recent years, these concerns have escaped from specialized publications and are increasingly reaching wider audiences. See, e.g., David Orentlicher, "Genetic Screening by Employers," *Journal of the American Medical Association*, 263 (1990), 1005, 1008; Robert Wright, "Achilles' Helix: The End of Insurance," *New Republic*, July 9, 1990, p. 26; David Stipp, "Genetic Testing May Mark Some People as Undesirable to Employers, Insurers," *Wall Street Journal*, July 9, 1990, p. B1; Dorothy Nelkin and Laurence Tancredi, *Dangerous Diagnostics: The Social Power of Biological Information* (New York: Basic Books, 1989); Neil Holtzman, *Proceed with Caution: Predicting Genetic Risks in the Recombinant DNA Era* (Baltimore: Johns Hopkins University Press, 1989); and Jerry E. Bishop and Michael Waldholz, *Genome: The Story of the Most Astonishing Scientific Adventure of*

Our Time—The Attempt to Map All the Genes in the Human Body (New York: Simon and Schuster, 1990), pp. 285–306.

Legal analysis of this issue, though valuable, is surprisingly limited. It seems to comprise one very useful book, Mark A. Rothstein, *Medical Screening and the Employee Health Cost Crisis* (Washington, D.C.: Bureau of National Affairs, 1989), and four insightful articles. Three of the articles are from one symposium: Lance Liebman, "Too Much Information: Predictions of Employee Disease and the Fringe Benefit System"; Mark A. Rothstein, "Medical Screening and Employment Law: A Note of Caution and Some Observations"; and Richard E. A. Epstein, "AIDS, Testing and the Workplace" (focusing solely on predicting AIDS costs), *University of Chicago Legal Forum* (1988), 57–92, 1–32, 33–56.

2. I have discussed the issues in this chapter in substantially more detail than current space allows in Henry T. Greely, "The Death of Health Insurance: Employment-Related Health Coverage and the Increasing Predictability of Individual Health" (forthcoming).

3. All estimates of health coverage have serious weaknesses. Official governmental records are kept only of Medicare and Medicaid patients. The other numbers rest mainly on survey evidence. All categories suffer uncertainties because many people are covered by more than one source. The best estimates on employment-related coverage come from Jon Gabel et al., "Employer-Sponsored Health Insurance in America," *Health Affairs*, 8 (Summer 1989), 116–128. The estimate for individually underwritten insurance is from the United States Office of Technology Assessment, *AIDS and Health Insurance*, 5 (February 1988). This estimate excludes those people covered by Medicare who also purchase individually underwritten "medigap" policies to cover some of the expenses that Medicare does not pay. The Medicare and Medicaid estimates are from Office of National Cost Estimates, Health Care Financing Administration, "National Health Expenditures, 1988," *Health Care Financing Review*, 11 (Summer 1990), 1–41. The number of people without coverage is controversial, but most sources estimate it at between 31 and 37 million at any one time. See M. Eugene Moyer, "A Revised Look at the Number of Uninsured Americans," *Health Affairs*, 8 (Summer 1989), 102–110, and Gail R. Wilensky, "Filling the Gaps in Health Insurance," *Health Affairs*, 7 (Summer 1988), 133–149. In addition to these sources, small numbers of people are covered exclusively through the Veterans Administration, the Indian Health Service, or other governmental sources.

4. For useful discussions of how medical underwriting of individually purchased insurance works, see United States Office of Technology Assessment, *AIDS and Health Insurance*; and Rothstein, "Medical Screening and the Employee Health Cost Crisis," pp. 205–210.

5. The first rigorous statement of the theory of adverse selection in the insurance context was Michael Rothschild and Joseph Stiglitz, "Equilibrium in Competitive Insurance Markets: An Essay on the Economics of Imperfect Information," *Quarterly Journal of Economics*, 90 (1976), 629–649. The concepts had been presaged to some extent in George A. Akerlof,

"The Market for 'Lemons': Quality, Uncertainty and the Market Mechanism," *Quarterly Journal of Economics*, 84 (1970), 488–500, in the context of the market for used cars. Some of the possible effects of adverse selection on the health insurance market were discussed in Mark V. Pauly, "Is Cream-Skimming a Problem for the Competitive Medical Market?" *Journal of Health Economics*, 3 (1984), 87–95.

6. For a general discussion of the increase in employment-related health insurance, see Paul Starr, *The Social Transformation of American Medicine* (New York: Basic Books, 1982), pp. 311–320.

7. For 1990, her marginal federal income tax rate would have been 28 percent and her marginal California income tax rate would have been 9.3 percent. She would also have paid 7.65 percent of her income in FICA contributions. Her combined federal and state marginal income tax rate will therefore be 44.95 percent. In addition, her employer would have to make a separate contribution to FICA on her behalf, totaling another 7.65 percent of her salary, or $2,295. Assuming the employee takes the standard deduction, the employer would give her a $929 raise, with another $71 going to its 7.65 percent FICA contribution. Her posttax share of that raise would be $511. The details of these rates have changed because of new legislation, but the overall picture remains the same.

8. The text focuses on insurance plans, but health maintenance organizations (HMOs) require special mention. Unlike insurance companies, which either reimburse the people they insure or pay health care providers chosen by the insureds for treatments provided to them, HMOs provide their members with health care directly, through specified doctors and hospitals. Federal law originally required federally qualified HMOs to practice community-rating. Amendments through 1988 have allowed them to experience-rate, but only on a prospective and somewhat limited basis. Nearly 20 percent of the people covered through employers are now members of HMOs. It is likely that HMOs will make the same move to experience-rating as insurance companies have, although, by their nature, they will not be displaced by employer self-insurance.

9. This seemingly paradoxical and inadvertent result was confirmed by the U.S. Supreme Court's 1985 decision in *Metropolitan Life Insurance Co. v. Massachusetts*, 471 U.S. 724 (1985). *Metropolitan Life* confirmed lower court decisions that had begun with *Wadsworth v. Whaland*, 562 F.2d 70 (1st Cir. 1977), *cert. den.* 435 U.S. 980 (1978). ERISA excepts from preemption state laws regulating the business of insurance or banking, but it further provides that an employer's benefit plan shall not be "deemed" to be insurance or banking. The effect of this complex drafting on health benefits has appropriately been called "semipreemption." Daniel M. Fox and Daniel C. Schaffer, "Health Policy and ERISA: Interest Groups and Semipreemption," *Journal of Health Policy, Policy, and Law*, 14 (1989), 239–260. This fascinating article on the provision's legislative history deserves attention for its illustration of the haphazard nature of much legislation concerning health issues.

10. The first surveys mentioned were conducted by the Health Insurance Association of America (HIAA) and are reported in several locations. The 1987 survey is reported in Jon Gabel et al., "The Changing World of Group Health Insurance," *Health Affairs,* 7 (Fall 1988), 48–65, and in Steven DiCarlo and Jon Gabel, "Conventional Health Insurance: A Decade Later," *Health Care Financing Review,* 10 (Spring 1989), 77–89. The 1988 survey is reported in Gabel et al., "Employer-Sponsored Health Insurance in America."

    The second survey was performed by Foster Higgins, an insurance brokerage and benefits consulting firm. Foster Higgins, *Health Care Benefits Survey 1988: Report of the Survey Findings* (Princeton, N.J.: Foster Higgins, 1988), p. 23.

11. Because of a federal requirement called COBRA continuation coverage, the employer may have to bear an employee's costs for years after he leaves. The Consolidated Omnibus Budget Reconciliation Act of 1985 requires employers who offer health benefits to offer covered beneficiaries—employees, spouses, or dependents—a continuation of that coverage for between 18 and 36 months after the coverage would otherwise have ended because of loss of employment, death, divorce, or otherwise. The details of this coverage are somewhat complicated. See Greely, "The Death of Health Insurance."

12. Except in Hawaii. Hawaii passed a statute requiring employers to provide health coverage the same year Congress passed ERISA. The courts quickly held that ERISA's preemption provisions overruled Hawaii's statute. After eight years of lobbying, Hawaii finally got a narrow amendment to ERISA preemption, permitting it to enforce its statute, but only as originally passed. Current attempts to have states require employers to cover employees (or to twist their arms to do so through special taxes) are likely to run afoul of ERISA.

13. These two methods also have some application to an employee's biological children, who will share half of her genes.

14. For more detail on AIDS and the implications of its predictability for health coverage, see Greely, "AIDS and the American Health Care Financing System," *University of Pittsburgh Law Review,* 51 (1989), 73–166.

15. For human papillomavirus, see William C. Reeves et al., "Human Papillomavirus and Cervical Cancer in Latin America," *New England Journal of Medicine,* 320 (1989), 1437–1441; Jan M. McDonnell, Anton J. Mayr, W. John Martin, "DNA of Human Papillomavirus Type 16 in Dysplastic and Malignant Lesions of the Conjunctiva and Cornea," *New England Journal of Medicine,* 320 (1989), 1442–46; and Ronald L. Moy et al., "Human Papillomavirus Type 16 DNA in Periungual Squamous Cell Carcinomas," *New England Journal of Medicine,* 261 (1989), 2669–2673. For Lyme disease, see Allen C. Steere, "Lyme Disease," *New England Journal of Medicine,* 321 (1989), 586–596. Interestingly, a person's risk of long-term damage from Lyme disease seems to have a genetic connection. Allen C. Steere, Edward Dwyer, and Robert Winchester, "Association of Chronic Lyme Arthritis with HLA-DR4 and HLA-DR2 Alleles," *New England Journal of*

Medicine, 323 (1990), 219–223. For a discussion of diabetes, see William J. Riley et al., "A Prospective Study of the Development of Diabetes in Relatives of Patients with Insulin-Dependent Diabetes," *New England Journal of Medicine,* 323 (1990), 1167–1172.

16. For a general discussion of the AAPCC problem and some of the research concerning it, see Jonathan Howland et al., "Adjusting Capitation Using Chronic Disease Risk Factors: A Preliminary Study," *Health Care Financing Review,* 9 (Winter 1987), 15–23; Gerald F. Anderson et al., "Capitation Pricing: Adjusting for Prior Utilization and Physician Discretion," *Health Care Financing Review,* 8 (Winter 1986), 27–34; and Joseph P. S. Newhouse et al., "Adjusting Capitation Rates Using Objective Health Measures and Prior Utilization," *Health Care Financing Review,* 10 (Spring 1989), 41–54. This research is discussed in more detail in Greely, "The Death of Health Insurance."

17. Newhouse et al., "Adjusting Capitation Rates." Among other virtues, this study is the only one based on a non-Medicare population similar to that covered by employment-related health coverage. Newhouse's work is based on data from the very important RAND Corporation health insurance experiment, which ran from 1974 to 1982. The experiment "enrolled" families in six parts of the country into insurance plans that had different cost-sharing characteristics. The purpose of the experiment was to study whether people got different amounts of health care and ended up with different levels of health depending on the terms of their health insurance coverage. The experiment also collected a great deal of data on the medical costs and health status of nearly 4,000 people, each one over 14 and under 65 and each one studied for either three or five years.

18. Newhouse, "Adjusting Capitation Rates," p. 41. This is true even though only 56 percent of those it enrolled, all of whom it predicted would have lower-than-average costs, would actually have lower-than-average costs. The remaining 44 percent would have higher-than-average costs, but their extra expenses would be more than overcome by the majority's lower costs. The method of turning the known predictive abilities of the various adjusters, plus the mean and standard deviation of the group's overall annual expenditures, into the dollar savings for predictions better than the AAPCC is quite technical. See ibid., pp. 41, 52–53.

19. One possible solution is the creation of government-sponsored "high-risk pools," recently adopted by at least 15 states. See generally Randall R. Bovbjerg and Christopher F. Koller, "State Health Insurance Pools: Current Performance, Future Prospects," *Inquiry,* 23 (1986), 111–121. These pools are unlikely to provide a broad answer to the problem because of problems with state funding of them and the remaining high costs of the insurance. See the discussion in Greely, "The Death of Health Insurance."

20. See, e.g., Joseph P. Newhouse et al., "Some Interim Results from a Controlled Trial of Cost Sharing in Health Insurance," *New England Journal of Medicine,* 305 (1981), 1501–1507 (total use of medical services is

inversely and strongly correlated to the copayments required by the health insurance plan); Robert H. Brook et al., "Does Free Care Improve Adults' Health? Results from a Randomized Controlled Trial," *New England Journal of Medicine*, 309 (1983), 1426–1434; Paula Braveman et al., "Adverse Outcomes and Lack of Health Insurance Among Newborns in an Eight-County Area of California, 1982 to 1986," *New England Journal of Medicine*, 321 (1989), 508–513. See Donald M. Berwick and Howard H. Hiatt, "Who Pays?" *New England Journal of Medicine*, 321 (1989), 541–542 (discussing this study); Mark B. Wenneker, Joel S. Weissman, and Arnold M. Epstein, "The Association of Payer with Utilization of Cardiac Procedures in Massachusetts," *Journal of the American Medical Association*, 264 (1990), 1255–1260.

21. Society as a whole would also suffer, at least in theory, from distortions in the labor market as more-skilled employees at higher risk for illness are shunted aside for less-skilled but lower-risk employees. The relationship of future health to expected productivity and the possible offsetting effects of, in some circumstances, lower pension costs make this cost more speculative and, in any event, its magnitude is unclear.

22. The seminal work on this point is Lionel Charles Robbins, *An Essay on the Nature and Significance of Economic Science*, 2nd ed. (London: Macmillan, 1940), pp. 136–143 (impossible to make interpersonal utility comparisons in a "scientific way"). In legal literature, the attack on utilitarian justifications for equalizing wealth and income distributions was nicely summarized many years ago by Walter J. Blum and Harry Kalven, Jr., *The Uneasy Case for Progressive Taxation* (Chicago: University of Chicago Press, 1953), pp. 56–63.

23. The Act takes effect for employers with more than 25 workers in July 1992 and for employers with more than 15 workers in July 1994. No covered employer may discriminate in hiring, firing, compensation, or any terms or conditions of employment on the basis of disability, including the disability of one of the applicant's or employee's relatives. For a more complete discussion of the Act as it applies to this problem, see Greely, "The Death of Health Insurance."

24. Other legislation is relevant. Most states have their own bans on discrimination on the basis of handicap and the earlier federal law remains effective and may confer stronger rights in a few cases. ERISA itself protects existing employees (though not job applicants) from actions taken against them for exercising rights to which they are entitled under benefits plans. This section has been used to provide damages for a man found to have been fired after his diagnosis with multiple sclerosis because of the self-insured employer's concern about his medical expenses. See *Folz v. Marriott Corporation*, 594 F. Supp. 1007 (W.D. Mo. 1984).

25. Section 501(c)(1) of the ADA allows insurers and other health organizations to "underwrite risks, assign risks, or administer risks," in any way not inconsistent with State law. Section 501(c)(2) allows employers to sponsor, observe, or administer a plan that meets Section 501(c)(1). Section 501(c)(3) goes farther and provides that an employer may establish,

sponsor, or observe or administer the terms of any "bona fide benefit plan that is not subject to State laws that regulate insurance."

The same is true of ERISA. Although ERISA has detailed provisions requiring the "vesting" of pension rights, nonpension employee benefits are not subject to vesting. ERISA §§201 and 301, 29 U.S.C. §§1051 and 1081 (1985). Except as bound by the terms of the plan, a collective bargaining agreement, or any other contracts, the employer can change the terms of those plans at will.

26. In addition, by September 1990 at least one bill had been introduced in Congress to provide expressly for the privacy of individual genetic information gathered by the federal government, its contractors, or grantees. H.R. 5612, 101st Cong., 2d Sess. The bill has been strongly supported by both Jeremy Rifkin, a vociferous and litigious opponent of much genetic research, and W. French Anderson, a pioneering scientist in genetic therapy. Rifkin has already advocated broadening the legislation to include employers and insurers. Susan K. Miller, "Genetic Privacy Makes Strange Bedfellows," *Science*, 249 (1990), 1368.

27. For a proposal that the United States adopt something similar to the Canadian system, see David U. Himmelstein and Steffie Woolhander, "A National Health Program for the United States: A Physicians' Proposal," *New England Journal of Medicine*, 320 (1989), 102–107.

28. Such a proposal is laid out in Alain C. Enthoven and Richard Kronick, "A Consumer-Choice Health Plan for the 1990s: Universal Health Insurance in a System Designed to Promote Quality and Economy," *New England Journal of Medicine*, 320 (1989), 29–37 (Part I) and 94–107 (Part II).

29. See Greely, "The Death of Health Insurance."

30. See the critique of the Canadian system in Enthoven and Kronick, "A Consumer-Choice Health Plan for the 1990s," pp. 99–100.

13. Nature, Nurture, and the Human Genome Project

1. U.S. Congress, Office of Technology Assessment, *Mapping Our Genes* (Washington, D.C.: Government Printing Office, 1988), p. 85; Daniel Koshland, "Sequences and Consequences of the Human Genome," *Science*, 146 (1989), 189.

2. Office of Technology Assessment, *Mapping Our Genes*, p. 79.

3. Koshland, "Sequences and Consequences," p. 189.

4. Daniel Koshland, "Nature, Nurture, and Behavior," *Science*, 235 (1987), 1445.

5. Quoted in Robert Wright, "Achilles' Helix," *New Republic*, July 9 & 16, 1990, 21–31.

6. Robert Plomin, "The Role of Inheritance in Behavior," *Science*, 248 (April 13, 1990), 187.

7. Diane Paul, "Eugenics and the Left," *Journal of the History of Ideas*, 45 (1984), 574; H. J. Muller, "Social Biology and Population Improvement," *Nature*, 144 (1939), 521–522.

8. Warren Weaver, Transcript of Oral History Memoir, Oral History Office, Butler Library, Columbia University (Record No. 343), 1962, pp. 282–283.

9. Warren Weaver, Progress Report, the Natural Sciences, Rockefeller Foundation, 915.1.7, February 14, 1934; for further discussion, see, e.g., Evelyn Fox Keller, "Physics and the Emergence of Molecular Biology," *Journal of the History of Biology,* 23 (Fall 1990), 389–409.

10. Paul, "Eugenics and the Left," p. 589.

11. See Daniel J. Kevles, *In the Name of Eugenics: Genetics and the Uses of Human Heredity* (New York: Alfred A. Knopf, Inc., 1985), for an extensive review of the history of eugenics.

12. In 1969, Arthur Jensen provoked widespread outrage with his diatribe against the "zeitgeist of environmental egalitarianism" and his attempt to resurrect a connection between race and IQ. See Kevles, *In the Name of Eugenics.*

13. Philip Handler, ed., *Biology and the Future of Man* (Oxford: Oxford University Press, 1970), p. 926.

14. Ibid., p. 928.

15. Robert Sinsheimer, "The Prospect of Designed Genetic Change," *Engineering and Science,* 32 (1969), 8–13; reprinted in Ruth Chadwick, ed., *Ethics, Reproduction, and Genetic Control* (London: Croom Helm, 1987), p. 145.

16. Ibid.

17. Ibid.

18. By 1989, the number of human genes mapped had risen to over 1,450— this, out of total number of human genes estimated at somewhere between 50,000 and 100,000. Victor McKusick, "Mapping and Sequencing the Human Genome," *New England Journal of Medicine,* 320 (1989), 910–915.

19. Edward J. Yoxen, "Constructing Genetic Diseases," in Troy Duster and Karen Garett, eds., *Cultural Perspectives on Biological Knowledge* (Norwood, N.J.: Ablex, 1984).

20. Yoxen, "Constructing Genetic Diseases," p. 41.

21. P. A. Baird, "Genetics and Health Care," *Perspectives in Biology and Medicine,* 33 (1990), 203–213.

22. Yoxen, "Constructing Genetic Disease," p. 49.

23. Ibid., p. 50.

24. Ibid., p. 48.

25. See the chapter by Walter Gilbert in this volume.

26. National Research Council, *Mapping and Sequencing the Human Genome* (Washington, D.C.: National Academy Press, 1988), pp. 1, 12–13, 45.

27. Ibid., p. 11.

28. The quotations are taken from a lecture that Watson gave at the California Institute of Technology, May 9, 1990. See his chapter in this volume.

29. Sinsheimer, "Prospect of Designed Genetic Change," p. 146.

30. Office of Technology Assessment, *Mapping Our Genes,* p. 84.

31. Ibid., p. 86.

32. James D. Watson, "The Human Genome Project—Past, Present, and Future," *Science*, 248 (April 6, 1990), 44–49.
33. Charles Cantor, informal lecture at the University of California, Berkeley, 1990.
34. Miranda Robertson, "False Start on Manic Depression," *Nature*, 342 (November 18, 1989), 222.
35. Watson lecture, May 9, 1990.
36. Ian Hacking, *The Taming of Chance* (Cambridge: Cambridge University Press, 1990), p. 168.
37. Watson, "The Human Genome Project," p. 46.

14. Reflections

1. Michael Syvanen et al. to Colleagues, electronic mail notice, April 16, 1990, printout in our possession; Martin C. Rechsteiner letter, February 1990, attached to Rechsteiner to Norman Davidson, February 23, 1990, copy in our possession; Bernard Davis et al., "The Human Genome and Other Initiatives," *Science*, 249 (1990), 342–343; John M. Barry, "Cracking the Code," *The Washingtonian*, February 1991, p. 183; Jeffrey Mervis, "On Capitol Hill: One Day in the Hard Life of the Genome Project," *The Scientist*, August 20, 1990, p. 1. Rechsteiner's letter was published in *The FASEB Journal*, 4 (1990), 2941–2942. James Wyngaarden, a former director of NIH and a supporter of the genome project, noted, "If you took a vote in the biological sciences on the project, it would lose overwhelmingly." Barry, "Cracking the Code," p. 63. See also Robert Wright, "Achilles' Helix," *The New Republic*, 203 (July 9 & 16, 1990), 30.
2. Barry, "Cracking the Code," p. 183; Leslie Roberts, "Plan for Genome Center Sparks a Controversy," *Science*, 246 (October 13, 1989), 204; Davis et al., "The Human Genome and Other Initiatives," p. 342. See also Bernard D. Davis, "Some Problems with a Crash Program," *FASEB Journal*, 5 (January 1991), 76. Rechsteiner, ms of article, April 22, 1991, p. 1, forthcoming in *TIBS*. In the spring of 1990, a student of the project noted, "This widespread fear among biologists of 'big science,' and the accompanying doubts about the feasibility of the project, especially in the context of a biological science community that remains unconvinced about the benefits of contributing to such a project, will undoubtedly continue to haunt administrators of the project over the next fifteen years (or more)." Tracy Friedman, "The Science and Politics of the Human Genome Project," Senior Thesis, Woodrow Wilson School of Public and International Affairs, Princeton University, April 1990, p. 82.
3. Rechsteiner letter, Feb. 1990, pp. 1–2; Natalie Angier, "Vast 15-Year Effort to Decipher Genes Stirs Opposition," *New York Times*, June 5, 1990, p. B5.
4. Tracy L. Friedman, "The Science and Politics of the Human Genome Project," pp. 93–94, 99–101; Davis et al., "The Human Genome and

Other Initiatives," p. 342; Barry, "Cracking the Code," p. 183; Mervis, "On Capitol Hill," p. 1.

5. Friedman, "Science and Politics of the Human Genome Project," pp. 99–101, 104–105; Rechsteiner ms, April 22, 1991, p. 2.

6. Rechsteiner ms, April 22, 1991, p. 2; Rechsteiner, letter of February 1990, p. 2; Davis et al., "The Human Genome and Other Initiatives," p. 342.

7. Data supplied by Eric Lander, private communication, September 22, 1991.

8. National Science Board, NSF, *Science Indicators 1972* (Washington, D.C.: U.S. Government Printing Office, 1972), p. 128; National Science Board, NSF, *Science and Engineering Indicators—1989* (Washington, D.C.: U.S. Government Printing Office, 1989), p. 224.

9. R. Keith Wilkinson, *Science and Engineering Personnel: A National Overview* (Washington, D.C.: National Science Foundation, Surveys of Science Resources Series, Special Report NSF 90–310, 1989), p. 62; Friedman, "Science and Politics of the Human Genome Project," pp. 93–94.

10. Barry, "Cracking the Code," p. 183.

11. Eric Lander, "The Human Genome Project," talk at "Scientist-to-Scientist Colloquium," Keystone, Colorado, August 1991.

12. "Physiology or Medicine 1954: Presentation Speech by Professor S. Gard, member of the staff of professors of the Royal Caroline Institute," *Nobel Lectures . . . Physiology or Medicine, 1942–1962* (Amsterdam: Elsevier, 1964), p. 447. Patricia Hoben, a molecular biologist and staff member at the Office of Technology Assessment, has remarked that the human genome project "goes against the whole culture of biomedical research" and that few biologists focus "on the kind of technical aspects of research and are really wizards in that, and there are few who would want to be." Friedman, "Science and Politics of the Human Genome Project," pp. 69–70.

13. John L. Heilbron and Robert W. Seidel, *Lawrence and His Laboratory: A History of the Lawrence Berkeley Laboratory,* vol. 1 (Berkeley and Los Angeles: University of California Press, 1990), pp. 156–157, 189–190, 219; Daniel J. Kevles, *The Physicists: The History of a Scientific Community in Modern America* (Cambridge: Harvard University Press, 1987), pp. 274–275; Robert E. Kohler, *Partners in Science: Foundations and Natural Scientists, 1900–1945* (Chicago: University of Chicago Press, 1991), pp. 358–391; Lily E. Kay, *Cooperative Individualism and the Growth of Molecular Biology at the California Institute of Technology, 1928–1953* (Ann Arbor: UMIT Dissertation Services, 1987), especially chaps. 2, 4–6.

14. Friedman, "Science and Politics of the Human Genome Project," pp. 1–2; Daniel E. Koshland, Jr., "Sequences and Consequences of the Human Genome," editorial, *Science,* 246 (October 13, 1989), 189; *The New York Times,* May 27, 1990, p. 1; September 4, 1990, p. B7.

15. The model of the Geological Survey affected the organization of scientific research in the private sector, particularly in the debates over how Andrew Carnegie's munificent gift to establish the Carnegie Institution of Washington should best be used. See Kevles, *The Physicists,* pp. 82–83.

16. Lander, "The Human Genome Project." The 1991 budget figures are estimates and were provided to us by Erin Burgess, Budget Officer, National Center for Human Genome Research.

17. For the high cost of high-technology biology in its early days, see Kay, *Cooperative Individualism and the Growth of Molecular Biology at the California Institute of Technology,* pp. 113–187, 226–240.

18. Leslie Roberts, "New Game Plan for Genome Mapping," *Science,* 245 (September 29, 1989), 1438–1440; Friedman, "Science and Politics of the Human Genome Project," pp. 26–27.

19. Wright, "Achilles' Helix," p. 23.

20. "Gene Maps That Guide Biological Explorers," *Independent,* [London], December 4, 1989, BioDoc collection, European Economic Community, DG-XII, Brussels (hereafter cited as BioDoc); Francis S. Collins, "The Genome Project and Human Health," *FASEB Journal,* 5 (January 1991), 77; Victor A. McKusick, "Mapping and Sequencing the Human Genome," *New England Journal of Medicine,* 320 (April 6, 1989), 914–915.

21. Paul Berg, "All Our Collective Ingenuity Will Be Needed," *FASEB Journal,* 5 (January 1991), 75.

22. McKusick, "Current Trends in Mapping Human Genes," *FASEB Journal,* 5 (January 1991), 19.

23. G. Christopher Anderson, "Genome Database Booms as Journals Take the Hard Line," *The Scientist,* October 30, 1989, p. 4; Alessio Vassarotti et al., "Genome Research Activities in the EC," *Biofutur,* October 1990, pp. 1–4.

24. Author's (DJK's) discussion with Giorgio Bernardi, a molecular geneticist at the Institut Jacques Monod, Paris, in Valencia, Spain, November 11, 1990.

25. Hugo Johnson, "Human Genome et Revolution PCR," *Biofutur,* December 1989, p. 23; "American Know-How for Paid-Up Members Only," *Independent* [London], December 11, 1989, p. 13; *The Economist,* December 16, 1989, p. 94; Michael Cross, "Japan Drags Its Feet on Project to Map the Human Genome," *New Scientist,* January 6, 1990, p. 25, copies in BioDoc; "Q&A with James Watson, Genome Project Chief," *Science & Government Report,* March 15, 1990, pp. 1–2; Yoji Ikawa, "Human Genome Efforts in Japan," *FASEB Journal,* 5 (January 1991), 68–69. The Japanese government proposed to increase its human-genome budget by 66 percent for 1992, raising it to $11 million a year. David Swinbanks, "Japanese Science Agency Targets Space, Genome," *Nature,* 353 (September 5, 1991), 3.

26. "American Know-How for Paid-Up Members Only," p. 13; John Maddox, "The Case for the Human Genome," *Nature,* 352 (July 4, 1991), 13.

27. A business in commercial cyclotrons did develop in the 1950s. Perhaps a thousand of them now operate in the United States, doing radiopharmacy and radiotherapy and also things undreamed of by its original promoters, such as microlithography for computer circuits. U.S. Government Accounting Office, *DOE's Physics Accelerators: Their Costs and Benefits* (Washington, D.C.: GAO, RCED-85-96, April 1, 1985), p. 45. For a

fuller discussion of these points and what follows about high-energy physics, see J. L. Heilbron and Daniel J. Kevles, "Finding a Policy for Mapping and Sequencing the Human Genome: Lessons from the History of Particle Physics," *Minerva*, 26 (Fall 1988), 299–314.

28. Atomic Energy Act of 1946, Secs. 4, 6; Atomic Energy Act of 1954, Sec. 152. Executive Order 10096, 23 January 1950, gave the government rights to all inventions made by government employees during working hours or while using government facilities. Case law originating in implementation of the Order is reviewed by John O. Tresansky, "Patent Rights in Federal Employee Relations," Patent and Trademark Society, *Journal*, 67 (1985), 451–488; DOE, *Annual Report*, 1983, p. 61.

29. F. D. Gault, "Physics Databases and Their Use," *Computer Physics Communications*, 22 (1981), 125–132, and "The Particle Physics Data Group in the UK," *Computer Physics Communications*, 33 (1984), 217–219.

30. E.g., Yoshio Yamaguchi, speaking on behalf of the International Union of Pure and Applied Physics, in International Conference on High-Energy Accelerators, XII (1983), *Proceedings*, p. xviii, and on behalf of the International Committee for Future Accelerators, in International Symposium on Lepton and Photon Interactions at High Energies, *Proceedings*, ed. M. Konuma and M. Takahashi (IUPAP, 1985), pp. 826–827.

31. William J. Martin and R. Wayne Davies, "Automated DNA Sequencing: Progress and Prospects," *Biotechnology*, October 4, 1986, pp. 890–895.

32. Ibid.; Statement of Susan C. Rosenfeld, Science and Law Committee, Association of the Bar of the City of New York, at the Office of Technology Assessment Workshop on Issues of Collaboration for Human Genome Projects, June 26, 1987, pp. 6–7; *Science*, 240 (July 11, 1986), p. 157.

33. Rebecca S. Eisenberg, "Patenting the Human Genome," *Emory Law Journal*, 39 (Summer 1990), 726–728.

34. Ibid.

35. Craig Venter, a senior NIH scientist and a pioneer in the scientific exploitation of the method of expression site tags, personal communication, September 25, 1991.

36. Commission of the European Communities, *Modified Proposal for a Council Decision, Adopting a Specific Research and Technological Development Programme in the Field of Health: Human Genome Analysis (1990–1991)*, COM (89) 532 final–SYN 146, Brussels, November 13, 1989), pp. 12–14.

37. We are grateful to Eric Lander for helpful and illuminating discussion on the issue of patenting cDNAs.

38. *La Vie en Kit: Éthique et Biologie* (Paris: L'Arche de la Defense, 1991), p. 25.

39. Philip R. Reilly, *The Surgical Solution: A History of Involuntary Sterilization in the United States* (Baltimore: The Johns Hopkins University Press, 1991), pp. 91–93. The last state eugenic sterilization law was passed in 1937, in Georgia, partly in response to conditions of overcrowding in the state's institutions for the mentally handicapped. Edward J. Larson, "Breeding Better Georgians," *Georgia Journal of Southern Legal History*, 1 (Spring/ Summer 1991), 53–79.

40. Steven Jay Gould, *The Flamingo's Smile: Reflections in Natural History* (New York: W. W. Norton, 1985), pp. 292–295, 301–303; *The New York Times*, August 15, 1991, p. 1.

41. Wright, "Achilles' Helix," p. 27; Joseph Bishop and Michael Waldholz, *Genome: The Story of the Most Astonishing Scientific Adventure of Our Time— The Attempt to Map All the Genes in the Human Body* (New York: Simon and Schuster, 1990), pp. 310–322.

42. Jane E. Brody, "Personal Health," *The New York Times*, November 8, 1990, p. B7; Barry Werth, "How Short Is Too Short?" *The New York Times Magazine*, June 16, 1991, pp. 15, 17, 28–29; European Parliament, Committee on Energy, Research, and Technology, *Report Drawn up on Behalf of the Committee on Energy, Research and Technology on the Proposal from the Commission to the Council (COM/88/424-C2-119/88) for a Decision Adopting a Specific Research Programme in the Field of Health: Predictive Medicine: Human Genome Analysis (1989–1991)*, Rapporteur Benedikt Härlin, European Parliament Session Documents, 1988–89, 30.01.1989, Series A, Doc A2–0370/88 SYN 146, pp. 25–26 (hereafter cited as Härlin Report); John Hodgson, "Editorial: Geneticism and Freedom of Choice," *Trends in Biotechnology*, September 1989, p. 221.

43. "Resolutions Adopted by the European Parliament on 16 March 1989," in European Parliament, Committee on Legal Affairs and Citzens' Rights, Rapporteurs: Mr. Willi Rothley and Mr. Carlo Casini, *Ethical and Legal Problems of Genetic Engineering and Human Artificial Insemination* (Luxembourg: Office for Publications of the European Communities, 1990), pp. 15, 38–39. Carlo Casini, from Italy, the rapporteur for the section of the report concerned with human artificial insemination, is known in the circles of the Parliament as virtually a papal representative to the legislative body. Embryo research and germ-line engineering is also opposed by many adherents of Islamic religion and by many Protestants, most recently in a 1989 report by the World Council of Churches. Lectures by Azeddine Guessos and Jack Stotts, "II Workshop on International Cooperation for the Human Genome Project: Ethics," Valencia, Spain, November 12, 1990.

44. "Resolutions Adopted by the European Parliament on 16 March 1989," in European Parliament, Committee on Legal Affairs and Citizens' Rights, *Ethical and Legal Problems of Genetic Engineering and Human Artificial Insemination*, p. 12; Daniel J. Kevles, "Unholy Alliance," *The Sciences*, September/October 1986, pp. 25–30.

45. Bishop and Waldholz, *Genome*, pp. 314–316; Sharon Kingman, "Buried Treasure in Human Genes," *New Scientist*, July 8, 1989, p. 37.

46. *Consumer Reports*, July 1990, pp. 483–488; Bishop and Waldholz, *Genome*, pp. 268–270.

47. Thomas H. Murray, "Ethical Issues in Human Genome Research," *FASEB Journal*, 5 (January 1991), 56; Benjamin S. Wilfond and Norman Fost, "The Cystic Fibrosis Gene: Medical and Social Implications for Heterozygote Detection," *Journal of the American Medical Association*, 263 (May 23/30 1990), 2777.

48. Bishop and Waldholz, *Genome*, pp. 291–294; Wilfond and Fost, "The Cystic Fibrosis Gene," p. 2777.
49. Bishop and Waldholz, *Genome*, p. 274.
50. Ibid., pp. 300–302; G. W. de Wit, "Gentechnology, Insurance, and the Future," paper delivered at "II Workshop on International Cooperation for the Human Genome Project: Ethics," Valencia, Spain, November 11–14, 1990.
51. Härlin Report, pp. 25–26.
52. Daniel J. Kevles, *In the Name of Eugenics: Genetics and the Uses of Human Heredity* (New York: Alfred A. Knopf, 1985), p. 278.
53. "The Genetic Age," *Business Week*, May 28, 1990, p. 69; Bishop and Waldholz, *Genome*, pp. 282–283; Paul Billings, "Genetic Discrimination: An Ongoing Survey," *geneWatch*, 6, nos. 4–5 (n.d. [1991]), 7, 15.
54. Wright, "Achilles' Helix," p. 26; Editorial, "More Genome Ethics" (p. 2), and Peter Aldhous, "California Tackles Insurance" (p. 5), *Nature*, 353 (September 5, 1991); Assembly Bill 1888, California State Legislature, "Legislative Counsel's Digest," September 1991, copy in our possession.
55. Bishop and Waldholz, *Genome*, pp. 299–300.
56. Judy Payne, senior policy analyst at the Health Insurance Association of America, has said: "Just because somebody else is paying the bill doesn't mean the bill doesn't exist . . . most people are so used to having their health care paid for that they just think you can shift the cost and the cost goes away." Judy Berlfein, "Genetic Testing: Health Care Trap," *Los Angeles Times*, April 30, 1991, p. B2.
57. Gary Taylor, "Houston Conference Explores Human Genome Project's Legal and Ethical Considerations," *Genetic Engineering News*, May 1991, p. 16.
58. Wright, "Achilles' Helix," p. 26; Alexander Morgan Capron, "Which Ills to Bear?: Reevaluating the 'Threat' of Modern Genetics," *Emory Law Journal*, 39 (Summer 1990), 694–695. Rate discrimination has become increasingly widespread in medical insurance on the basis of, for example, age. *New York Times*, March 24, 1991, p. 1.
59. de Wit, "Gentechnology, Insurance, and the Future," pp. 8, 14; G. W. de Wit, "The Politics of Rate Discrimination: An International Perspective," *Journal of Risk and Insurance*, 53 (1986), 660. The European Parliament's Committee on Legal Affairs, less sanguine than de Wit, warns that private insurers in Europe might exclude high-risk people, leaving them to the care of the public social security system. European Parliament, Committee on Legal Affairs and Citizens' Rights, *Ethical and Legal Problems of Genetic Engineering and Human Artificial Insemination*, pp. 55–56.
60. Wright, "Achilles' Helix," p. 26; Berlfein, "Genetic Testing," p. B2.
61. *New York Times*, July 4, 1991, p. 12; European Parliament, Committee on Legal Affairs and Citizens' Rights, *Ethical and Legal Problems of Genetic Engineering and Human Artificial Insemination*, p. 29.
62. Bishop and Waldholz, *Genome*, pp. 226–234, 243–244, 251–262; "Mental Illness Theory—Study Raises Doubts," *Los Angeles Times*, November 17,

1989, p. 29. The confirmation study for manic depression failed even though it used the same cell line employed in the initial investigation. Marcia Barinaga, "Manic Depression Gene Put in Limbo," *Science*, 246 (November 17, 1989), 886.

63. Kathleen Doheny, "Researchers Find Genetic Link to Shyness, Hay Fever," *Los Angeles Times*, November 27, 1990, p. E4.

64. Lawrence K. Altman, "Scientists Say a Specific Gene May Foreshadow Alcoholism," *New York Times*, April 18, 1990, p. 1; Gina Kolata, "Researchers Cannot Confirm a Genetic Link to Alcoholism," *New York Times*, December 26, 1990, p. 10. Even Robert Plomin, a geneticist at Penn State University who has noted enthusiastically that "the role of inheritance in behavior has become widely accepted" has remarked that "acceptance of genetic influence has begun to outstrip the data in some cases, such as alcoholism." Robert Plomin, "The Role of Inheritance in Behavior," *Science*, 248 (April 13, 1990), 187–188.

65. European Parliament, Committee on the Environment, Public Health, and Consumer Protection, *Opinion for the Committee on Energy, Research and Technology on the Proposal from the Commission of the European Communities for a Council Decision Adopting a Specific Research Programme in the Field of Health: Predictive Medicine: Human Genome Analysis (1989–1991)* (COM (88) 424 final–SYN 146 Doc. C2–119/88, 1989), Draftsman: Mrs. Lentz-Cornette, pp. 3, 5–8; Telephone conversation with Gordon Lake, staff member, European Parliament, April 24, 1991.

66. Claude Cheysson, foreword to the catalogue of *La Vie en Kit*, p. 1.

Selected Bibliography

Science, Technology, and Medicine

Anderson, G. Christopher. "Creation of Linkage Map Falters, Posing Delay for Genome Project." *The Scientist*, (January 8, 1990), 1, 10, 12–13.

Ayala, Francisco. "Two Frontiers of Human Biology: What the Sequence Won't Tell Us." *Issues in Science and Technology*, 3 (Spring 1987), 51–56.

Baltimore, David. "Genome Sequencing: A Small-Science Approach." *Issues in Science and Technology*, 3 (Spring 1987), 48–50.

——— "RNA-dependent DNA Polymerase in Virions of RNA Tumour Viruses." *Nature*, 226 (1970), 1209–1211.

Bishop, Joseph, and Michael Waldholz. *Genome: The Story of the Most Astonishing Scientific Adventure of Our Time—The Attempt to Map All the Genes in the Human Body*. New York: Simon and Schuster, 1990.

Botstein, David, et al. "Construction of a Genetic Linkage Map in Man Using Restriction Fragment Length Polymorphisms." *American Journal of Human Genetics*, 32 (1980), 314–331.

Burke, D. T., G. F. Carle, and M. V. Olson. "Cloning of Large Segments of Exogenous DNA into Yeast Artificial Chromosome Vectors." *Science*, 236 (1987), 806–808.

Cohen, S. N., et al. "Construction of Biologically Functional Bacterial Plasmids in Vitro." *Proceedings of the National Academy of Sciences*, 70 (1973), 3240–3244.

DeLisi, Charles. "Computers in Molecular Biology: Current Applications and Emerging Trends." *Science*, 240 (April 1, 1988), 47–52.

Dulbecco, Renato. "A Turning Point in Cancer Research: Sequencing the Human Genome." *Science*, 231 (1986), 1055–1056.

Gallo, R. C., et al. "Isolation of Human T-cell Leukemia Virus in Acquired Immune Deficiency Syndrome (AIDS)." *Science*, 220 (1983), 865–867.

Geever, R. F., et al. "Direct Identification of Sickle Cell Anemia by Blot Hybridization." *Proceedings of the National Academy of Sciences,* 78 (1981), 5081–5085.

Gordon, J. W., et al. "Genetic Transformation of Mouse Embryos by Microinjection of Purified DNA." *Proceedings of the National Academy of Sciences,* 77 (1980), 7380–7384.

Gusella, James, et al. "A Polymorphic DNA Marker Genetically Linked to Huntington's Disease." *Nature,* 306 (1983), 234–238.

Hall, Stephen S. "James Watson and the Search for Biology's 'Holy Grail.'" *Smithsonian,* February 1990, pp. 41–49.

Hozumi, N., and S. Tonegawa. "Evidence for Somatic Rearrangement of Immunoglobulin Genes Coding for Variable and Constant Regions." *Proceedings of the National Academy of Sciences,* 73 (1976), 3628–3632.

The Human Genome Organization (HUGO). Montreaux, Switzerland: HUGO, 1988.

Jacobson, C. B., and R. H. Barter. "Intrauterine Diagnosis and Management of Genetic Defects." *American Journal of Obstetrics and Gynecology,* 99 (1967), 796–807.

Judson, Horace Freeland. *The Eighth Day of Creation: Makers of the Revolution in Biology.* New York: Simon and Schuster, 1979.

—— "Reflections on the Historiography of Molecular Biology." *Minerva,* 18 (Autumn 1980), 369–421.

Kan, Y. W., M. S. Golbus, and A. M. Dozy. "Prenatal Diagnosis of Alpha-thalassemia. Clinical Application of Molecular Hybridization." *New England Journal of Medicine,* 295 (1976), 1165–1167.

Kay, Lily E. *Cooperative Individualism and the Growth of Molecular Biology at the California Institute of Technology, 1928–1953.* Ann Arbor: UMI Dissertation Service, 1987.

Kerem, B. S., et al. "Identification of the Cystic Fibrosis Gene: Genetic Analysis." *Science,* 245 (1989), 1073–1080.

Lejeune, J., M. Gautier, and R. Turpin. "Etudes des chromosomes somatique de neuf enfants mongoliens." *Comptes Rendus de l'Academie des Sciences* (Paris), D, 248 (1959), 1721–1722.

Littlefield, J. W. "Selection of Hybrids from Matings of Fibroblasts In Vitro and Their Presumed Recombinants." *Science,* 145 (1964), 709–710.

Mandel, M., and A. Higa. "Calcium-dependent Bacteriophage DNA Infection." *Journal of Molecular Biology,* 53 (1970), 159–162.

Maxam, A. M., and W. Gilbert. "A New Method for Sequencing DNA." *Proceedings of the National Academy of Sciences,* 74 (1977), 560–564.

McKusick, Victor. "Mapping and Sequencing the Human Genome." *New England Journal of Medicine,* 320 (April 6, 1989), 910–915.

—— "Current Trends in Mapping Human Genes." *The FASEB Journal,* 5 (January 1991), 12–20.

McKusick, V. A., C. A. Francomano, and S. E. Antonarakis. *Mendelian Inheritance in Man. Catalogs of Autosomal Dominant, Autosomal Recessive, and X-linked Phenotypes.* 9th edition. Baltimore: The Johns Hopkins University Press, 1990.

Meselson, M., and F. W. Stahl. "The Replication of DNA in *Escherichia coli*." *Proceedings of the National Academy of Sciences*, 44 (1958), 671–682.

Myerowitz, R., and F. C. Costigan. "The Major Defect in Ashkenazi Jews with Tay-Sachs Disease Is an Insertion in the Gene for the ∝-Chain of β-Hexosaminidase." *Journal of Biological Chemistry*, 263 (1988), 18587–18589.

National Research Council, Committee on Mapping and Sequencing the Human Genome, Board on Basic Biology, Commission on Life Sciences. *Mapping and Sequencing the Human Genome.* Washington, D.C.: National Academy Press, 1988.

Nelson, D. L., et al. "*Alu* Polymerase Chain Reaction: A Method for Rapid Isolation of Human-specific Sequences from Complex DNA Sources." *Proceedings of the National Academy of Sciences*, 86 (1989), 6686–6690.

Olby, Robert. *The Path to the Double Helix.* Seattle: University of Washington Press, 1974.

Palca, Joseph. "Hard Times at NIH." *Science*, 246 (November 24, 1989), 988–990.

Parker, R. C., H. E. Varmus, and J. M. Bishop. "Cellular Homologue (*c-src*) of the Transforming Gene of Rous Sarcoma Virus: Isolation, Mapping, and Transcriptional Analysis of *c-src* and Flanking Regions." *Proceedings of the National Academy of Sciences*, 78 (1981), 5842–5846.

Pines, Maya. *Mapping the Human Genome.* Bethesda, Md.: Howard Hughes Medical Institute, 1987.

Roberts, Leslie. "Whatever Happened to the Genetic Map?" *Science*, 247 (January 19, 1990), 281–282.

Robertson, Miranda. "False Start on Manic Depression." *Nature*, 342 (November 18, 1989), 222.

Rosenberg, S. A. "Gene Transfer into Humans—Immunotherapy of Patients with Advanced Melanoma, Using Tumor-infiltrating Lymphocytes Modified by Retroviral Gene Transduction." *New England Journal of Medicine*, 323 (1990), 570–578.

Saiki, R. K., et al. "Enzymatic Amplification of β-Globin Genomic Sequences and Restriction Site Analysis for Diagnosis of Sickle Cell Anemia." *Science* 230 (1985), 1350–1354.

Saiki, R. K., et al. "Primer-directed Enzymatic Amplification of DNA with a Thermostable DNA Polymerase." *Science*, 239 (1988), 487–491.

Sanger, F., and A. R. Coulson. "A Rapid Method of Determining Sequences in DNA by Primed Synthesis with DNA Polymerase." *Journal of Molecular Biology*, 94 (1975), 441–448.

Schwartz, D. C., and C. R. Cantor. "Separation of Yeast Chromosome–Sized DNAs by Pulsed Field Gradient Gel Electrophoresis." *Cell*, 37 (1984), 67–75.

Smith, Lloyd M. et al. "Fluorescence Detection in Automated DNA Sequence Analysis." *Nature*, 321 (June 12, 1986), 674–679.

Southern, E. M. "Detection of Specific Sequences among DNA Fragments Separated by Gel Electrophoresis." *Journal of Molecular Biology*, 98 (1975), 503–517.

Stephens, J. C., et al. "Mapping the Human Genome: Current Status." *Science*, 250 (1990), 237–244.

Temin, H. M., and S. Mizutani. "RNA-dependent DNA Polymerase in Virions of Rous Sarcoma Virus." *Nature*, 226 (1970), 1211–1213.

U.S. Congress, Office of Technology Assessment. *Mapping Our Genes—The Genome Project: How Big, How Fast?* OTA-BA-373. Washington, D.C.: U.S. Government Printing Office, April 1988.

Villa-Komaroff, L., et al. "A Bacterial Clone Synthesizing Proinsulin." *Proceedings of the National Academy of Sciences*, 75 (1975), 3727–3731.

Volkow, N., and L. Tancredi. "Neural Substrates of Violent Behavior: A Preliminary Study with Positron Emission Tomography." *British Journal of Psychiatry*, 151 (1987), 668–673.

Watson, James D. "The Human Genome Project—Past, Present, and Future." *Science*, 248 (April 6, 1990), 44–49.

Watson, J. D., and F. H. C. Crick. "Molecular Structure of Nucleic Acid: A Structure for Deoxyribonucleic Acid." *Nature*, 171 (1953), 737–738.

Weinberg, Robert A. "The Human Genome Sequence: What Will It Do for Us?" *BioEssays*, 9 (August–September 1988), 91–92.

Wilson, E. O. *Sociobiology: The New Synthesis*. Cambridge: Belknap Press, Harvard University Press, 1975.

Wingerson, Lois. *Mapping Our Genes: The Genome Project and the Future of Medicine*. New York: Dutton, 1990.

Wolff, J. A., et al. "Direct Gene Transfer into Mouse Muscle in Vivo." *Science*, 247 (1990), 1465–1468.

Ethics, Law, and Society

Adams, Mark, ed. *The Wellborn Science: Eugenics in Germany, France, Brazil, and Russia*. New York: Oxford University Press, 1990.

Allen, Garland. "The Misuse of Biological Hierarchies: The American Eugenics Movement, 1900–1940." *History and Philosophy of the Life Sciences*, 5 (May 2, 1983), 105–128.

Badouin, Jean-Louis, and Catherine Labrusse-Riou. *Produire l'homme de quel droit?: Etude juridique et éthique des procréations artificielles*. Paris: Press Universitaire de France, 1987.

Baird, P. A. "Genetics and Health Care." *Perspectives in Biology and Medicine*, 33 (1990), 203–213.

Ballantyne, John, et al., eds. *DNA Technology and Forensic Science*. Banbury Report: No. 32. Cold Spring Harbor, N.Y.: Cold Spring Harbor Press, 1989.

Bayles, Michael. *Reproductive Ethics*. Englewood Cliffs, N.J.: Prentice-Hall, 1984.

Bennett, Neil G., ed. *Sex Selection of Children*. New York and London: Academic Press, 1983.

Bernard, Jean. *De la Biologie à l'Éthique: Nouveau Pouvoirs de la Science, Nouveaux Devoirs de l'Homme*. Paris: Éditions Buchet/Chastel, 1990.

Bioéthique, Pouvoirs: Revue Française d'Études Constitutionelles et Politiques, 56 (1991).

Blanc, Marcel. *L'Ère de la Génétique.* Paris: Éditions la Découverte, 1986.

Brock, D. J. H. *Early Diagnosis of Fetal Defects.* Edinburgh and London: Churchill Livingstone, 1982.

Canguilhem, George. *The Normal and the Pathological.* Cambridge: MIT Press, 1989.

Capron, Alexander Morgan. "Which Ills to Bear?: Reevaluating the 'Threat' of Modern Genetics," *Emory Law Journal,* 39 (Summer 1990), 665–696.

Ciba Foundation. *Human Genetic Information: Science, Law and Ethics.* Ciba Foundation Symposium 149. New York: John Wiley, 1990.

Coles, Gerald. *The Learning Mystique.* New York: Pantheon, 1987.

Commission of the European Communities. *Modified Proposal for a Council Decision, Adopting a Specific Research and Technological Development Programme in the Field of Health: Human Genome Analysis (1990–1991).* (COM (89) 532 final–SYN 146), Brussels, November 13, 1989.

Cook-Deegan, Robert Mullan. "The Human Genome Project: The Formation of Federal Policies in the United States, 1986–1990." In Kathi E. Hanna, ed., *Biomedical Politics,* pp. 99–168. Washington, D.C.: National Academy Press, 1991.

Dagognet, François. *La maîtrise du vivant.* Paris: Hachette, 1988.

Davis, Joel. *Mapping the Code: The Human Genome Project and the Choices of Modern Science.* New York: John Wiley, 1990.

de Wit, G. W. "The Politics of Rate Discrimination: An International Perspective." *Journal of Risk and Insurance,* 53 (1986), 644–661.

Douglas, Mary. *How Institutions Think.* Syracuse: Syracuse University Press, 1986.

Duster, Troy. *Back Door to Eugenics.* New York: Routledge, 1989.

Eisenberg, Rebecca S. "Patenting the Human Genome." *Emory Law Journal,* 39 (Summer 1990), 721–745.

Enthoven, Alain C. *Health Plan: The Only Practical Solution to the Soaring Cost of Medical Care.* Reading, Mass.: Addison-Wesley, 1980.

European Parliament, Committee on Energy, Research, and Technology. *Report Drawn up on Behalf of the Committee on Energy, Research and Technology on the Proposal from the Commission to the Council (COM/88/424-C2–119/88) for a Decision Adopting a Specific Research Programme in the Field of Health: Predictive Medicine: Human Genome Analysis (1989–1991).* Rapporteur: Benedikt Härlin. European Parliament Session Documents, 1988–89, 30.01.1989, Series A, Doc. A2–0370/88 SYN 146, Brussels.

European Parliament, Committee on the Environment, Public Health, and Consumer Protection. *Opinion for the Committee on Energy, Research and Technology on the Proposal from the Commission of the European Communities for a Council Decision Adopting a Specific Research Programme in the Field of Health: Predictive Medicine: Human Genome Analysis (1989–1991) (COM (88)424 final–SYN 146, Doc. C2–119/88).* Draftsman: Mrs. Lentz-Cornette. Brussels.

European Parliament, Committee on Legal Affairs and Citizens' Rights. *Ethi-*

cal and Legal Problems of Genetic Engineering and Human Artificial Insemination. Rapporteurs: Mr. Willi Rothley and Mr. Carlo Casini. Luxembourg: Office for Official Publications of the European Communities, 1990.

Evans, Mark I., Alan O. Dixler, John C. Fletcher, and Joseph D. Schulman, eds. *Fetal Diagnosis and Therapy: Science, Ethics and the Law.* Philadelphia: Lippincott, 1988.

Fischer, Jean-Louis, ed. *Confidences d'un biologiste: Jean Rostand.* Paris: Éditions La Découverte, 1987.

Fletcher, John C., and Dorothy Wertz. "Ethics, Law, and Medical Genetics: After the Human Genome Is Mapped." *Emory Law Journal,* 39 (Summer 1990), 747–809.

Foucault, Michel. *Discipline and Punish: The Birth of the Prison.* New York: Vintage Books, 1979.

Fox, Daniel M., and Daniel C. Schaffer. "Health Policy and ERISA: Interest Groups and Semipreemption." *Journal of Health Policy, Policy, and Law,* 14 (1989), 239–260.

Friedman, Tracy L. "Science and Politics of the Human Genome Project." Senior thesis, Woodrow Wilson School of Public and International Affairs, Princeton University, April 1990.

Gaudilliere, Jean-Paul. "French Strategies in Molecular Biology." Paper delivered at conference on the Human Genome Project, History of Science Department, Harvard University, June 1990.

German Bundestag. *Report of the Commission of Enquiry on Prospects and Risks of Genetic Engineering.* 10th Legislative Period; Paper 10/6775. Bonn: German Bundestag, June 1, 1987.

Gilligan, Carol. *In a Different Voice: Psychological Theory and Women's Development.* Cambridge: Harvard University Press, 1982.

Goldstein, Robert. *Mother Love and Abortion: A Legal Interpretation.* Berkeley: University of California Press, 1988.

Graham, Loren R. "Science and Values: The Eugenics Movement in Germany and Russia in the 1920s." *American Historical Review,* 83 (1978), 1135–1164.

Greely, Henry T. "AIDS and the American Health Care Financing System." *University of Pittsburgh Law Review,* 51 (1989), 73–166.

—— "The Death of Health Insurance: Employment-Related Health Coverage and the Increasing Predictability of Individual Health." Forthcoming.

Hacking, Ian. *The Taming of Chance.* Cambridge: Cambridge University Press, 1990.

Handler, Philip, ed. *Biology and the Future of Man.* New York: Oxford University Press, 1970.

Harris, Irving. *Emotional Blocks to Learning: A Study of the Reasons for Failure in School.* Glencoe, Ill.: Free Press, 1962.

Harwood, Jonathan, ed. "Genetics, Eugenics, and Evolution." Special Issue of *The British Journal for the History of Science,* 22 (1989), 257–265.

Henry, Robin Marantz. "High-Tech Fortunetelling." *New York Times Magazine,* December 24, 1989, pp. 20–22.

Hermitte, Marie-Angèle. "Le Sequencage du Genome Humain: Liberté de la Recherche et Demarche Democratique." Unpublished manuscript in possession of the author.

Hoffman, Elaine Baruch, Amadeo F. D'Amado, and Joni Seager, eds. *Embryos, Ethics and Women's Rights: Exploring the New Reproductive Technologies*. New York and London: Harrington Park Press, 1988.

Holtzman, Neil A. *Proceed with Caution: Predicting Genetic Risks in the Recombinant DNA Era*. Baltimore: Johns Hopkins University Press, 1989.

"The Human Genome Initiative." A series of articles in *The FASEB Journal*, 5 (January 1991), 1–78.

Institut National de la Santé et de la Recherche Médicale, Comité Consultatif National d'Éthique. *Avis de Recherches sur L'embryon*. La Fabrique du Corps Humain. Paris: Actes Sud et INSERM, 1987.

—— *Comités d'Éthiques a travers le Monde: Recherches en Cours 1987*. Paris: Éditions Tierce, 1988.

—— *Éthique Médicale et Droits de L'Homme*. La Fabrique du Corps Humain. Paris: Actes Sud, 1988.

Jauvert, Vincent, ed. *Va-t-on modifier l'espèce humaine?: Les prodiges et les menaces de la révolution bio*. Paris: *Le Nouvel Observateur*, Documents No. 10, 1990.

Jeffrey, C. Ray. *Criminology: An Interdisciplinary Approach*. Englewood Cliffs, N.J.: Prentice Hall, 1990.

Jeffrey, C. Ray, in collaboration with R. V. Del Carmen and J. D. White. *Attacks on the Insanity Defense: Biological Psychiatry and New Perspectives on Criminal Behavior*. Springfield, Ill.: Charles C. Thomas, 1985.

Jensen, Arthur. "How Much Can We Boost IQ and Scholastic Achievement?" *Harvard Educational Review*, 33 (1969), 159–179.

Keller, Evelyn Fox. "Physics and the Emergence of Molecular Biology." *Journal of the History of Biology*, 23 (Fall 1990), 389–409.

Kevles, Daniel J. *In the Name of Eugenics: Genetics and the Uses of Human Heredity*. Berkeley: University of California Press, 1986.

Koshland, Daniel. "Nature, Nurture, and Behavior." *Science*, 235 (March 20, 1987), 1445.

—— "Sequences and Consequences of the Human Genome." *Science*, 246 (October 13, 1989), 189.

Lander, Eric S. "DNA Fingerprinting on Trial." *Nature*, 339 (1989), 501–505.

Larson, Edward J. "Belated Progress: The Enactment of Eugenic Legislation in Georgia." *Journal of the History of Medicine and Allied Sciences*, 46 (January 1991), 44–64.

Lecourt, Dominique. *Contre La Peur: De la science à l'éthique, une aventure infinie*. Paris: Hachette, 1990.

"A Legal Research Agenda for the Human Genome Initiative." *Jurimetrics Journal*, forthcoming.

Lemain, Gérard, and Benjamin Matalon. *Hommes supérieurs, hommes inférieurs*. Paris: Armand Colin, 1985.

Lewis, Ricki. "Genetic-Marker Testing: Are We Ready for It?" *Issues in Science and Technology*, 4 (Fall 1987), 76–82.

Liebman, Lance. "Too Much Information: Predictions of Employee Disease and the Fringe Benefit System." *University of Chicago Legal Forum*, 1988, 57–92.

Lindee, Mary Susan. "Mutation, Radiation, and Species Survival: The Genetics Studies of the Atomic Bomb Casualty Commission in Hiroshima and Nagasaki, Japan." Ph.D. diss., Department of History, Cornell University, 1990.

Lucas, Philippe. *Dire L'Éthique: Éthique Biomédicale: Le Débat.* La Fabrique du Corps Humain. Paris: Actes Sud et INSERM, 1990.

Ludmerer, Kenneth M. *Genetics and American Society: A Historical Appraisal.* Baltimore: Johns Hopkins University Press, 1972.

Macklin, Ruth. "Mapping the Human Genome: Problems of Privacy and Free Choice." In Aubrey Milunsky and George J. Annas, eds., *Genetics and the Law III: Proceedings of the Third National Symposium on Genetics and the Law Held in Boston 2–4 April 1984*, pp. 107–114. New York: Plenum Press, 1985.

Müller-Hill, Benno. *Murderous Science: Elimination by Scientific Selection of Jews, Gypsies, and Others, Germany, 1933–1945.* New York: Oxford University Press, 1988.

National Institute of Mental Health. *Approaching the 21st Century: Opportunities for NIMH Neurosciences Research.* Report to Congress on the Decade of the Brain. Washington, D.C.: U.S. Department of Health and Human Services, January 1988.

Nelkin, Dorothy. *Selling Science: How the Press Covers Science and Technology.* New York: W. H. Freeman, 1988.

Nelkin, Dorothy, and Michael S. Brown. *Workers at Risk.* University of Chicago Press, 1984.

Nelkin, Dorothy, and Laurence Tancredi. *Dangerous Diagnostics: The Social Power of Biological Information.* New York: Basic Books, 1989; paperback, 1991.

Newhouse, Joseph P., et al. "Adjusting Capitation Rates Using Objective Health Measures and Prior Utilization." *Health Care Financing Review*, 10 (Spring 1989), 41–54.

Paul, Diane. "Eugenics and the Left." *Journal of the History of Ideas*, 45 (October 1984), 567–590.

———— "The Rockefeller Foundation and the Origins of Behavioral Genetics." In Keith R. Benson, Jane Maienschein, and Ronald Rainger, eds., *The Expansion of American Biology*, pp. 262–283. New Brunswick: Rutgers University Press, 1991.

Petchesky, Rosalind Pollack. *Abortion and Woman's Choice: The State, Sexuality and Reproductive Freedom.* New York: Longman, 1984.

Plomin, Robert. "The Role of Inheritance in Behavior." *Science*, 248 (April 13, 1990), 183–188.

President's Commission for the Study of Ethical Problems in Medicine and Biomedical Research. *Screening and Counseling for Genetic Conditions: A Report on the Ethical, Social and Legal Implications of Genetic Screening, Counseling and Education Progams.* Washington, D.C.: The Commission, 1983.

Proctor, Robert N. *Racial Hygiene: Medicine Under the Nazis.* Cambridge: Harvard University Press, 1988.

Quéré, France. *L'Éthique et La Vie.* Paris: Éditions Odile Jacob, 1991.

"The Randolph W. Thrower Symposium: Genetics and the Law." *Emory Law Journal,* 39 (1990).

Reich, William. "Diagnostic Ethics: The Uses and Limits of Psychiatric Explanation." In Laurence Tancredi, ed., *Ethical Issues in Epidemiologic Research,* pp. 37–69. New Brunswick: Rutgers University Press, 1986.

Reilly, Philip R. *The Surgical Solution: A History of Involuntary Sterilization in the United States.* Baltimore: Johns Hopkins University Press, 1991.

Roll-Hansen, Nils. "Eugenics before World War II: The Case of Norway." *History and Philosophy of the Life Sciences,* 2 (1980), 269–298.

——— "The Progress of Eugenics: Growth of Knowledge and Change in Ideology." *History of Science,* 26 (1988), 295–331.

Rothman, Barbara Katz. *Recreating Motherhood: Ideology and Technology in a Patriarchal Society.* New York: W. W. Norton, 1989.

——— *The Tentative Pregnancy: Prenatal Diagnosis and the Future of Motherhood.* New York: Viking Penguin, 1986.

Rothstein, Mark A. *Medical Screening and the Employee Health Cost Crisis.* Washington, D.C.: Bureau of National Affairs, 1989.

——— *Medical Screening of Workers.* Washington, D.C.: Bureau of National Affairs, 1984.

Rowley, Peter. "Genetic Discrimination." *American Society of Human Genetics,* 43 (July 1988), 105–106.

Schneider, William H. *Quality and Quantity: The Quest for Biological Regeneration in Twentieth-Century France.* Cambridge: Cambridge University Press, 1990.

Scriver, Charles R. "Presidential Address." *American Journal of Human Genetics,* 40 (1987), 199–211.

Searle, G. R. *Eugenics and Politics in Britain, 1900–1914.* Leyden: Noordhoff International Publishing, 1976.

Sensabaugh, G., and J. Witkowski, eds. *DNA Technology and Forensic Science.* Banbury Report: No. 34. Cold Spring Harbor, N.Y.: Cold Spring Harbor Press, 1989.

Shaw, Margery W. "Conditional Prospective Rights of the Fetus." *Journal of Legal Medicine,* 63 (1984), 63–116.

——— "Presidential Address: To Be or Not to Be? That Is the Question." *American Journal of Human Genetics,* 36 (1984), 1–9. (Based on presidential address delivered at 33rd Annual Meeting of American Society of Human Genetics, Detroit, September 30, 1982.)

Sinsheimer, Robert. "The Prospect of Designed Genetic Change." *Engineering and Science,* 32 (1969), 8–13.

Stanworth, Michelle, ed. *Reproductive Technologies: Gender, Motherhood and Medicine.* Minneapolis: University of Minnesota Press, 1987.

Starr, Paul. *The Social Transformation of American Medicine.* New York: Basic Books, 1982.

Stemerding, Dirk. "Political Decision-Making on Human Genome Research in Europe." Paper delivered at conference on the Human Genome Project, History of Science Department, Harvard University, June 1990.

Tétry, Andrée. *Jean Rostand: Un homme du future.* Lyon: La Manufacture, 1988.

U.S. Congress. *Genetic Screening in the Workplace.* Washington, D.C.: Government Printing Office, 1990.

U.S. Congress, Office of Technology Assessment. *Medical Testing and Health Insurance.* Washington, D.C.: Government Printing Office, 1988.

Walsh, Diane Chapman. *Corporate Physicians: Between Medical Care and Management.* New Haven: Yale University Press, 1987.

Weatherall, D. J. *The New Genetics and Clinical Practice.* 2d edition. Oxford: Oxford University Press, 1985.

Weindling, Paul. *Health, Race and German Politics between National Unification and Nazism, 1870–1945.* Cambridge: Cambridge University Press, 1990.

Weiss, Sheila Faith. *Race Hygiene and National Efficiency: The Eugenics of Wilhelm Schallmayer.* Berkeley: University of California Press, 1987.

Wertz, Dorothy, and John G. Fletcher, eds. *Ethics and Human Genetics: A Cross-Cultural Perspective.* New York: Springer-Verlag, 1989.

Wess, Ludger. *Die Träume der Genetik: Gentechnische Utopien von Sozialem Fortschritt.* Hamburger Stiftung für Sozialgeschichte des 20. Jahrhunderts. Nördlingen: Delphi Politik, 1989.

White, B. "Biological Causes for Violent Behavior: Research Could Affect Legal Decisions." *Texas Bar Journal,* 50 (1987), 446.

Woodhead, Avril D., and Benjamin J. Barhart, eds., and Katherine Vivirito, technical ed. *Biotechnology and the Human Genome: Innovations and Impact.* Basic Life Sciences, vol. 46. Based on the Science Writers Workshop on Biotechnology and the Human Genome, held September 14–16, 1987, at Brookhaven National Laboratory, Upton, N.Y. New York: Plenum Press, 1988.

Wright, Robert. "Achilles' Helix." *The New Republic,* 201 (July 9 & 16, 1990), 21–31.

Yoxen, Edward J. "Constructing Genetic Diseases." In Troy Duster and Karen Garett, eds., *Cultural Perspectives on Biological Knowledge,* pp. 41–62. Norwood, N.J.: Ablex, 1984.

—— *Unnatural Selection: Coming to Terms with the New Genetics.* London: Heinemann, 1986.

Glossary

address, chromosomal A site characterized by a DNA sequence (greater than 16 base pairs) which occurs only once in the genome.

adenine (A) See **base pair**

allele One of several alternative forms of a gene occupying a given locus on the chromosome. A single allele for each locus is inherited separately from each parent, so every individual has two alleles for each gene.

Alu A set of 500,000 closely related genetic sequences, each about 300 base pairs long, in the human genome. An example of a repetitive gene family. *Alu* sequences are widely dispersed among all 46 chromosomes.

amino acid Any of a class of twenty molecules that combine in linear arrays to form proteins in living organisms. The sequence (order) of amino acids in a protein (and, hence, the function of the protein) is determined by the order of coding triplets in a gene. The order of amino acids in a protein dictates how the molecules fold into a particular shape to generate a three-dimensional molecular machine with one or more functions.

antibody A large defense protein synthesized by the immune system to neutralize or eliminate pathogens in the body.

antigen Any pathogen whose entry into an organism's body provokes synthesis of an antibody.

autoradiography A technique for producing a visual image of the distribution of DNA fragments separated by length by gel electrophoresis. Each strand of DNA is labeled with a radioactive marker, which is recorded on X-ray film. The result is a series of bands that shows the different fragment sizes and allows comparison of one DNA sample with another.

autosome A chromosome not involved in sex determination. The diploid human genome consists of 46 chromosomes: 22 pairs of autosomes and 1 pair of sex chromosomes (the X and Y chromosomes). Each parent contributes one haploid set of chromosomes (22 autosomes and 1 sex chromosome) to each offspring.

bacteriophage See **phage**

band, chromosomal A narrow portion of the chromosome darkened by interaction with a dye. Each human chromosome displays a unique pattern of bands and can be identified by its pattern.

base pair (bp) Two bases—adenine and thymine or guanine and cytosine—held together by weak bonds. A base is just one of the subunits (see nucleotide) that make up DNA, but it is the sequence of the bases that encodes the instructions for the production of different proteins. Two strands of DNA are held together in the shape of a double helix by the bonds between base pairs.

carrier An individual with one disease form of a gene (allele) and one normal form of the gene.

cDNA Complementary or copy DNA is a man-made copy of the coding sequences of a gene; cDNA is produced in a test tube—it is not a natural product. In a living cell, the protein-coding sequences of DNA are transcribed as mRNA (see messenger RNA). Molecular biologists use reverse transcriptase, an enzyme that makes DNA copies from RNA, to make copies of the mRNA. The resulting cDNA— a copy of a copy, so to speak—may then be analyzed by various methods.

cell, sex A gamete (eggs and sperm) and its precursors. In humans, normal sex cells have only one copy of the 22 autosomes and one sex chromosome—either an X or a Y—so they are referred to as "haploid cells." When an egg and a sperm combine, the resulting zygote therefore has a diploid number of chromosomes. See also **cell, somatic**

cell, somatic Any cell in the body except gametes (or sex cells) and their precursors. In the human body, each normal somatic cell has a full complement of 46 chromosomes, two copies of the 22 autosomes and two sex chromosomes. Somatic cells are thus also called "diploid cells."

centimorgan A unit of measure of recombination frequency. One centimorgan is equal to a 1 percent chance that a genetic locus will be separated from another marker because of recombination in a single generation. In human beings, 1 centimorgan is equivalent, on average, to 1 million base pairs.

chromosome A rod-like structure composed of proteins and the cellular DNA that bears in its nucleotide sequence the linear array of genes. The backbone of the chromosome is a very long molecule of DNA. Chromosomes can be seen in the light microscope only during certain stages of cell division, when they are in condensed form. The 24 different chromosomes in the human genome (22 autosomes plus the X and Y sex chromosomes) are believed to contain approximately 100,000 genes.

cloning The process of asexually producing a group of cells (clones), all genetically identical, from a single ancestor. In recombinant DNA technology, the use of various procedures to produce multiple copies of a single gene or segment of DNA is referred to as "cloning DNA."

coding region See **exon**

codon See **genetic code**

cosmid An artificially constructed cloning vector containing the *cos* gene of phage lambda as well as the DNA segment to be cloned. Cosmids can be packaged in lambda-phage particles for infection in *Escherichia coli;* this permits cloning of larger DNA fragments (up to 45 kb) than can be introduced into bacterial hosts by other kinds of plasmid vector.

cosmid map A physical map comprising a collection of bacteria containing cosmids that carry the DNA fragments under study.

crossing-over See **recombination, chromosomal**

cytosine (C) See **base pair**

cytogenetics The study of genetic variation through an examination of differences in chromosomal structure.

deoxyribonucleic acid See **DNA**

diagnostics, DNA The deployment of different techniques to identify variations in genes or chromosomes that are associated with disease. The techniques include use of molecular complementarity between probe and target DNAs to test for matches and mismatches between the two.

DNA Deoxyribonucleic acid; the molecule that encodes genetic information. DNA is a double-stranded chain of nucleotides held together by weak bonds between base pairs. In nature, base pairs form only between adenine (A) and thymine (T) and between guanine (G) and cytosine (C); thus the sequence of each single strand can be deduced from that of its partner.

DNA fingerprinting Analysis of the DNA from one individual by the molecular techniques used in DNA diagnostics to create a unique DNA profile.

DNA synthesis The chemical joining of nucleotides to create an artificial DNA molecule.

dominant Pertaining to the form of a gene (see **allele**) that exerts its effect when present in the individual in just a single copy. Huntington's disease is an example of a disease caused by a dominant gene.

double helix The natural shape of DNA; the coiled conformation of two complementary, antiparallel chains of nucleotides.

electrophoresis A method of separating large molecules (such as DNA fragments or proteins) from a mixture of similar molecules. An electric current is passed through a medium containing the mixture, and each kind of molecule travels through the medium at a different rate, depending on its electrical charge and size. The fragments are therefore separated, or fractionated, according to their size. Agarose and acrylamide gels are the media commonly used for electrophoresis of proteins and nucleic acids.

embryo An organism in its earliest stages of development.

enzyme A protein that acts as a catalyst, speeding the rate at which a biochemical reaction proceeds but not altering the direction or nature of the reaction.

enzyme, restriction A protein that recognizes specific, short nucleotide sequences and cuts DNA at those sites.

exon The protein-coding DNA sequences of a gene. See also **intron**

expression See **gene expression**

flow cytometry The analysis of biological material by detection of the light-absorbing or fluorescing properties of cells or subcellular bodies (such as chromosomes) passing in a narrow stream through a laser beam. An absorption or fluorescence profile of the sample is produced. Automated sorting devices analyze successive droplets of the stream and sort them into different fractions according the fluorescence emitted by each droplet. Individual cells or chromosomes may be isolated and characterized in such droplets.

folding, protein   The arrangement of the chain of amino acids that make up a protein molecule in a nonlinear, or "bunched-up," form. The order of the amino-acid subunits dictates how the chain folds in three dimensions to form a molecular "machine." The protein-folding problem—that is, determining the rules by which amino acids form different shapes—is an important question in biology because a protein's shape helps determine its function.

gene   The fundamental physical and functional unit of heredity. A gene is an ordered sequence of nucleotides located in a particular position on a particular chromosome. Each gene encodes a specific functional product, such as a protein or RNA molecule. See also **allele**

gene expression   The process by which a gene's coded information is converted into the structures present and operating in the cell. Expressed genes include those that are transcribed into messenger RNA and then translated into protein and those that are transcribed into RNA but not translated into protein (transfer and ribosomal RNAs, for example).

gene regulation   The DNA and protein interactions in a gene that determine the temporal and spatial modes of expression as well as the amplitude of expression.

genetic code   The sequence of nucleotides, coded in triplets along messenger RNA, that determines the sequence of amino acids in protein synthesis. The DNA sequence of a gene can be used to predict the mRNA sequence, and the genetic code can in turn be used to predict the amino acid sequence. The four letters of the DNA alphabet form 64 triplets, or codons, which specify the 20 different amino-acid subunits and the stop signals that end the production of the protein. Hence, most amino acids are coded for by more than one triplet.

genetic mapping   Determination of the relative positions of genes on a DNA molecule (chromosome or plasmid) and of the distance, in linkage units or physical units, between them.

genome   All the genetic material in the chromosomes of a particular organism. The size of a genome is generally given as its total number of base pairs.

genotype   The total genetic, or hereditary, constitution that an individual receives from his or her parents. See also **phenotype**

germ line   The DNA of sex cells.

guanine (G)   See **base pair**

heterozygote   An individual organism having different alleles of a particular gene on each member of a pair of chromosomes.

homologue   One member of a related pair of human chromosomes.

homozygote   An individual organism having identical alleles of a particular gene on each member of a pair of chromosomes. A person may be homozygous for one gene and heterozygous for another.

hybridization, cellular   The fusion of cells from two different organisms into one cell that combines the chromosomes of both.

inborn error of metabolism   A defect in one of the many genes that regulate metabolism.

intron   DNA sequence interrupting the protein-coding sequences of a gene; these sequences are transcribed into nuclear RNA but are cut out of the message (mRNA) before it is translated into protein.

in vitro   Pertaining to procedures carried out in a test-tube.

in vivo   Pertaining to procedures carried out in the living organism.

kilobase (kb)   Unit of length for DNA fragments on physical maps (equal to the distance spanned by 1,000 base pairs).

lambda phage   See **phage**

library, genetic   An unordered collection of DNA clones from a particular organism.

linkage   The proximity of two or more genetic markers—a gene or some other polymorphism—on a chromosome; the closer together the markers are, the lower the probability that they will be separated during DNA repair, replication, or recombination and hence the greater the probability that they will be inherited together.

linkage map   A genetic map determined by an analysis of the linkage patterns of genes and markers on chromosomes.

locus The position on a chromosome of a gene or other chromosome marker; also, the DNA sequence at that position. Some restrict use of the word to regions of DNA that are expressed.

maps See **cosmid map, genetic mapping, linkage map, physical map, restriction fragment length polymorphism,** and **sequence tagged site**

marker An identifiable physical location on a chromosome whose inheritance can be monitored. Markers can be expressed regions of DNA (genes), a sequence of bases that can be identified by restriction enzymes, or a segment of DNA with no known coding function. Genetic maps are maps of the relative positions of markers and genes on the chromosomes.

megabase One million base pairs.

messenger RNA (mRNA) A class of ribonucleic acid (RNA) whose function is to carry the genetic code from the chromosome (in the nucleus) to the ribosome (in the cytoplasm) and direct protein synthesis there.

mutation Any replicatable change in DNA sequence.

nucleotide A subunit of DNA or RNA consisting of a nitrogenous base (adenine, guanine, thymine, or cytosine in DNA; adenine, guanine, uracil, or cytosine in RNA), a phosphate molecule, and a sugar molecule (deoxyribose in DNA and ribose in RNA). Thousands of nucleotides are linked to form DNA or RNA molecules.

nucleic acid A natural polymer, either single- or double-stranded, comprising a sugar phosphate backbone to which different bases are attached.

oligonucleotide A short DNA sequence, often generated by chemical synthesis.

oncogene A gene that in one or more of its forms is associated with cancer. In their normal form as proto-oncogenes, many oncogenes are involved, directly or indirectly, in controlling the rate of cell growth.

PCR See **polymerase chain reaction**

phage A virus for which the natural host is a bacterial cell. Phi-X and lambda are types of phages.

phenotype The appearance and other physical characteristics of an organism, a result of the interaction of an individual's genetic constitution with the environment. Phenotype differs from genotype in that it includes only the outward manifestations of genes.

physical map   An overlapping collection of DNA fragments which span a particular chromosomal region. A common type of physical map comprises DNA fragments contained in cosmids.

plasmid   An extrachromosomal entity containing nucleic acid and replicating independently of the chromosome. Foreign DNA can be inserted into a plasmid and made to replicate with it.

polygenic   Pertaining to the combined action of alleles of more than one gene. Height is an example of a polygenic trait, as are predispositions to different types of heart disease.

polymerase, DNA   An enzyme that acts as a catalyst in the replication of DNA.

polymerase chain reaction (PCR)   A method for making a great many copies of a particular DNA sequence. The procedure requires two kinds of primers—primers are essential for DNA synthesis— each kind complementary to just one end of the DNA fragment to be amplified; a thermostable DNA polymerase; and a supply of nucleotides. First a solution containing the DNA fragment, the primers, and the nucleotides is heated, and the two strands of the DNA come apart; the primers then anneal to the appropriate ends. After the solution is cooled, the polymerase is added, and the enzyme effects the replication of the DNA fragment between the two primers on the ends. Each newly synthesized strand of DNA subsequently serves as a template for yet another strand, so the supply doubles with each repetition of the procedure. PCR can also be used to detect the existence of a defined sequence in a DNA sample (see **sequence tagged site**).

polymorphism   Any difference in DNA sequence among individuals. Genetic variations occurring in more than 1 percent of a population would be considered useful polymorphisms for genetic linkage analysis.

primer   Short, preexisting polynucleotide chain to which new deoxyribonucleotides can be added by DNA polymerase.

probe   Single-stranded DNA or RNA molecule of a known sequence that is, labeled either radioactively or immunologically. Probes are used to detect complementary base sequences by hybridization.

protein   A large molecule composed of one or more chains of amino acids in a specific sequence; the sequence is determined by the sequence of nucleotides in the gene coding for the protein. Proteins are required for the structure, function, and regulation of the body's cells, tissues, and organs, and each protein has unique functions.

proto-oncogene   A normal cellular gene that can become a cancer-producing oncogene. The change may occur via different mechanisms, including point mutation, chromosome translocation, insertional mutation, and amplification.

pulsed-field gel electrophoresis (PFGE)   The use of alternating electric fields in electrophoresis; the pulsing causes large DNA molecules to collapse and hence permit them to migrate through gel. While standard gel electrophoresis can separate DNA fragments up to 25,000 base pairs in size, PFGE can separate fragments ranging in size from 100 to 10 million base pairs.

radiolabel   To label or mark a molecule with a radioactive substance.

recessive   Pertaining to an allele that is expressed only when present in two copies, one on each member of a pair of chromosomes.

recombinant DNA   The hybrid DNA produced in the laboratory by joining pieces of DNA from different sources.

recombination, chromosomal   The interchange of chromosomal segments (by breaking and rejoining) between two homologues during replication. Since a piece of one chromosome now resides on the other chromosome in the pair, and vice versa, the recombination is said to be heterologous. Also called "crossing-over."

recombination, homologous   The insertion of genes or segments of DNA into their proper positions in a chromosome.

replication   Duplication of a molecule by following the pattern of a template. Replication can occur *in vitro* as well as *in vivo*.

restriction fragment length polymorphism (RFLP)   The variation, between individuals, in DNA fragment sizes cut by specific restriction enzymes; polymorphic sequences that result in RFLPs are used as markers on both physical maps and genetic linkage maps. RFLPs are usually caused by a mutation at a cutting site.

RNA   Ribonucleic acid; a chemical found in the nucleus and cytoplasm of cells; it plays an important role in protein synthesis and other chemical activities of the cell. The structure of RNA is similar to that of DNA, except that it includes uracil, in place of thymine, as one of its bases. There are several classes of RNA molecules—messenger RNA (mRNA), transfer RNA (tRNA), ribosomal RNA, and other small RNAs—each serving a different purpose.

RNA splicing pattern The combination of DNA sequences copied from a gene by messenger RNA. The mRNAs transcribed from a single gene may splice together different parts of the sequence of the gene.

sequence The order of nucleotides in a nucleic acid or the order of amino acids in a protein.

sequence tagged site (STS) A short DNA sequence, delineated by two PCR primers, that uniquely identifies a mapped gene or other chromosomal region. The order and spacing of these sequences constitute an STS map.

thymine See **base pair**

transgenic animal An animal whose cells contain genetic material originally derived from another animal. For example, transgenic mice may contain genetic material from humans.

uracil See **RNA**

vector The agent used (by researchers) to carry new genes into cells. Plasmids currently are the vectors of choice, although viruses and bacteria are increasingly being used for this purpose.

X chromosome, Y chromosome The sex chromosomes. Normal human females have a pair of X chromosomes in each somatic cell; normal human males have one X and one Y.

X linkage The location of a gene or marker on the X chromosome. Recessive, abnormal genes located on the X chromosome will be expressed in more males than females, because males lack the second copy of the X that might have carried the dominant, normal copy of the gene (which, being dominant, would have exerted its effect). Examples of X-linked genetic diseases are hemophilia and Duchenne muscular dystrophy; color blindness is a much less serious trait that is determined by an X-linked gene.

yeast artificial chromosome (YAC) The three components of a yeast chromosome that are necessary for replication joined to a large fragment of foreign DNA. A YAC is capable of maintaining and replicating itself and the foreign DNA.

Contributors

Charles Cantor is professor of biochemistry at Berkeley and chief scientist of the DOE genome project. He was trained in physical biochemistry and has written a major textbook on that subject. Together with David Schwartz, he developed pulsed-field gel electrophoresis, an important technique for separating large DNA fragments. An advocate of the genome project since its early days, he was formerly director of the Genome Center at the Lawrence Berkeley Radiation Laboratories. Cantor's interests now focus on chromosome structure and organization.

C. Thomas Caskey, the Henry and Emma Meyer Professor and the Director of the Institute for Molecular Genetics at the Baylor College of Medicine, trained as a physician and later at NIH as a molecular biologist with a focus on protein synthesis. He is currently interested in somatic-cell therapy, somatic-cell genetics, medical genetics, and chromosome mapping and sequencing. He has established an NIH Genome Center focused on these topics. Caskey has been a pioneer in applying molecular biology to medical genetics and recently has isolated the gene for the fragile X syndrome, the most common cause of mental deficiency in newborns.

Ruth Schwartz Cowan is a historian of science and technology in the nineteenth and twentieth centuries. Her publications include the prize-winning *More Work for Mother: The Ironies of Household Technology from the Open Hearth to the Microwave* (Basic Books, 1983) and a variety of studies in the history of biology and eugenics. She is a professor of history at the State University of New York at Stony Brook, where she has directed the Women's Studies Program and where her teaching has included courses in social and ethical issues in reproductive technology. Currently, she is at work on a book about the history of prenatal diagnosis.

Walter Gilbert, the Carl Loeb University Professor at Harvard, was trained as a physicist, switched to molecular biology, and carried out pioneering work on protein DNA interactions. He developed, together with Allan Maxam, one of the major techniques for DNA sequencing, an accomplishment for which he was awarded the Nobel Prize in Chemistry, in 1980. Gilbert attended the first genome meeting in Santa Cruz in 1985 and since that time has been an eloquent advocate for the genome project. He is currently establishing a laboratory to carry out large-scale DNA sequencing focused on the analysis of the one-megabase genome of the microorganism mycoplasma. He has broad interests including gene regulation and molecular evolution.

Henry T. Greely is professor of law at Stanford University, specializing in health law and policy. A 1977 graduate of Yale Law School, he served as a law clerk to Judge John Minor Wisdom and Justice Potter Stewart and as a staff assistant to the Secretary of Energy. He entered private law practice in 1981 and joined the Stanford faculty in 1985.

Leroy Hood, Bowles Professor of Biology at California Institute of Technology, is internationally known for his research in immunology and for his pioneering innovation of technologies for biological science, particularly in the analysis of the human genome. He has won numerous honors for his work, including the Lasker Award and membership in the National Academy of Sciences. In September 1992 he will move to the University of Washington as the William Gates III Professor of Molecular Biology.

Horace Freeland Judson is Senior Research Associate in the Program in History of Science at Stanford University. Among his books are *The Eighth Day of Creation, The Search for Solutions,* and a collection of critical essays on the sciences and arts, soon to be published, titled *"Give It Mouth!"*

Evelyn Fox Keller, a professor at the University of California, Berkeley, brings to her work in the history and philosophy of science first-hand experience as a researcher in theoretical physics as well as in mathematical and molecular biology. She has written extensively on a variety of social issues in science from a perspective that is not only knowledgeable but provocative. She is the author of *A Feeling for the Organism: The Life and Work of Barbara McClintock* (W. H. Freeman, 1983) and *Reflections on Gender and Science* (Yale University Press, 1985). Her newest books are *Keywords in Evolutionary Biology* (Harvard University Press, 1992), co-edited with Elisabeth Lloyd, and *Secrets of Life/Secrets of Death* (Routledge, 1991).

Daniel J. Kevles is the Koepfli Professor of the Humanities at California Institute of Technology. He has written extensively about the social and political relations of science for a variety of publications, including the *New Yorker* and the *New York Review of Books*. He is author of *The Physicists: The History of a Scientific Community in Modern America* and *In the Name of Eugenics: Genetics and the Uses of Human Heredity*.

Eric S. Lander is a geneticist and mathematician whose research interests include human, mouse, and population genetics as well as computational and mathematical methods in biology. A member of the Whitehead Institute for Biomedical Research, he is an associate professor of biology at MIT and the director of the MIT Center for Genome Research. He was awarded a MacArthur Foundation Prize Fellowship, in 1987, for his work in genetics. He has developed methods for the genetic analysis of complex inherited traits and has worked on the construction of complete genomic maps in connection with the human genome project. A member of the National Research Council Committee on DNA Technology in Forensic Science, he has played a major role in assessing the merits of DNA fingerprinting as a method of determining the identification of criminal suspects.

Dorothy Nelkin holds a University Professorship at New York University, where she teaches in the Department of Sociology and the School of Law. Formerly a professor at Cornell University, she has been a Guggenheim Fellow and a Visiting Scholar at the Russel Sage Foundation. She has served on the Board of Directors of the American Association for the Advancement of Science and is currently a member of the National Advisory Council of the NIH Genome Project. She is the author of many books on the relationship of science to society, including *The Creation Controversy*, *Selling Science: How the Press Covers Science and Technology*, *Dangerous Diagnostics: The Social Power of Biological Information* (with L. Tancredi), and, most recently, *The Animal Rights Crusade* (with J. Jasper).

James Watson shared in the 1962 Nobel Prize in Physiology or Medicine for his role in the discovery of the structure of DNA. An ornithologist before becoming a molecular biologist, he has maintained a broad spectrum of biological interests and has contributed to research in the control of gene expression. A member of the Harvard faculty for many years, Watson became director of the Cold Spring Harbor Laboratory in 1968 and has created an outstanding research and educational environment there. In 1989, he assumed the directorship of NIH's genome project and has since that time been perhaps the most influential single figure in promoting this endeavor with Congress, NIH, and the scientific community.

Nancy Wexler trained in clinical psychology and is professor of clinical psychology, neurology and psychiatry at Columbia University College of Physicians and Surgeons. She has pioneered an attack on Huntington's disease that has led her to analyze a large population of afflicted individuals in Venezuela, a project that has made the cell lines of the study population available for analysis by contemporary techniques in molecular biology. Together with James Gusella, she localized the gene for Huntington's disease to the short area of human chromosome 4. She has been very active in the Hereditary Disease Foundation, which her father, Milton Wexler, a psychoanalyst, established and which has played a pivotal role in attracting outstanding scientists to work on the problems of Huntington's disease. Nancy Wexler is chairperson of the ethics working group for the committee on the human genome at the National Institutes of Health. Her current work focuses on more detailed mapping of the Huntington's gene.

Index